LABORATORY   *   *   *

EPISTEMOLOGIES

Experimental Futures
Technological Lives, Scientific Arts, Anthropological Voices
A series edited by
Michael M. J. Fischer and Joseph Dumit

LABORATORY ✳ ✳ ✳ EPISTEMOLOGIES

✳ ✳ ✳ A HANDS-ON PERSPECTIVE ✳ ✳

JENNY ✳ ✳ ✳ ✳ ✳ ✳ ✳ ✳ ✳ BOULBOULLÉ

DUKE UNIVERSITY PRESS   DURHAM AND LONDON   2024

© 2024 DUKE UNIVERSITY PRESS
All rights reserved

Project Editor: Livia Tenzer
Designed by A. Mattson Gallagher
Typeset in Minion Pro and Source Code Pro
by Westchester Publishing Services

Library of Congress Cataloging-in-Publication Data
Names: Boulboullé, Jenny, author.
Title: Laboratory epistemologies : a hands-on perspective / Jenny Boulboullé.
Other titles: Experimental futures.
Description: Durham : Duke University Press, 2024. | Series: Experimental
futures | Includes bibliographical references and index.
Identifiers: LCCN 2024005738 (print)
LCCN 2024005739 (ebook)
ISBN 9781478030966 (paperback)
ISBN 9781478026754 (hardcover)
ISBN 9781478059981 (ebook)
Subjects: LCSH: Knowledge, Theory of. | Philosophy and the life sciences. |
Life sciences—Philosophy. | Science—Philosophy. | Science—Experiments—
Philosophy. | BISAC: SOCIAL SCIENCE / Anthropology / General |
SCIENCE / Life Sciences / General
Classification: LCC Q175.32.K45 B68 2024 (print) | LCC Q175.32.K45 (ebook) |
DDC 001.401—dc23/eng/20240630
LC record available at https://lccn.loc.gov/2024005738
LC ebook record available at https://lccn.loc.gov/2024005739

Cover art: Herwig Turk, *hands-on (version 3)*, 2014. Video still,
two-channel video installation. Courtesy of the artist.

For Robert, Noor, and Pim,
and my parents, Anja and Guido

CONTENTS

The philosopher and experimenter René Descartes (1596–1650) wrote his treatises and corresponded in Latin and in French. His masterwork, *Meditations on First Philosophy*, was first published in Latin in 1641 but soon translated into French and issued during his lifetime with his approval. See "Works by René Descartes" in the bibliography for the editions of Descartes's treatises and letters cited in this book and the modern English-language translations consulted. The citation system is explained here.

For Descartes's works and correspondence, I generally cite the standard modern edition: René Descartes, *Oeuvres de Descartes*, 11 vols., edited by Charles Adam and Paul Tannery (Paris: Vrin, 1996 [1896–1901]), abbreviated as AT I–XI. Citations show the shortened title of Descartes's work with the appropriate volume number of AT, as in this example: Descartes, *Les méditations métaphysiques*, AT IX. Page numbers appear at the end in all citations. AT is accessible online at History of Philosophy: Texts Online, http://philosophyfaculty.ucsd.edu/faculty/ctolley/texts/descartes.html.

When I cite directly from the first print editions of individual works by Descartes, I give the shortened title of the work, followed by the year of publication, as in this example: Descartes, *Les méditations métaphysiques* (1647). Select first editions are accessible online at Corpus Descartes: Édition en ligne des oeuvres et de la correspondance de Descartes, https://www .unicaen.fr/puc/sources/prodescartes/presentation.html.

For all English-language translations of Descartes's writings, I have whenever possible consulted existing translations. As needed, I have adapted and modified them to keep them as close as possible to the original French texts. I relied primarily on: René Descartes, *The Philosophical Writings of Descartes*, 2 vols., translated by John Cottingham, Robert Stoothoff, and Dugald Murdoch (Cambridge: Cambridge University Press, 1985), abbreviated as CSM 1–2; and René Descartes, *The Philosophical Writings of Descartes*, Vol. 3, *The Correspondence*, translated by John Cottingham, Robert Stoothoff, Dugald Murdoch, and Anthony Kenny (Cambridge: Cambridge University Press, 1991), abbreviated as CSMK. Citations show the short title of Descartes's work or details of the correspondence, followed by the appropriate volume number, as in these examples: Descartes, *Meditation on First Philosophy*, CSM 2; Descartes, letter to Mersenne, June 1632, CSMK.

Other translations I consulted are: René Descartes, *Discourse on Method, Optics, Geometry, and Meteorology*, translated and edited by Paul J. Olscamp, (Indianapolis, IN: Bobbs-Merrill, 1976); and René Descartes, *The World and Other Writings*, translated and edited by Stephen Gaukroger (Cambridge: Cambridge University Press, 1998). These are cited as in this example: *Optics*, in Descartes, *Discourse on Method*.

★   ★   ★

One of the most critical passages of Descartes's *Meditations on First Philosophy* for the present book is his detailed account of a "wax *experience*," described in the Second Meditation. The relevant pages from the first French edition of the work, *Les méditations métaphysique* (1647), 26–27, follow here (see also figs. 1.2 and 1.3).

bride, afin que venant cy-apres à la retirer doucement
& à propos , nous le puiſſions plus facilement regler
& conduire.

Commençons par la conſideration des choſes les
plus communes, & que nous croyons comprendre le
plus diſtinctement, à ſçauoir les corps que nous touchons & que nous voyons. Ie n'entens pas parler des
corps en general, car ces notions generales ſont d'ordinaire plus confuſes, mais de quelqu'vn en particulier. Prenons pour exemple ce morceau de cire qui
vient d'eſtre tiré de la ruche, il n'a pas encore perdu
la douceur du miel qu'il contenoit , il retient encore
quelque choſe de l'odeur des fleurs dont il a eſté recueilly ; ſa couleur, ſa figure, ſa grandeur, ſont apparentes : il eſt dur, il eſt froid, on le touche, & ſi vous
le frappez, il rendra quelque ſon. Enfin toutes les
choſes qui peuuent diſtinctement faire connoiſtre
vn corps, ſe rencontrent en celuy-cy.

Mais voicy que cependant que ie parle on l'apro-
che du feu, ce qui y reſtoit de ſaueur s'exale , l'odeur
s'éuanoüit, ſa couleur ſe change, ſa figure ſe perd , ſa
grandeur augmente, il deuient liquide, il s'échauffe,
à peine le peut-on toucher , & quoy qu'on le frappe il
ne rendra plus aucun ſon : La meſme cire demeure-
t'elle aprés ce changement ? Il faut auoüer qu'elle demeure , & perſonne ne le peut nier. Qu'eſt-ce donc
que l'on connoiſſoit en ce morceau de cire auec tant
de diſtinction ? Certes ce ne peut eſtre rien de tout ce
que i'y ay remarqué par l'entremiſe des ſens, puis que

toutes les chofes qui tomboient fous le gouft, ou l'o-
dorat, ou la veuë, ou l'attouchement, ou l'ouye fe
trouuent changées, & cependant la mefme cire demeure. Peut-eftre eftoit-ce ce que ie penfe maintenant, à fçauoir que la cire n'eftoit pas, ny cette douceur du miel, ny cette agreable odeur des fleurs, ny
cette blancheur, ny cette figure, ny ce fon, mais feu-
lement vn corps qui vn peu auparauant me paroiffoit
fous ces formes, & qui maintenant fe fait remarquer
fous d'autres. Mais qu'eft-ce precifément parlant que
i'imagine, lors que ie la conçoy en cette forte? Confi-
derons-le attentiuement, & éloignant toutes les chofes qui n'appartiennent point à la cire, voyons ce qui
refte. Certes il ne demeure rien que quelque chofe
d'eftendu, de flexible & de muable: Or qu'eft-ce que
cela flexible & muable? n'eft-ce pas que i'imagine que
cette cire eftant ronde eft capable de deuenir quarrée,
& de paffer du quarré en vne figure triangulaire? non
certes ce n'eft pas cela, puis que ie la conçoy capable
de receuoir vne infinité de femblables changemens,
& ie ne fçaurois neantmoins parcourir cette infinité
par mon imagination, & par confequent cette conception que i'ay de la cire ne s'accomplit pas par la faculté d'imaginer.

Qu'eft-ce maintenant que cette extenfion? n'eft-
elle pas auffi inconnuë? Puifque dans la cire qui fe fond
elle augmente, & fe trouue encore plus grande quand
elle eft entierement fonduë, & beaucoup plus encore
quand la chaleur augmente dauantage; & ie ne con-

★ ★ ★ ★ ★ ★ ★ ★ ★ ★ ★ ★ ★ ★

Introduction

A FEELING FOR THE LIFE SCIENCES

Looking back at my first day in a teaching laboratory for life sciences and chemistry students, I remember how unfamiliar it felt in the beginning to hold a pipette, to pick up the sterile tips without touching them, and to fill them without sucking up air bubbles. Microliter pipettes are delicate and expensive instruments adapted for exact measurements. They form an indispensable part of innumerable procedures in technoscientific research and production laboratories all over the world. Ensuring that these manually operated precision tools are used accurately and function properly is not trivial but is part and parcel of good laboratory practice in the life sciences. Today's automatic microliter pipettes are adapted for accuracy and exactness to measure and transfer volumes of liquids as small as 2 microliters (μl), swiftly with one hand—if calibrated correctly and handled

2

I.1 * "Experiment 1: IJken van een pipet" (translation, *opposite*). Instructions for biochemistry practicum, Gorlaeus Laboratories, Leiden University. From Erik Vijgenboom et al., "Handleiding Introductie & Biochemie Practicum I, Sept–Okt 2005," p. 11.

properly and with care. Functioning, well-calibrated pipettes are, indeed, vital to life science laboratories' routines. Perhaps not surprisingly, then, the first experiment I performed in the biochemistry practicum at Leiden University was the calibration of a microliter pipette (figure I.1).[1] Indeed, it is no coincidence that we as novices spent our first day in the laboratory calibrating pipettes: setting measurements, fitting plastic tips onto pipettes, sucking up tiny amounts of fluids, transferring and releasing fluids into plas-

Experiment 1.

Calibration of a Pipette

Introduction
The functioning and the use of the pipette can be tested in a simple manner. By means of a set of weight experiments, it can be determined whether the set volume is delivered indeed. . . .

Protocol
Perform the following measurements and process the data using Excel. Automatic pipette 2–20 µl:
Fill a beaker with Milli-Q water.
Put an empty plastic medicine cup on the analytical balance and set to zero.
Set the pipette to 2 µl, draw water with the pipette, and pipette the volume into the plastic cup.
Record the weight.
Set the balance back to zero.
Pipette 5 µl water into the cup and record the weight.
Repeat this for 7.5 µl, 10 µl, 15 µl, and 20 µl.

Likewise, repeat the procedure with the 20–200 µl pipette (use 20, 40, 60, 80, 100, 125, 150, 200 µl) and the 200–1,000 µl pipette (use 100, 200, 400, 600, 800, 1,000 µl).

Calculate from the recorded weights and the relative density of water (see Table 1) the (average) weight volume. Plot the weight volume against the set volume and make a statement about the precision with which you pipetted. The deviation should not exceed 2 percent. If the deviation is larger, you should repeat the measurements. Pay close attention to handling the pipette correctly during all steps of the procedure. If you believe the pipette is not functioning properly (the deviation is larger than 2 percent, in spite of working correctly), discuss with the assistant the next step(s).[2]

Table 1

Specific Gravity of Water	
Temperature (°C)	g/mL
4	1.0000
10	0.9997
15	0.9991
20	0.9982
25	0.9971
30	0.9957

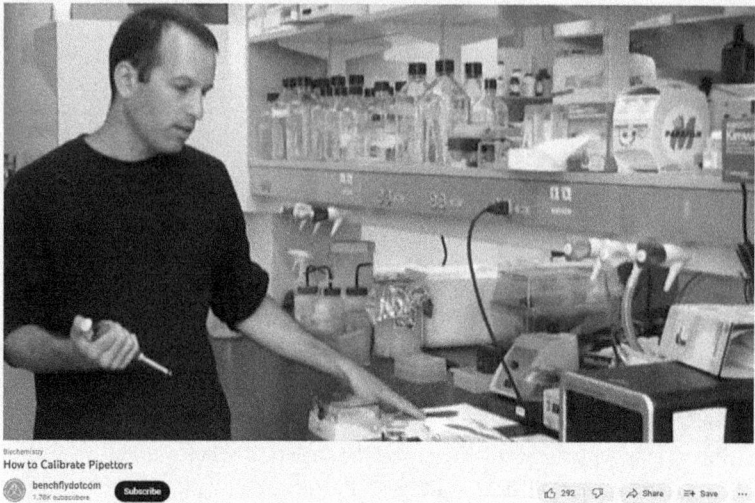

I.2 ★ Still from "How to Calibrate Pipettors," YouTube, posted by bench-flydotcom, October 23, 2009, https://www.youtube.com/watch?v=ImFy3tBC -8o&feature=youtu.be.

tic cups, weighing them with high-precision scales and minutely recording the weights, entering data into Excel tables, calculating average weights and statistical errors, and finally producing digital graphs and interpreting them.

In retrospect, our calibration experiment turned out to be much more than a repetitious, dull, and tedious mechanical procedure of weighing mi-nuscule water droplets. In fact, calibrating an everyday lab tool on our first day introduced us to indispensable skills needed for successful experimen-tation. Moreover, when pipettes are used conscientiously—in conjunction with disposable sterile plastic tips—experienced pipettors can handle tiny amounts of liquids with hardly any risk of contamination. Numerous on-line instructions and how-to videos on the proper use and maintenance of pipettes show that good pipetting is not only a basic skill but also a matter of ongoing concern in contemporary laboratories (figure I.2).

Life sciences research and instruction sites are devoted to the study of life processes on cellular and molecular levels—mostly invisible to the naked eye. In retrospect, measuring microscopic amounts of transparent and odorless fluids epitomized my experience of doing molecular biologi-cal experiments. Only rarely could I perceive the life processes that we were investigating with my ordinary senses—that is, by sight or touch—though we definitely smelled the specific odor of the *Escherichia coli* bacteria that naturally inhabit our lower intestines. Most of the life-forms we worked with were cultivated in the lab; as such, we would not generally be able to encounter these "model organisms" in nature, outside the confines of the laboratory. Most important, I learned that it is crucial in a life science lab to handle barely visible amounts of media, living and nonliving, and that it was critical to avoid any unnecessary physical disturbance of these media and particularly to not bring them directly, or indirectly, into contact with any surfaces or parts of my body. Proper handling of manual precision tools is hereby of vital importance.

In hindsight, our calibration experiment can perhaps best be described as a complex process of "enskillment," "skilled practice," and cultivation of specific "bodily techniques" that involves a training of manual as well as cognitive abilities—minds and hands.[3] In practice, our calibration experiment made visible the entangled activities of manual benchwork, material engagement, observational documentation, and statistical data processing. Calibration understood *as practice* makes us attentive to the interdependency of mind, hand, instruments, and materials in scientific experimentation. Moreover, performing the experiment made plain that calibration is an *embodied* practice that fine-tunes body-instrument interactions and fosters a mindful engagement with materials for scientific analysis.[4]

Hands-On Knowledge Making in the Life Sciences

The initializing calibration experiment turned out to be programmatic for the book that you now hold in your hands. This seemingly simple experiment brings us right to the core of what this book is about: developing a better understanding of apparently mundane, yet on closer view complex, knowledge-making practices in the life sciences that remain resistant to categorical distinctions between mind and hand. What is at issue is the very idea that we can envision a neat dividing line between the cognitive and the manual in doing life sciences.

Instead, I argue that we can best describe the work of life scientists in instruction and research laboratories with a continuum approach that does not privilege mind over hand and rationalization over material engagement. However, while life science handbooks are full of scientific descriptions of complex life processes, they lack any descriptions of the complexities of benchwork or idiosyncrasies of bodily work in microbiological labs that underpin the wealth of information that life science students have to get

5

into their heads—and the bits they have to be able to reproduce by hand. Paradoxically, the manual knowledge that comes with skilled practice—and forms an integral part of lab-based learning and experimental research in the life sciences—appears to have been written out of life science handbooks. And not only out of books: manual training and research are done in the lab but, generally, are not topics of discussion in life science lecture halls and research meetings. In this book, I put the manual center stage. To better understand how it functions as an integral part of knowledge making in the life sciences, I investigate hands-on laboratory practices from an ethnographic, historical, and philosophical perspective. I argue that the manual should be acknowledged not only as a fundamental part of daily life science research but also as a vital feature of epistemologies concerned with life processes on cellular and molecular levels—that is, the theoretical discourses of *how* we can gain knowledge of life processes on (sub)cellular levels. As knowledge makers, we should be able not only to manually produce knowledge but also to talk about and understand—culturally, historically, and philosophically—*how we make knowledge*. In this sense, this book functions as a how-to book for life sciences research, that is, as *a manual* for making *the manual* explicit in experimental life sciences.

Fundamentally, this is an epistemological question that requires recalibration from a hands-on perspective. This book thus situates reflections on today's experimental life sciences within broader debates on knowledge making in history and philosophy of science. It provides a *long durée* philosophical perspective with a radical rereading of a key source text in epistemology: at its heart lies an invitation to read René Descartes's (1596–1650) famous *Meditations* against the grain of an outdated but persistent Cartesianism that has traditionally located capacities of knowing in our minds, not our hands. Based on participatory hands-on experiences in molecular biology, I offer a critical reading of laboratory ethnographies that fundamentally recalibrates anti-Cartesian discourses in science studies.

Recalibrations

Learning to work accurately with a micropipette and becoming habituated to proper handling of pipettes and tips takes a while. It is a skill that is learned through continuous practice. Moreover, becoming attuned to a manual precision tool is a reciprocal process that affects both instrument and operator.

6

During our first-year calibration experiment, no information was given on how to proceed in case an inaccuracy of our tested pipette would have been established, nor were we given access to calibration tools to make such adjustments ourselves. I suppose our teachers and tutors generally assumed that any significant deviations in the accuracy of the instrument were more likely due to our poor performances during the process of calibration than to dysfunction of our tested pipette. Most likely, our first-day experiment was just as much geared toward calibrating the bodily skills of us novices as toward calibrating our new tools. In retrospect, I like to think of this first day of weighing water droplets with my bench partner, a first-year chemistry student, as an initiation rite into the art of pipetting with an automatic microliter pipette—an indispensable skill in experimental life sciences research. The *Oxford English Dictionary* defines *pipettor* (= *pipetter*, n.) as an instrument that allows for automated operation of one or more pipettes.[5] Yet *pipettor* cannot refer only to a machine that can automatically operate several pipettes. *Pipettor* is grammatically and etymologically analogous to terms like *instructor*, *lecturer*, or *practitioner*. In all these examples the suffix *-er* or *-or* is added to the stem of a verb. The endings *-er* and *-or* historically served to designate persons according to their profession or occupation: someone who experiments becomes an *experimenter*. Derivatives from action verbs are also called agent nouns. Agent nouns, like *instructor*, *teacher*, and *vlogger*, are nowadays used to denote not only human but also material agents, or even instruments. *Pipettor* is such a multivalent term, which can in fact denote both—an instrument (i.e., a material agent) and the operator of an instrument (i.e., a human agent who operates a pipette). This semantic ambivalence aptly conveys the intimate relationship of instrument and instrument operator that is key to good scientific benchwork. So we might wonder what it was that was calibrated during our first-day experiment—the pipettes or us, the instrument users? Perhaps, we can best say that the calibration of a *pipettor* teaches us—albeit implicitly—a lesson in the reciprocal and fine-tuned body-instrument actions of laboratory work and that the semantic ambiguity of *pipettor* testifies to the intimately entangled user-instrument relations and skillful mind/hand/tool/material interactions in scientific practices.

What motivated me to write this book was experiencing how knowledge making in life science labs requires as much training in dexterity and care as cognitive capacities for operating with biological concepts in the abstract realms of genes, proteins, molecules, deoxyribonucleic acid (DNA) codes,

7

and Mendel's laws. Yet, when interning with life scientists in the molecular genetics laboratories of the Leiden Institute of Chemistry, I wondered why successful careers in the life sciences are rarely acclaimed for the expertise that researchers gained during years-long or even lifelong training in highly skilled hands-on practices. The aim of this book is to write researchers' hands and bodies back into the sciences of life and to make them explicit in theories of scientific knowledge production. The book draws on historical and ethnographic accounts of past and present hands-on practices to explicate the role of the manual and embodied practices in knowledge-formation processes since the rise of the experimental life sciences in the seventeenth century.

Becoming a Life Scientist

You let the material tell you where to go, and it tells you at every step what the next has to be because you are integrating with an overall brand new pattern in your mind. You are not following an old one; you are convinced of a new one.

Barbara McClintock, quoted in Evelyn Fox Keller, *A Feeling for the Organism*

The biography *A Feeling for the Organism* traces the academic career of Barbara McClintock (1902–1992), a plant biologist and a pioneer in the field of cytogenetics in the context of the emergent life sciences.[6] The conversations on which the publication is based are a testimony to an encounter between two extraordinary women of science. The interviewee, McClintock, was the first woman to receive an unshared Nobel Prize in Physiology and Medicine (1983). The interviewer, Evelyn Fox Keller (1936–2023), was an eminent scholar known for her important oeuvre on gender and science and the history and philosophy of the life sciences that shape our understanding of genetics and molecular biology.[7] Keller interviewed not only McClintock but also McClintock's peers and former students, to give voice to a researcher whose work on plant genetics has long remained in the shadow of more prominent male-dominated narratives of the rapidly unfolding "molecular revolution" in twentieth-century biology. Keller's portrayal of McClintock and her lifelong research on innumerable minute details of genetic organization in plants and fungi shows how this life scientist, who could see what others could not see, contributed to the rise of a "new biology" that came to focus on genetics and cellular and molecular processes.[8] Yet, McClintock's groundbreaking research on mobile genetic elements in maize was strikingly slow to win acclaim from her peers. When her biography was published in 1983—the same year that she would receive the highest possible scientific recognition for the research she conducted in the 1940s—the biologist had reached the eminent age of eighty-two.

In one of the interviews, McClintock recalls how she became fascinated with biological research, first as a student and later as a scientist specializing in plant genetics. When Keller probed her about the capacities that made her into an outstanding scientist, the biologist expressed her wish to be "free of the body." "The body," she told her biographer, "was something you dragged around." To her, it was not something that enabled her to experience and perceive what interested her but something that stood in the way, a burden that appeared to hold her back from fulfilling her scientific ambition. As she explained: "I always wished that I could be an objective observer, and not be what is known as 'me' to other people." Her own "bodily me" was first and foremost an impediment that others mistakenly associated her with and which she preferred to ignore.[9]

Keller chronicles McClintock's ambivalent feelings about her body throughout her academic career. When the biologist talks about her passion for learning and how she enjoyed writing an exam in her favorite subject while still a high school student, she recalls how she experienced her own body as something irritating and annoyingly irrelevant, yet persistently present: "I think it had to do with the body being a nuisance. What was going on, what I saw, what I was thinking about, and what I enjoyed seeing and hearing was so much more important."[10] McClintock intimates a longing for a transparent body that could not block her view and would not get in the way of all the more important things she wanted to explore and understand. This wish appears at once sensible and puzzling in light of her own research on our bodies' innermost life processes, which we share with other living organisms. Her biographer presents these recollections as moments of special concentration, highlighting McClintock's remarkable ability to be fully absorbed in learning and later in research: they provide a glimpse "of the characteristics that would be so important in defining her as a scientist."[11] We see here how McClintock's extraordinary "feeling for the organism" is framed as a disembodied affair. What marks her out as a successful scientist is her capacity to forget her "bodily me"—or, as she puts it, an ability to make sense of a living being despite the nuisance of having a body that appears to obscure rather than illuminate the clarity of her exceptional vision of a body's innermost life processes.

9

In the course of the first half of the twentieth century, biology and the field of classical Mendelian genetics saw a rapid transformation when the concept of the gene became more and more materially defined, culminating in the mid-twentieth century in Oswald Avery's identification of DNA as the bearer of genetic information and its reconceptualization as the "master molecule of life."[12] Mendel's fuzzy concept of a gene had now become a *molecular object*, a term used by James Watson, one of the three male laureates who won the Nobel Prize in 1962 for providing a molecular structure for DNA.[13] Noticeably, as Mark Lawler posted in *The Conservation* on April 24, 2018, these researchers' female collaborator, Rosalind Franklin, has never received the proper recognition for her scientific contribution to these groundbreaking publications.[14] Franklin (1920–1958), whose work on experimental visualizations of DNA molecules with X-ray diffraction photography was instrumental in discovering the double helix structure of deoxyribonucleic acid in the 1950s, never made it onto the list of Nobel Prize nominees in the 1960s, as the prize is not awarded posthumously.[15] Despite her untimely, early death, this fact may still raise questions about gender biases in scientific committees and also might suggest that a scientist's skilled and innovative mastery of new imaging technologies did not speak as much to the imagination of the Nobel Prize Committee as did the idea of a marvelous "meeting of minds" that was later promoted by Watson and Crick.[16] As we will see, this emphasis on ingenious minds was omnipresent and went hand in hand with a conspicuous disregard for the virtuoso experimental handiwork involved in writing the history of the life sciences in twentieth-century laboratories.

The first half of the twentieth century saw a growing interest in genetics and the mechanisms of inheritance among physicists, who brought, in the words of the famous theoretical physicist Max Delbrück (1906–1981), a "new intellectual approach to biology."[17] McClintock's work on maize genetics must be seen in the context of a larger shift in focus from multicellular plants and animals to quickly reproducing single-celled microorganisms and minuscule bacterial viruses—too small to see under a light microscope—as preferred objects for gene studies. A new biology was on the rise that aimed at bringing life phenomena back to the most basic mechanisms and to study these as much as possible in isolation from nature's diversity, complexity, and the time-consuming cultivation dependent on annual seasons. The bold reductionism that arrived in 1945 along with Delbrück at Cold Spring Harbor Laboratory, on Long Island, New York, "was steeped in a tradition that seeks understanding in simplicity rather than complexity."[18] For the "new biologists" this quest for simplicity entailed—"naturally," so to speak—a search for the smallest living organism that could provide the simplest possible system for the study of fundamental life processes.[19] In a lecture delivered in 1946, Delbrück introduced the rhetorical figure of an "imaginary physicist" equipped with the powerful vision needed to solve the secret of life with an irresistibly reductionist model: a bacterial virus that was hailed as the smallest and simplest organism for the study of life's basic mechanisms.[20] Delbrück envisioned this as follows: "We will do a few experiments at different temperatures, in different media, with different viruses, and we will know. Perhaps we may have to break into the bacteria at intermediate stages between infection and lysis. Anyhow, the experiments can only take a few hours each, so the whole problem cannot take long to solve."[21]

In the brilliant mind of the imaginary physicist, the laborious lab life of experimental biologists and biochemists is reduced to a task that can be solved within a matter of hours. Somewhat chastened by his own experimental experience, Delbrück curbed this boldly optimistic time estimate, acknowledging in the same lecture that it might take more time—and trained hands—to conduct the laboratory experiments.[22] But the main message of Delbrück's carefully crafted recollections remains unaltered: to meet the challenges of a new biology, what was needed most was a new generation of brilliant minds who commit themselves to understanding the basic mechanisms of life. In Delbrück's fast-paced vision, there is no room for manual know-how and skilled expertise acquired in years-long hands-on training at the laboratory bench or in fields of plants. What remains unvoiced in his narrative is any kind of material and technical mastery in working with organic and biochemical material realities. Delbrück's well-known disdain for biochemistry's messiness is apparent in his airy, disembodied vision of an imaginary physicist.[23] For the bright-minded new life scientists the practical and material challenges of daily experimental laboratory life and its demands for bodily know-how, technical skills, and material literacy are—ideally—of no major concern. For the theoretical physicist Delbrück, the new science of life was essentially a matter of minds, not hands. Yet, at the same time, McClintock attended to her maize crops at the Cold Spring Harbor Laboratory and immersed herself in plant experiments that ultimately would challenge the appealing simplicity of reductionist approaches to the study of (sub)cellular life processes.

"The mind's eye" became the hallmark of a rapidly evolving and highly successful "molecularization" of genetics and later of biology,

11

medicine, and life.[24] Keller's historical account illustrates how an intellectually biased reductionism goes hand in hand with a theoretical attitude vis-à-vis the pursuit of scientific knowledge that profoundly marginalizes forms of manual know-how and makes redundant a biologist's hard-won intimate acquaintance with living organisms and the material literacy that comes with years-long bench training. The preeminence of a Cartesian mind-body dualism in this discourse is not coincidental but was deeply ingrained in the rising new science of genetics: "The theory of the gene was quintessentially mechanistic and the methodology of genetics inherently quantitative; geneticists were entranced with the power of numbers."[25] It was partly in opposition to such a disembodied and bloodless understanding of life processes that McClintock and other cytologists came to define their own practice and approach in the mid-twentieth century.

In 1964, a decade after Crick and Watson's groundbreaking publication in *Nature* on the DNA molecule structure that so profoundly contributed to the transformation of twentieth-century biological sciences, giving rise to the new field of molecular genetics, the First Chromosome Conference was convened in Oxford.[26] Some thirty years after the publication of C. D. Darlington's seminal *Recent Advances in Cytology*, which was hailed by his peers for introducing the chemists' and physicists' reductionist and unifying attitude into the biological sciences, the British cytologist opened the conference with a noticeable cautionary note.[27] Darlington voiced a shared concern among cytologists about the limitations of the approach of the chemist, who reduces the chromosome to the model of a chemical structure, and that of the mathematical geneticist, who formalizes the genetic constituents of the chromosome in mechanical models.[28] Looking back on the rapid developments of the preceding ten years and their shift of focus toward biochemical formalizations and molecular model constructions, Darlington reminded his audience: "We must applaud the success achieved by our colleagues on the basis of these assumptions [the theoretically deduced properties of chromosomes and their genetic constituents]. But they see the chromosomes through the mind's eye. We, who believe we see actual chromosomes through the microscope, must explain what we have seen, and point out that it is not always what our friends expect. For us, neither the chemical code, nor the linkage map of the chromosome, nor the genes embodied in it, are enough."[29]

Darlington's speech expressed a critique of a "mind's eye" approach that "shifted attention away from the organism as a whole" and led to a deeply

12

problematic divide between the mechanistic and quantitative explanations of geneticists, on the one hand, and the qualitative science of cytologists, embryologists, and developmental biologists, on the other.[30] The latter, including McClintock, remained much more intimately engaged with biological materialities, and many shared the belief that the particular realities of bio matter ought to undergird a robust heredity theory and provide the material basis for the conceptual entities of more quantitative approaches.[31] Keller emphasizes that McClintock's work in cytogenetics required "physical, active participation with the material itself."[32] The recalcitrant nature of McClintock's experimental research on plant genetics in maize—a complex, higher organism demanding a particular kind of enduring and season-dependent attendance—apparently did not fit very well into the brisk pace of the "new intellectual approach to biology."[33]

Skilled Vision

I found that the more I worked with them [the chromosomes of *Neurospora*, a red mold on bread] the bigger and bigger [they] got, and when I really was working with them I wasn't outside, I was down there. I was part of the system. I was right down there with them, and everything got big. I even was able to see the internal part of the chromosomes—actually everything was there. It surprised me because I actually felt as if I were right down there and these were my friends.

Barbara McClintock, quoted in Evelyn Fox Keller, *A Feeling for the Organism*

The cell's innermost secrets don't reveal themselves to a bright mind in the blink of an eye. Seeing and making sense of what one sees on a microscopic and submicroscopic level is a laborious process. Skillful hands-on work with materials, living and nonliving, played a crucial part in McClintock's experimental exploration of life processes. If we listen closely to the oral accounts documented in her biography, we hear how the biologist gave voice to embodied experiences of doing her research.[34] In McClintock's experiential account, the scale of the chromosomes is reenvisioned in relation to her own body: the visual encounter is narrated as an embodied experience in which she can relate spatially and even socially to the invisibly and intangibly small subjects of inquiry that, in her own perception, "got big" and became "her friends."[35] Such a friendship cannot be built within mere hours but grows over time. Her biographer notes that in the process

13

of embodied and materially engaged knowledge-making practices, "the objects of her study become subjects in their own right."[36] And she calls attention to the fact that her "intimate knowledge" was "made possible by years of close association with the organism she studies."[37] Keller learns during their conversations that a "motivated observer develops faculties that a casual spectator may never be aware of," and that it takes years of intimate engagement to bring forth the "heightening powers of discernment" that McClintock was known for among her peers.[38]

Indeed, McClintock's mastery of observational tools and experimental techniques was legendary.[39] Among the peers who acknowledged McClintock's "skilled vision" was the biochemist Rollin Hotchkiss (1911–2004), a "brilliant, analytical mind" who gained fame for determining the genetic material in bacteria and is remembered today for his pioneering work on the isolation and characterization of new antibiotics from soil bacteria.[40] Hotchkiss remembers McClintock as an "expert microscopist" and calls to mind that she was admired for her "masterful control" of dye and staining techniques and her cunning use of biochemical tools for her cytological studies of chromosome structures and behavior.[41]

Many of the processes McClintock studied were too small to be observed with the microscope but could be inferred from related, observable features or studied with analytical tools from biochemistry. She recalled that when she started to work with George Beadle at Stanford University on the cytology of the mold *Neurospora*, she "couldn't see anything well with the light microscope," but, as Keller points out, she could draw on her many years of experience in experimental observation.[42] This enabled McClintock, as Beadle put it, to do "more to clean up the cytology of *Neurospora* than all other cytological geneticists had done in all previous time on all forms of mold."[43]

Her biographer coins the term *virtuoso technique* to describe the ways with which McClintock attained an intimate knowledge of the chromosomal behavior of the organisms she worked with.[44] Scattered through the biography, we find mention of McClintock's hands-on expertise: her improvements of techniques of staining plant cells, and her careful and meticulous preparations of samples.[45] "Everything," Keller writes, "depends on the care and ingenuity with which cells are fixed and stained on the slides"—which is done by hand, of course.[46] The knowing touch and gaze with which McClintock could assess the quality of the crops in the cornfields marked her out. Throughout her scientific career, she continued to engage in the "most laborious parts of her investigations herself, leaving none of the labor, however onerous or routine, to others"; such a hands-on attitude is not exceptional but expected in making a career in science. As Keller put it, "In this she did as almost all beginning scientists."[47]

A former student recalled how long it took her to learn the art of looking that McClintock practiced. Evelyn Witkin, then a student at Cold Spring Harbor and later the holder of the Barbara McClintock Chair at Rutgers University, described how rewarding it was to become apprenticed in the skilled vision of this cytogeneticist: "Just looking over her shoulder, looking at the spots, you could visualize what was going on—she made you see it." Under McClintock's supervision, Witkin learned to discern chromosomal movements and structural changes that McClintock studied intensively in her laboratory until Witkin herself "became competent at reading the patterns so well, that she, too, could 'actually see genes turning on and off at definite times.'"[48]

Keller also interviewed the influential cytogeneticist Marcus Rhoades (1903–1991), who recalled a conversation with McClintock in which he marveled at her capacity to "look at a cell under the microscope and see so much!" To this, McClintock replied, "Well, you know, when I look at a cell, I get down in that cell and look around."[49] Looking into the depth of the corn cells to understand how cellular and chromosomal processes on a (sub)microscopic scale relate to observable traits in the full-grown plants demands "due attentiveness."[50] McClintock's "special blend of observational and cognitive skills" and her "heightened powers of discernment" allowed her to get in touch with life on a scale that for many of us would be unimaginably small.[51] Although many of the indirect methods of observation she worked with later became common practice in life science labs for the experimental study of cellular and molecular processes, McClintock was in particular remembered for being the only one who could "learn quite so many of the cell's secrets simply by close observation."[52] On closer inspection, *simply* is perhaps not the right word to account for the skilled vision of this seasoned biologist who could make sense of the "wealth of new patterns of color and texture in the tissues of the maturing plants that could be seen with the naked eye" and through the microscope.[53] Based on observations through a "cytological window" of how the genome functioned, the groundbreaking findings did not come easily but required an "extensive training of the eye."[54] In due course, McClintock acquired an extraordinary material literacy and observational competency that did not simply spring

15

to the brilliant mind of a smart scientist but came with many years of diligent hands-on experience.

Knowledge Cultures

I would say that the best part of working here [at Cold Spring Harbor Labora-
tory] is being a small part of all the research that goes on here. Even though it's
a small part of that research, it's still probably the most important part because
you can't understand DNA in plants if you don't have any plants.

Tim Mulligan, "Uplands Farm"

Corn genetics is hard work.

Evelyn Fox Keller, *A Feeling for the Organism*

The experimental life sciences call for sensitive and sensible finger work as
well as bright and well-trained minds, but the hands-on skills that are involved in the life sciences' cultures of knowledge often go unnoticed. This
is quite surprising, given that life scientists undergo years-long hands-on
training at the bench. Through bench training they learn how what they do
with their hands relates to complex theoretical understandings and cognitive operations. This is obvious to experimental scientists, but the constitutive role of manual expertise in the experimentalization of life processes has
rarely been acknowledged as part of scientific achievements.

   In 1983 the Nobel Prize Committee praised McClintock for her experimental work "carried out with great ingenuity and intellectual stringency."[55]
The committee portrayed McClintock as a solitary genius "working completely
on her own" while "her observations received very little attention" from her
contemporaries.[56] Indeed, it took her a lifetime to have her academic peers
"realize the generality and significance of her findings."[57] The committee
lauded McClintock for "carrying out experiments of great sophistication"
and for her "immense perseverance and skill."[58] At the same time, the committee described her work at the award ceremony as especially "encouraging" because it "shows that great discoveries can still be made with simple
tools."[59] Similarly, the Nobel Prize organization claimed in its press release
that maize, the model organism to which McClintock dedicated almost her
entire career, has the great advantage that its individual chromosomes "are
easily studied" and mutations "easily observed."[60] McClintock's groundbreaking achievements were presented by the prize committee as the feat of
an extraordinary scientific mind, while her experimental handiwork with

16

plants and tools was made to appear surprisingly "simple" and "easy." Implicit in the committee's laudation is a dismissive gesture toward manual labor and the embodied practice of trained observation. In addition, conspicuously absent are any references to the biologist's extraordinarily skilled expertise in cultivating, handling, and examining the corn plants that were vital to her research on cytogenetics. The account, though, that McClintock imparted to her biographer sheds a different light on her work with maize: "Over and over again, she tells us one must have the time to look, the patience to 'hear what the material has to say to you,' the openness to 'let it come to you.'"[61] Keller's oral history of McClintock's extraordinary attentive capacity might suggest that McClintock had a particular gift or inborn talent. She quotes McClintock's own words: "Above all, one *must have* 'a feeling for the organism.'"[62] However, if we listen closely to McClintock's oral account, we get a glimpse of the ways in which McClintock *developed* such a capacity through on-the-spot observation and a caring engagement with the plants over long periods of time. Motivated by the drive to genuinely get to know another organism, the biologist wanted to understand

> how it grows, understand its parts, understand when something is going wrong with it. [An organism] isn't just a piece of plastic, it's something that is constantly being affected by the environment, constantly showing attributes or disabilities in its growth. You have to be aware of all that. . . . You need to know those plants well enough so that if anything changes . . . you [can] look at the plant and right away you know what this damage you see is from—something that scraped across it or something that bit it or something that the wind did.[63]

One needs to develop an awareness for every individual plant, she explains to her interviewer: "No two plants are alike. They are all different, and as a consequence, you have to know that difference."[64] The commitment this work demands is not only immensely time- and labor-intensive but also—in her words—deeply rewarding: "I start with the seedling, and I don't want to leave it. I don't feel I really know the story if I don't watch the plant all the way along. So I know every plant in the field. I know them intimately, and I find it a real pleasure to know them."[65]

Keller here reaches the intriguing conclusion that McClintock's "intimate knowledge, made possible by years of close association with the organism, is a prerequisite for her extraordinary perspicacity."[66] The observational skills that formed the basis of McClintock's scientific discoveries were honed through many years of committed and caring work with maize plants. In addition, her scientific peers acknowledged the "unique virtuosity" that she had developed in practice.[67] Rollins A. Emerson (1873–1947), who was instrumental in promoting the maize plant as a major research tool for genetic studies, described her as "the best trained and most able person in this country on the cytology of maize genetics."[68] Recently, these intimate aspects of her work attracted more attention in academia.[69]

The Nobel Prize Committee also omitted any mention of collaborative work with corn farmers that we can assume must have been part of McClintock's daily work. It is a widely held assumption that her work has long remained unnoticed because "her results were reported in publications that were not widely read, such as the annual report of the institute where she worked and in special newsletters exchanged by plant breeders working with maize."[70] Yet the fact that she published for a community of plant breeders also shows the relevance of expert communities for her own research that have mostly remained unacknowledged or ignored by academia and scientific institutions. McClintock not only supervised researchers but also closely engaged with farming staff and convinced the committee to hire a farm manager she had mentored. In her heyday at Cold Spring Harbor, McClintock used to have her own acres and was known to do everything herself. She continued to work at Uplands Farm, the agricultural field station that the Cold Spring Harbor Laboratory maintains for its plant research group, in the last years of her career.[71] Tim Mulligan, the farm manager, recalls how McClintock, who had persuaded the hiring committee to give him his first job at Uplands Farm, taught him how to drive the tractor to properly plow a maize field when he was managing his first crop in 1989:

> When you turn the soil over you create furrows and you are supposed to level them out in the field and I was not doing this correctly with my equipment. As a researcher doing crosses, you walk constantly, maybe there are twenty plants in a row and you have 200 rows to cross, so to make it easier the field has to be flat with no stones. So here I am turning this thing out that I guess was pretty terrible and she said, "You know you have to do a better job." She talked and told me how to drive the tractor and she gave me my marching orders. I turned out a much better field that next time. They were kidding me about this because she had short legs and she always liked her field to be flat and immaculate.[72]

The knowledge that Barbara McClintock harvested from her corn plants depended on her extraordinary capacity to link the highly abstract language of genetics to her hands-on work in the cornfields and in her lab. Another of McClintock's mentees, the biochemist and later Nobel laureate Elizabeth Blackburn (born in 1948), recalls her first encounter and provides a vivid impression of the intrinsically entangled practices of intellectual discourse and material engagement that characterized McClintock's research practice: "She just absolutely enchanted and enraptured me, because she immediately got deeply into a scientific discussion. She had all her ears of corn all around. She was showing us all the different stocks."[73] McClintock's "mind's eye" was profoundly shaped by manual engagement with the organisms she studied. Blackburn recalls a conversation they had on odd experimental results in 1977 in which McClintock encouraged her "to go with your intuition, really trust what you see."[74] Blackburn explains, "[When] I had told her about my unexpected findings with the rDNA end sequences, she urged me to trust my intuition about my scientific research results. This advice was surprising to me then, because intuitive thinking was not something that at the time I allowed myself to admit might be a valid aspect of being a biology researcher. I think her advice recognizes an important and sometimes overlooked aspect of the intellectual processes that underlie scientific research, and for me it had a liberating aspect to it."[75]

The intuition that McClintock refers to here had been cultivated over years of working intimately with maize both in and outside the lab. McClintock's acute perceptiveness did not derive from mental effort but instead was a certainty that grows from knowing by hand how plant genetics work. Her "seeing" is not the gaze of an uninvolved observer but the skilled vision of a caring practitioner who comes to know the object of study through intimate embodied and sensory engagement. To be in touch with plant life is a highly sensitive experience:

> Animals can walk around, but plants have to stay still to do the same things, with ingenious mechanisms. . . . Plants are extraordinary. For instance, . . . if you pinch a leaf of a plant you set off electric pulses. You can't touch a plant without setting off an electric pulse. . . . There is no question that plants have [all] kinds of sensitivities. They do a lot of responding to their environment. They can do almost anything you can think of. But just because they sit there, anybody walking down the road considers them just a plastic area to look at, [as if] they are not really alive.[76]

19

Hands-on expertise and care are of vital importance to experimental life sciences, and not only in plant cytogenetics. Conducting experiments with model organisms and cell cultures in today's molecular biology laboratories depends on highly skilled manual training and expertise in culturing and maintaining model organisms in the lab. Knowledgeable hands-on work is constitutive for these epistemic cultures. However, Keller's discussion of McClintock's research practices also remains indebted to a theory-laden, immaterial, and disembodied history of science.[77] References to McClintock's skillful ways of engaging with the materialities of her experimental daily practices are scarce and scattered throughout her biography, and a more substantial discussion of her "virtuoso technique" and hands-on engagements with organic materials and tools is mainly lacking. Keller's epistemological analyses of McClintock's research practice tend to quickly turn from experiential accounts to the immaterial realm of "qualitative and quantitative reasoning," and multisensory embodied practices are reduced to the domain of the visual.[78] Any forms of nonpropositional, embodied, or material ways of knowing are readily delegated to nonworldly realms of "mysticism."[79] The fine motoric and material practice of close microscopic inspection is described as requiring "an extensive training of the eye," but eventually, we need to turn inward toward essentially "internal visions" to understand how this outstanding scientist makes cytogenetic knowledge.[80] Also, here the mind's eye takes over when the embodied act of seeing becomes knowing: "For her the eyes of the body were the eyes of the mind."[81] In this description, Keller echoes the famous assertion "I become a transparent eye-ball; I am nothing; I see all" by the American transcendentalist Ralph Waldo Emerson.[82] Keller interprets McClintock's attentive absorption as a way to transcend her "bodily me": McClintock apparently found a way to overcome her wish to be freed of her body, and it is precisely this capacity that defines her as a successful scientist.[83]

Hence, the specialized and embodied acts of seeing of the trained cytologist and expert microscopist are turned into an essentially disembodied, mental activity or cognitive achievement: Keller portrays the Nobel Prize–winning biologist here, too, primarily as a knowing mind that eclipses the image of the hands-on experimenter who can only see "what others can't see" through an embodied practice of careful attention and a laborious process of skillful manual operations and preparations. In Keller's narrative the mental and the manual are hierarchically ordered: in the act of knowing, the mental takes primacy over the body, and the latter recedes into the background. The epistemic bias for the mind at the cost of the body has a long history that dates back to the ancient philosophies of Plato and Aristotle. Its importance for modern Western thinking was reaffirmed with René Descartes's influential account of the knowing subject (the Cartesian cogito) that has absolute epistemic agency and is categorically distinguished from objects and bodies. Interestingly, Keller makes an effort to conceptualize McClintock's fully absorbed self not in terms of a Cartesian "knowing mind" that is self-consciously affirmed through the act of reasoning. In her attempt to circumvent contestable subject-object and mind-body dichotomies, Keller interprets this "loss of self" not as an affirmation of a modern knowing I, but as the disappearance of a self-conscious subject in a "state of subjective fusion with the object of knowledge."[84] For Keller, a self-forgetting absorption and an extraordinary feeling for the organism are defining characteristics of Barbara McClintock's research experiences, which cannot be told in terms of a rational subject alone. However, despite Keller's attempt to push beyond a Cartesian mind-body dualism, we see that her descriptions remain indebted to a language and epistemic discourse that readily gravitates toward the knowing mind and easily loses sight of the manual and material.

The Disembodied Knower

The contested yet persistent image of knowers as "thinkers without hands" underpins the modern rationalistic and positivistic ideal of an uninvolved observer who can view natural processes under controlled conditions but without human intervention or manipulation. This ideal of noninterventionist yet controlled observation chimes conspicuously well with McClintock's desideratum of a transparent body that could not block her vision and would go unnoticed by others. The biologist's recollections conjure up a contested model of a disembodied observer, a knower with virtually no hands and an all-pervasive gaze who seeks knowledge through detached observation and pure contemplation. McClintock's wish to get rid of her body to become the scientific persona she wishes to be feeds into long-standing debates in arts, sciences, and philosophy on the interrelation between praxis and theory, doing and knowing, material practices and immaterial concepts, laboring bodies and knowing minds, dating back to the ancient philosophies of Plato and Aristotle. The disembodied knower model is today commonly linked with René Descartes's epistemology that postulates a knowing subject as an immaterial entity that is distinguished from the material worlds of bodies and objects. The so-called Cartesian mind-body distinction has decisively shaped our thinking in the Western world on how cognition works and how (scientific) knowledge making is conceptualized and theorized. It is also intimately linked to an ongoing cultural discourse about the role of the human body in knowledge formation.

In art historical and aesthetic discourses, the disembodied knower model has its counterpart in the image of a "painter with no hands." The paradox of a Renaissance master painter as a man of ideas with no need for laboriously trained hands, hands-on knowledge of paint making, or pens and brushes to materialize his ideas proves illuminating for a better understanding of the problematic model of a bodiless maker of knowledge in scientific discourse. Renaissance scholars have shown how artists promoted drawing (*disegno*) as a superior artistic form. The celebration of *disegno* "as a purely abstract and mental moment in creation" was motivated by the wish to ascertain the intellectual nature of artistic production and to shift attention away from artists' and artisans' manual labor.[85] This mental turn in art theory aimed to elevate the status of painting from a mechanical art (*techne*) to one of the liberal arts (*arti liberali*), while at the same time obscuring the fact that Renaissance artists gained their artistic mastery in years-long hands-on workshop training.[86] The conceptualization of *disegno* as a mental achievement contributed to the devaluation of manual labor in visual arts discourse. For instance, due to a persistent tendency to consider Renaissance painting as a matter of the intellect, in-depth inquiries into the role of the hand in creating artworks had long remained understudied.[87]

The long tradition of devaluating manual aspects of art making was particularly celebrated in the Enlightenment era by the German poet and art critic Gotthold Ephraim Lessing, who introduced the famous trope of a "Raphael with no hands" in his play *Emilia Galotti* (1772). In a dialogue, the court painter Conti laments how much gets lost between the eye and hand of an artist in the act of painting; he asks his prince rhetorically: "Or do you think, Prince, that Raphael would not have been the greatest genie in painting, if he would have had the bad fortune to be born without hands?"[88] The reference to the famous Italian Renaissance painter Raffaello Sanzio da Urbino (1483–1520) as an artistic genius—even without hands—suggests that we can make a hierarchical distinction between the work of the mind and that of the hand in the process of artistic creation. Lessing's Raphael-without-hands conveys the idea that great art is first and foremost a mental achievement; it is a product of the artist's mind, whereas the handicraft involved in making the work of art should not be considered as a genuine part of the creative process and aesthetic value of the artwork. This alleged opposition between artists' mental and manual work has been challenged in art theory and practice, and there exists a long tradition in art scholarship of debating the impossibility of categorical distinctions between manual and material investments in and the intellectual content of artworks. The persistent occlusion of manual work and material engagement as constitutive elements of artistic meaning has been met with much criticism, and scholars have emphasized instead that handmade artifacts manifest the unique results of intractable collaborations between minds and hands, at least for art production up to the nineteenth century.[89] Similarly, scholars have started to criticize a long tradition in history and philosophy of science that tends to frame the early modern period in Western Europe and the rise of modern experimental sciences in terms of "autonomous ideas and disembodied mentalities."[90] This tendency manifested itself in a historiographical tradition that construed the Scientific Revolution in seventeenth-century Europe as a primarily conceptual revolution to be written in terms of mental operations with comparatively little attention to bodily issues or the innovative material and manual ways of *making* scientific knowledge that this era witnessed.[91] In the last century, beginning in the 1990s, we have observed a shift in perspective in history and philosophy of science from realms of ideas to embodied and material worlds of knowledge makers. Historians have begun to concentrate on the actual "work of science" and stress that "all ways of knowing involve crafts" and bodies.[92]

The rhetoric of McClintock's dismissive body talk reflects a theoretically skewed and heavily biased understanding of knowledge production and is strangely at odds with laboratory ethnographies that reveal the conspicuously manual character of epistemic practices in molecular genetics laboratories.[93] McClintock's desire to become a disembodied observer was driven by the idealized model of a knower who attains knowledge by means of detached and unobtrusive observation alone, as if experimental observation and natural inquiries could be done without intervention or manipulation, and without any recourse to technical know-how, material literacy, or skilled handiwork. Such a stance has been strongly criticized since the early 2000s with growing attention to embodied practices in the preceding decades by philosophers, historians of science, and social scientists whose work has contributed significantly to the reconceptualization of "knowing not as a faculty of the human mind, but as an activity of the human body."[94] This shift in attention also fed into the rising field of embodied cognition (e-cog) studies in the cognitive sciences, which has undergone a paradigm shift toward an understanding of the mind as intrinsically

23

embodied, embedded, and extended.[95] In many ways my book is a contribution to this important development, but it can also be read as a critique of universalistic mind theories and a profoundly ahistorical attitude that characterizes much of the e-cog debates.

The Art of Culturing Bacterial Life in the Lab

He travelled around and collected thousands of soil samples from all over Japan which he used for culturing bacteria. In one of Ōmura's bacteria cultures, which was sent to William Campbell's laboratory, a whole new strain of Streptomyces was discovered—one that would change the world.

Hans Forssberg, "Award Ceremony Speech: The Nobel Prize in Physiology or Medicine 2015"

Cultivating a feeling for life science's organisms, living and nonliving, and developing the necessary material literacy and technological dexterity to conduct experiments in a life science lab demands not only bright brains but also meticulous and sensible handiwork, executed with care, precision, and ingenuity. This is reflected in the Nobel Committee's announcement of the Nobel Prize in Physiology or Medicine in 2015, some thirty years after McClintock received the Nobel Prize. Satoshi Ōmura, who shared the prize with William C. Campbell for their collaborative work in developing a new antiparasitic drug, received the highest scientific acknowledgment for his successes in culturing new soil bacteria strains in his microbiological laboratory. During the award ceremony, the committee praised the microbiologist for his "extraordinary skill in developing unique methods for large-scale culturing and characterization," thus acknowledging the expertise and sophisticated practical knowledge that are needed to successfully identify, isolate, and maintain new strains of bacterial cultures under lab conditions.[96]

Ōmura is known for a habit that illustrates the importance of the seemingly mundane, often unnoticed manual work that frequently underpins Nobel Prize–winning discoveries. In an interview Ōmura imparted that he always carries a small plastic bag in his wallet, ready to hand so he can collect soil samples wherever he goes.[97] Let's imagine how Ōmura arrives at a conference, then takes a walk or visits the golf course, where he squats down to inspect the soil under his feet, perhaps assessing the quality of the earth between his fingertips, then taking out his small plastic bag. Maybe he has a spoon with which he can scoop some soil into a sample bag without dirtying his hands. Perhaps he uses bags that he can just slip over his hand—like a glove—to take a handful of soil, turn the bag inside out, seal it, and walk on. From the soil samples he collected over a lifetime, Ōmura and his team were able to identify promising bacterial strains and managed to find the right laboratory conditions to harvest their products. His research laboratory succeeded in isolating and culturing *Streptomyces avermitilis*, the producing organism of avermectin, an organic substance with excellent antiparasitic activity, from which William Campbell's laboratory developed the antiparasitic drug ivermectin.[98]

Already in 1986, Ōmura had received the Hoechst-Roussel Award, a prestigious prize from the American Society for Microbiology. As a recipient of the award, he was invited to publish on his research philosophy, an invitation he accepted "with great pleasure" as it gave him the opportunity to present his own "viewpoint and ideas on research work."[99] In the resulting paper, "Philosophy of New Drug Discovery," Ōmura comments on the art of microbiological research that depends on "believing in the great capabilities of microorganisms" and "the ability to devise various conditions for the successful isolation and cultivation of microorganisms."[100] On the complexities of creating efficient "screening systems" to identify bacteria that produce promising bioactive substances, he explains: "Sometimes we have to spend as much as a year or two to devise a satisfactory one."[101] Although this kind of work depends on diligent execution of various repetitive, manual tasks, Ōmura emphasizes that "screening is not just routine work," and he stresses the importance of participating in daily screening work "to improve one's research capabilities."[102]

Microbiological benchwork calls for hands-on commitment to the creation and maintenance of a thriving microbiological material culture. It depends on fruitful cohabitation of investigated microorganism and investigating researcher in the research lab—what we could call a living experimental system that depends on years-long working experience, the right tools and technologies, *and* intimate hands-on knowledge of culturing bacterial strains.[103] When Ōmura heard that he had been awarded the Nobel Prize for Physiology or Medicine in 2015, he echoed the main tenets of his philosophy in his first reaction: "I have learned so much from microorganisms and I have depended on them, so I would much rather give the prize to microorganisms"; in another interview he added that he "merely borrowed the power of microorganisms."[104] Ōmura understands his research practice as an intrinsically collaborative endeavor depending on successful cooperation that comes with lifelong training in working with another organism and becoming attuned to the organism's needs. He seems to allude

25

to what we might call "a feeling for the microorganism" as a requisite skill for successful scientific work in a microbiological lab.

* * *

Experiments fail most of the times. Things usually prove to be much more difficult than expected, or just fail. But, sometimes they go surprisingly well, and once you've experienced that, you will never get afraid of failure, no matter how often you fail. That's the fun part of research. Let's give it another try, another shot, or another night.

Satoshi Ōmura, in Mitsuko Nishikawa, "Japan's Latest Nobel Laureate"

Successful experimental life science laboratories excel in hands-on knowledge and techniques that have been acquired by research groups over years of failures, as well as innumerable trial-and-error explorations of means, methods, and materials. Ōmura's comment makes clear that laboratories build up a material culture and accumulate practical know-how over years of experimental work that provide the fertile grounds for the much rarer experimental successes, scientific breakthroughs, and prize-winning discoveries. Why, then, is there a tendency to omit the hands-on making and cultivation of such fertile grounds from knowledge-making theories?

My fieldwork brought me into cellular and molecular genetics laboratories, where I interned with a research group working on the complex processes of DNA damage recognition and repair in the genome of yeast, one of the first higher organisms, after the maize plants studied by McClintock, in which so-called jumping genes were observed.[105] After my first months working as an intern in a lab with life scientists, I realized I had quite happily been ignoring my own body during many years of university education in the humanities. Body issues had been of no specific practical concern in art history and philosophy curricula, even though bodies and bodily practices were matters of ongoing concern, for example, in phenomenology and in the theoretical interpretation of installation and performance art. In practical classes in the life sciences, however, it became impossible to ignore my bodily self. To *do* life sciences, it was not sufficient only to memorize all the new facts from the latest edition of Alberts's *Molecular Biology of the Cell*.[106] I had to pay at least as much attention to training my hands to perform all kinds of unfamiliar benchwork like pipetting, plating cells on agar plates, and pouring agarose gels. Seemingly simple tasks demanded acute bodily awareness, such as donning one glove and remembering to use only my gloved hand when washing carcinogenic

26

chemicals from an agarose gel that I had used for DNA electrophoresis. I quickly became especially intrigued by the strict sterile working procedures that I had to adopt during a gene technology course and a tissue-culturing experiment.

Learning to work in a sterile manner led me to conceive of my body not only in an instrumental sense as something that I avail myself of to conduct scientific experimentation but also as something I had to develop a special awareness of in order to avoid contamination of my samples. On a very practical level, I learned here that paying attention to my body was an important part of learning to conduct scientific research and performing experiments successfully. For example, I could avoid interference with results by polluting bacterial samples with microbes from my hands only by becoming a trained observer of my physical engagement with the media and instruments at the bench. Bodily issues appeared to be of daily concern for life scientists. In my training lab, making phenomena and processes that take place on a cellular and molecular level visible and interpretable was in the first place an embodied activity that involved as much manual know-how as brain work to understand what I was actually doing at the bench. Indeed, I learned quickly that it takes many hours of hands-on lab training to develop the skilled vision of a life scientist.[107] I was surprised to see nothing of this reflected in McClintock's deprecatory body talk. During my fieldwork I experienced a profound lack of literacy when it came to talking or writing about the bodywork of life scientists. This conspicuous discursive and conceptual lack motivated me to write the present book about hands-on knowledge-making practices in the sciences of life on cellular and molecular levels.

A Multisited Ethnography and Historical Epistemology

This book is an ethnographic, philosophical, and historical exploration of bodily ways of knowing in the life sciences. Moreover, it investigates why hands-on practices have rarely been acknowledged as part of scientific knowledge making in modern epistemology since the rise of the experimental sciences in early modern Europe. The study contributes to a growing body of scholarship that has started to laboriously develop—often using ethnographic methods—more sophisticated, mindful, and attentive ways of talking about science as a material and embodied practice. However, for better descriptions and conceptualizations of how life science is *done in practice*, we also need to understand *why* the exclusion of embodied aspects became so pervasive in Western thinking about scientific knowledge making in the first place. What are the philosophical underpinnings of this profound uneasiness or awkwardness in expressing and conceptualizing manual work in life science laboratory research? This is not an analytical problem that can be solved through reasoning; it demands a historical understanding of the conditions under which a particular philosophical attitude or project could arise and flourish. The conspicuous absence of hands-on notions in Western modern epistemologies has traditionally been explained by gesturing toward the intellectual legacy of the French seventeenth-century philosopher René Descartes, who has remained a celebrated figure in curricula of college programs and philosophy departments as the founding father of modern epistemology and rationalist philosophy. In modern philosophy the establishment of a separate realm for the human mind as the primary site of knowledge production that is independent of the body and the material world has traditionally been associated with a philosophical attitude that is grounded in a categorical distinction between mind and body. The Cartesian mind-body dualism posits a dichotomy between the knowing subject and the knowledgeable object that has profoundly shaped Western thinking. This Cartesian legacy has been as pervasive as it has been criticized in Western philosophy. Many scholars before me have, for example, convincingly shown that the conceptual erasure of the embodied experimenter from experimental scenes of knowledge is a theoretical desideratum that is impossible to achieve in practice.[108] However, in history and philosophy of science the recent shift in attention from ideas and theories to embodied and material practices of the sciences most commonly goes hand in hand with a call for anti-Cartesian approaches.

On a closer look, however, I found that anti-Cartesian approaches remain deeply flawed and that anti-Cartesian frameworks are actually not very helpful to describe the constitutive role of the manual in experimental life sciences; instead, they help to obscure the complexities and idiosyncrasies of life sciences hands-on practices. How, then, can we gain a better understanding of the embodied dimensions of experimental life science research? In this book I show that this requires a radically historicized understanding of the Cartesian epistemological project. Drawing on historical sources, I portray Descartes as a fervent anatomical experimenter and hands-on practitioner who cannot be reduced to the figure of a thinking meditator and rational epistemologist. The mind-body dualism is not the product of a philosophizing mind but took shape in the hands of an experimenting anatomist. Hence, the starting point for this epistemological discussion is not the "first philosophy" that Descartes sets forth in his *Meditations* but *the making of a first philosophy* in the context of Descartes's natural philosophical hands-on explorations. In short, I argue that we need to understand Descartes's epistemological project from *within* his own experimental practice in the early modern manual art of anatomy.

At the core of this study is a radical rereading of the famous "wax argument" in Descartes's *Meditations on First Philosophy* (*Meditationes de prima philosophia*; first Latin edition, Paris, 1641; second Latin edition, Amsterdam, 1642).[109] The Latin work was soon translated into the vernacular, and the first French edition, *Les méditations métaphysiques de René Des-Cartes touchant la première philosophie*, authorized by Descartes, was published during his lifetime, in 1647. My rereading of the French text in the context of other historical sources reveals a yet unacknowledged double movement in Descartes's doing and thinking: I trace how the natural philosopher and anatomical experimenter departs from hands-on embodied practices of knowledge making that are only in a subsequent philosophical operation obscured. I argue in this book that the rise of an epistemological attitude that deliberately obliterates modern epistemologies' grounding in hands-on experiences can, in fact, not be properly understood from the perspective of the thinking philosopher alone but needs to be understood in the context of Descartes's *manual* experimental practice. In chapters 1 and 2 of this book, I provide close readings of passages from the Second Meditation and of Descartes's correspondence on his anatomical experimentation in the context of the rise of the experimental sciences and recent scholarship on Descartes's involvement in the "new sciences." My analysis subverts the still pervasive reception of the early modern philosopher as a thinker without hands and instead places hands-on experiences and aesthetic reflection at the heart of Descartes's modern epistemological project. In chapters 3 to 5, I set forth a theory of knowledge making grounded in historical *and* ethnographic observations, for which I combined participatory hands-on observations in contemporary molecular genetics labs and microbiology cleanrooms with a *longue durée* philosophical account starting in the seventeenth century. Building on pioneering work of laboratory ethnographers, beginning at the end of the 1970s, my book demonstrates why we are in need of more embodied approaches to theorize knowledge-making practices in the life sciences. Yet, I disagree with pioneers such as Bruno Latour and Annemarie Mol, who declared that epistemology is a dead discipline. By contrast, the aim of this book is to revive epistemology

29

from within Descartes's philosophical undertakings in the role of meditator and experimenter as an inherently historical, aesthetic, and at the same
time pluralistic and embodied undertaking.

Central to this book is the notion of *hands-on* with which I draw together an argument that threads between seventeenth-century natural philosophy and experimental practices, laboratory practices in the present-day
life sciences, and contemporary artistic practices. I interweave observations taken from fieldwork with bio artists whose work was instrumental
for my reconceptualization of hands-on notions in life science research.
Throughout the book, these three heuristic methods—philosophical and
historical analysis, ethnographic accounts of lab work and cleanroom conditions, and bio art experiences and interpretations—are combined into a
multisited historioethnographic study.[110] *Multisited*, or *multilocale*, refers
here to studies across diverse historical and contemporary sites of inquiry,
drawing on historical sources and experiential accounts of life sciences and
bio art practices. My approach embraces the famous adage by the microbiologist Ludwik Fleck (1896–1961) "no epistemology without history" and
transforms it into the motto of this book: to reflect on knowledge making
in the life sciences, we need history *and* ethnography.

The Structure of This Book

In chapters 1 and 2 of this book, I describe shifts toward experimentation
in seventeenth-century natural philosophy against the background of my
participatory observations in life science laboratories. My multisited ethnographic approach takes me from the bench in an instruction lab, in which
students are trained to study life on the molecular level, to the specialized
contexts of experience where early modern natural philosophers investigate
phenomena that lie beyond the limits of human perception. I discuss the
work of exemplary experimenters, such as the innovative networker Marin
Mersenne (1588–1648), and then turn to the French philosopher and anatomist René Descartes, who engaged actively in the emerging experimental
scene in seventeenth-century Amsterdam. At the heart of these chapters
is a radical rereading of a famous epistemological argument, set forth by
Descartes in his *Meditations*. I provide a reading of Descartes's meditation
on a piece of wax from a hands-on perspective. Though Descartes plays
a prominent role in these chapters, my interpretation is not restricted to
Cartesian exegesis. Rather, I take Descartes and Mersenne as case studies
that make palpable how shifts toward experimentation and manipulation

30

in early modern experimental sciences brought about an epistemological dilemma.

Chapter 3 is dedicated to methodological reflections that reframe my historical inquiry as a philosophical project. Drawing on the work of the German philosopher Edmund Husserl (1859–1938), I first describe how the philosophical method of *epoché* made it possible to reread the famous wax passage in Descartes's Second Meditation in the context of his anatomical experiments. Second, I argue that we can understand the elimination of hands-on notions from the Cartesian epistemological project as the result of a philosophical operation that can be described as a form of substitution with the Husserlian concept of *Unterschiebung*.[111] The idea of a hands-on perspective as a point of departure for epistemological considerations is further elaborated with a phenomenological description of my own bodily experience. Drawing on recent interpretations of Husserl's and Maurice Merleau-Ponty's (1908–1961) phenomenological body, this chapter explicates how I experience my body both as mine and as a foreign or strange thing.[112] The chapter prepares the ground for further reflections on the resistance of experimenters' bodies to become transparent in a form of thinking that wants to ignore the persistent presence of bodies.[113]

In chapter 4, I discuss what the relevance of these findings is for today's discussion of experimental practices in the (life) sciences in the context of the practical and material turn in science studies. I critically explore how the authors of pioneering ethnographies of life science laboratories laid the foundations for thinking about *science as practice* but framed their participative bench approach as an intrinsically anti-Cartesian take on scientific knowledge formation and epistemology. My analysis of Bruno Latour and Steve Woolgar's *Laboratory Life* (first published in 1979) and Karin Knorr-Cetina's *Epistemic Cultures* (1999) demonstrates the shortcomings of anti-Cartesian frameworks in explicating the embodied aspects of life science laboratory work. In chapter 5, I turn to Don Ihde's and Shaun Gallagher's influential studies of body-instrument-relations and incorporation processes and argue that these embodiment philosophies are deficient in accounting for the peculiar hands-on/hands-off dynamic of sterile regimes practiced in today's technoscientific spaces.[114]

The epilogue is devoted to an exploration of contamination-controlled spaces that play a key role in life science research, such as isolator technologies and microbiological cleanrooms. These technoscientific phenomena give us insight into the conceptual complexities of embodied benchwork. In these final reflections, I introduce the concept of cleanroom aesthetics for which I draw on historical sources, in situ observations, and interviews with laboratory personnel. My "snapshot story" of a cleanroom visit provides a description, or perhaps better, a "practiography," of hands-on experiences and bodily practices in microbiological cleanroom facilities.[115] From this experiential grounding, I reassess the Cartesian legacy and argue that epistemology is not dead but needs to be revived and rethought as a historical, contextual, and pluralistic endeavor from the embodied standpoint of a hands-on practitioner—*with* and not *against* Descartes. This book is an interdisciplinary contribution to scholarship in historical epistemology, science studies, philosophy of embodied practices, and anthropological investigations into life sciences, biotechnologies, and artistic explorations of technoscientific life-forms.[116] It provides the philosophical and historical groundwork to study life science laboratories as idiosyncratic sites of embodied knowledge production processes.

Knowing by *Experience*

A Crisis in Perception: The New Sciences in Seventeenth-Century Europe

The emergence of modern experimental science is generally dated back to what has become known as the Scientific Revolution of the seventeenth century.[1] The rise of a "new science" of mechanics in that century posed a challenge to the Aristotelian tradition that dominated natural philosophers' inquiry into nature.[2] Inventors of and adherents to this new discipline, among them Galileo Galilei (1564–1642) and René Descartes (1596–1650), broke with central assumptions of Aristotelian epistemology when they based their inquiry into nature on mathematics and experiment. While such a schematic opposition between the Aristotelian tradition and the new science will be helpful for the following analysis, it should be noted that recent scholarship has provided a much more nuanced view of this period as one that marks a transition from Aristotelianism and Scholastic conceptions of science (*scientia*) toward natural philosophical inquiries for which experiment-based observations became more and more important. Simply opposing an "old Aristotelian empiricism" to the "new experimental sciences" obscures the varied ways in which complex and diverse ancient and medieval forms of knowledge, comprising practical as well as theoretical components such as geodetic surveys, the art of navigation, pharmacology, and surgery—to name but a few—were integrated into premodern natural inquiries.[3] Yet, we still can ask relevant questions about this period that saw a devaluation of Aristotelian empiricism and the rise of the new sciences in early modern Europe. What kind of epistemological dilemmas did the natural philosophers of the early seventeenth century face? How did new experimental methods at that time affect the thinking about sensual perception? What kind of bodily skills and spatial conditions gave shape to knowledge-making practices that increasingly relied on experimentation?

Space and place play a decisive role in understanding how experimental science introduced an empiricism that could no longer be reconciled with an Aristotelian sense-based epistemology. Aristotelian empiricism could be practiced by anyone anywhere. For natural philosophers in the Aristotelian tradition, it was most important to experience the events as they would normally happen in nature; thus, in principle, empirical observations did not depend on specialized observational tools or the proper operation of an apparatus.[4] An experimental setup, on the other hand, sets a "scene of knowledge" that is deliberately put into place to enable scientific observations.[5] The experiment, as a methodological device, thus redefined empirical observation as intrinsically dependent on particular spaces and places.[6]

We can observe a desire in the experimental work of natural philosophers, now often referred to as new scientists, to exclude disturbances and contingency as operating factors, so that valid observations and secure knowledge may be acquired in artificial spaces freed of contaminating influences.[7] With the emergence of artificially designed scenes of knowledge, empiricism lost its grounding in Aristotelian sense-based epistemology. The move from observing occurrences in the course of nature to observing phenomena resulting from experimental setups in specially constructed spaces brought with it a "crisis of perception."[8] Experiment-based observations challenged the new scientists to redefine the role of sensual perception in the production of knowledge. That being so, aesthetic issues became increasingly problematic

34

for an experimental epistemology. I use the term *aesthetics* here in the broad sense of a theory of perception. The term is etymologically derived from the Greek *aisthesthai* (perceive) and *aisthētikos* (perceptible by the senses). The term *epistemology* stems from the Greek word *epistēmē* (knowledge) and *epistasthai* (to know; to know how to do). The active and practical senses of the Greek verbal forms—perceiving and knowing how to do—are especially important for the following discussion of the ways in which aesthetics and epistemology come to interrelate in experimental empiricism.

The epistemological debate between the old and new empiricism and the role of experiment revolved around knowing how to *do* knowledge and theorizing about this *doing*. The epistemological problem can be described as a dilemma: practitioners of the new experimental empiricism adhered to the importance of observation; that is, they assigned the senses a key role in the practice of producing new knowledge, but at the same time sensual perception became newly problematic when observation became dependent on special facilities, preparations, and scientific instruments. The development of experimental science into a technology-dependent practice that shaped and was shaped by the invention of its scientific apparatus coincided with a growing disdain for ordinary perception. The primacy of the senses that marked out Aristotelian empiricism became problematic when new scientists started to question the epistemological relevance and validity of common observations. In the words of Francis Bacon (1561–1626): "For the subtlety of experiments is far greater than that of the sense itself, even when assisted by exquisite instruments; such experiments, I mean, as are skillfully and artificially devised for the express purpose of determining the point in question. To the immediate and proper perception of the sense therefore I do not give much weight."[9] Though the crisis in perception did not lead to a disqualification of sensual perception on the whole, it did pose a challenge to sense-based Aristotelian epistemology that relied on ordinary perceptions of everyday occurrences.[10]

Of course, Aristotelian epistemology itself must be understood as a response to the contested status of perception in ancient Greek philosophy. The mistrust of abstraction and idealization inherent in Aristotelian natural philosophy is a critical response to Platonic theories of perception and knowledge. Yet, the universalism of the new experimental empiricism is not a Platonic abstraction and is at odds with the Aristotelian tradition of sensory observation for an entirely different reason. It does not share the ontological distrust of sensual perception inherent in Platonic theories. Rather, it follows the Aristotelian tradition of placing epistemological primacy on

35

empirical evidence while, at the same time, fundamentally redefining the meaning of empirical observation. For Aristotelian natural philosophers the artificial nature and special environments of experiments could not qualify as suitable sources of knowledge about nature.[11] For the experimental empiricists, on the other hand, evidential observations of natural processes could only be established within the artificially designed contexts that characterized the experimental approach.[12] An example from the late sixteenth century illustrates how the term *experiment*, which commonly referred to everyday experiences in the Scholastic tradition, was redefined: Galileo deliberately uses *experiment* for situations that are brought about by deliberate skill and artifice; with this term he distinguishes contrived occurrences from naturally arising occurrences.[13] Experiment becomes intricately intertwined with a form of perception that is dependent on particular processes of aestheticization: for example, Galileo's experiments on the fall of bodies along an inclined plane involved careful preparations, such as extensive polishing of the bronze balls and careful smoothing of the parchment that lined the slope down which the balls rolled. Galileo's description of this experiment from 1638 reads as follows:

> In a wooden beam or rafter about twelve braccia long, half a braccio wide, and three inches thick, a channel was rabbeted in along the narrowest dimension, a little over an inch wide and made very straight; so that this would be clean and smooth, there was glued within it a piece of vellum, as much smoothed and cleaned as possible. In this there was made to descend a very hard bronze ball, well rounded and polished, the beam having tilted by elevating one end of it above the horizontal plane from one to two braccia, at will. As I said, the ball was allowed to descend along [*per*] the said groove, and we noted (in the manner I shall presently tell you) the time that it consumed in running all the way, repeating the same process many times.[14]

Galileo gives much attention to the ways in which he modified the "natural conditions" and reports how much care was given to the choice of certain materials and how they were manipulated (e.g., polished and smoothed) to produce a certain effect. The strategies of the new science diverged in a fundamental way from the Aristotelian tradition in alienating its objects of inquiry from their natural surroundings and in making phenomena visible and measurable by turning "nature into an aesthetic and experimentally reproducible artifact."[15]

Strategies of aestheticization gain importance in a double sense, given the apparent imperceptibility of the new scientists' objects of interest. Galileo's phenomena, for example, were "all too often not apparent to the untrained eye."[16] The requirement of a trained eye means that mere looking cannot qualify any longer as "scientific observation," and our sensual apparatus becomes subject to the demand for improvement. Whereas the old empiricism relates experience to the commonplace, the new empiricism is about a purposeful sophistication of the means of observation and observable phenomena.[17] With experiment arises the need to train our senses specifically, to develop strategies to sensitize them to particular appearances—to perceive, for example, the slightest changes in a vibrating string and the faintest tones resulting from its movement when investigating physical properties of sound and movement of air.

With the rise of modern science as an experimental research practice, sensual perception became differentiated into ordinary perception of everyday occurrences and an aestheticized mode of perception that became invested with epistemological significance at the expense of the former. Experimental empiricism redefined observation as an act dependent on the performance of an artificially contrived occurrence in a specific location for which an appropriate training of the investigator became indispensable. Faced with the problem of investigating phenomena that lie beyond the realm of ordinary perception, natural philosophers started to tackle this problem systematically with experiments. They turned the imperceptible into objects of inquiry that became measurable with artificial means and methods.

Making the Imperceptible Observable: Marin Mersenne's Experimental Study of Inaudible Tones

With the new experimental empiricism, perception became intricately interwoven with processes of making. The new scientists did not simply have impressions of phenomena; what they perceived was the outcome of a laborious process that involved the work of minds and hands, a skilled use of instruments, and a deliberate training of the senses. The French Franciscan monk Marin Mersenne (1588–1648) was influential in the development of seventeenth-century natural philosophy. As a "propagandist of the new experimental culture," he played a vital role in the rise of new science by sustaining a wide network of contacts with other innovative investigators, among whom were Galileo and Descartes.[18] Mersenne, himself a member of the Roman Catholic religious order of the Minim Friars, practiced a decidedly pragmatic approach that facilitated the support of scientific endeavors independent of confessional or political interests—he worked with Catholics, Protestants, and Jews on the Continent and on the other side of the English Channel.[19] This attitude is manifest in his comprehensive work *The Truth of the Sciences against the Sceptics* (1625), one of the earliest published textbooks that claimed to contain a list of all one could have known in Mersenne's time about mathematics and inquiries into natural phenomena.[20] Mersenne's pioneering work in mechanist natural philosophy and areas such as acoustics and musical theory is conspicuously anti-Aristotelian and anti-Scholastic in approach. Here I will take a closer look at his inquiry into vibrating musical strings, published in 1636 as *Harmonie universelle contenant la théorie et la pratique de la musique* (Universal harmony comprising the theory and practice of music), which exemplifies aesthetic issues that lay at the core of the new scientists' epistemological dilemma.[21]

In this case, Mersenne's object of inquiry is a phenomenon that manifests itself outside the range of our ordinary perception. He begins to study the inaudible and invisible, low- and high-frequency vibrations of musical strings. He tries to measure the upper and lower limits of audible frequency to bring an imperceptible realm under empirical observation. Mersenne builds his theory on calculations by himself and others, but his main contribution was that he succeeded in grounding a new theory of consonance not in numbers alone but based on empirical observation.[22] The main question of his acoustic inquiry was whether a musical tone can be described as a physical entity and, if so, how such a tone comes into existence.[23] Mersenne's objective was to push empirical observation with experiment into the unknown territory that lies beyond our ordinary perception. He comments in his *Universal Harmony* on the limits of sensory perception, raising the question of whether a tone exists that we cannot perceive: "It is therefore necessary to know, before we move on, if the tone, which is the subject or object of music and the sense of hearing, has a real existence, and what that is: because there are many who believe that the tone is nothing if it is not heard [by anyone], and that it is but a simple impression of the air that one cannot call a tone if there is no ear that hears it and that distinguishes it from other things."[24]

Mersenne's acoustic studies are a good example of the distinct interest among seventeenth-century natural philosophers in carrying out experimental observations on phenomena that in the *communis opinio* did not exist.[25] He gave very detailed and precise descriptions of his observations, though historical studies give little insight into the material and technical details of the experimental setups that Mersenne used to establish his theory of sound.[26] Nonetheless, his experiments illustrate how imperceptible phenomena became subject to experimental inquiry and how such an approach went hand in hand with an aestheticization of the objects of inquiry and the need for artificially constructed observation sites. We can assume that Mersenne's experiments must have been conducted at a site protected from interruptions by sound or uncontrollable airflows in order to observe phenomena that happen at and beyond the limits of our aural and visual perception. Moreover, Mersenne worked extensively with a pendulum as a measuring device, a tool that not only requires a steady hand but also implies a site of experimentation that is sheltered from unwanted disturbances. A divergence from the Aristotelian tradition becomes apparent in his mistrust of mere looking or hearing: in his experiments, he relied on mathematical calculations and had to draw on mathematical evidence as a corrective in those cases where the phenomena he describes evade perception.[27] His accounts of the difficulties he encountered testify to his precise attempts to explore the limits of experimental observation: "But the experiments are so difficult that with a chord of less than 1000 feet one cannot be sure; and one can never be so certain of the sites to which [the chord] returns every turn, that one cannot doubt whether [the chord] has not gone further, and whether [the chord] has finished its coming and goings exactly at the points that one has marked."[28]

Mersenne's experiments were dedicated to making phenomena perceivable and to revealing underlying universal laws that we are unable to grasp with our senses. His experimental practice has been understood as a struggle between empiricist and rationalist attitudes.[29] The loss of self-evidence that the perceptual or sensory experience undergoes when pushing empiricism into the realms of the imperceptible becomes apparent and has to be supplemented by rational calculations "in such a way that it is always necessary that the reason supplements something in the experiences that alone cannot provide the guidelines [or principles] for the sciences, which desire a perfect exactness that the senses cannot notice."[30] The reader becomes a witness to the experimenter's efforts to contrive an observable phenomenon. Mersenne included in his study reports of a series of operations he performed to reconstruct in his text the concrete moment of an experimental experience: the text has the function of creating a sense of immediacy, of documenting

39

a perceptual act; via the textual medium the experimenter shares with his readers an observation that has been made under exclusive and artificial conditions; the report has to make the experimental moment palpable for the readers, and it lets them imagine the physical exertion involved in making this observation.[31] This is also apparent in the headings Mersenne used to structure his experimental reports; it has been noted that they contain a conspicuous number of verbs that indicate doings or practical activities.[32]

This shift of attention from "natural phenomena" to artificially contrived phenomena that yield knowledge about nature's imperceptible secrets became an important trait of modern sciences. It has later been described as a "technoscientific productivity" inherent in the experimental sciences and reconceptualized as a tendency for the modern sciences to "produce," "engender," "enact," and "enhance" their own objects of inquiry.[33] I return to this problem in context of the experimental life sciences in chapters 3 and 4.

The preceding examples illustrate how scientific observation became inherently dependent on hands-on practices of constructing particular spatial settings and experimental setups, skillful handling, and the making and proper handling of instruments that either improve or alter our sensual apparatus. We can observe how experimental sciences now and then involve a practice of staging, a deliberate invention and application of aesthetic strategies.[34] Alienating objects of inquiry from their natural surroundings and recontextualizing them in experimental workshops, seventeenth-century laboratory work can best be understood as a practice of staging and shaping new spaces of experience. The concepts of alienation and engendering can be understood gradually, meaning that we can observe stronger and weaker forms of decontextualization and technoscientific productivity in experimental contexts.[35] The phase of transition away from an Aristotelian natural philosophy was marked by a crisis of perception and the call for an epistemology that could account for the necessary entanglement of action and reflection inherent in experimental knowledge practices.

Practices of generating new knowledge called for new vocabularies for sensation and new training in perception. With the experiment as a new methodological device, inquiries into nature necessarily turned perception into a problem that had to be addressed not only in theory but also in practice. The new sciences have not simply been pitted against "old" Aristotelian empiricism; they also are readily associated with the age of the Enlightenment and the rise of rationalist thought. We should note here that these kind of inquiries into nature, while driven by *ratio* (reason), inherently depended on skillful instrument makers, proficient practitioners, and knowledgeable hands.[36]

Engendering: An Epistemological Dilemma

Seventeenth-century experimental philosophy has been characterized by its "need to confine experiments to places very dissimilar from the natural world."[37] What an experimenter needed to do was to *make* objects of inquiry perceivable, and to this end aesthetic strategies had to be devised that would produce the phenomena to be investigated.[38] Natural philosophers working with experiments had to solve all sorts of aesthetic problems to create adequate conditions for the production of new knowledge. Steven Shapin and Simon Schaffer's seminal study of Robert Boyle's air pump experiments shows, for example, how Boyle (1627–1691) had to solve the problem of how to make concepts such as empty space visible and measurable in order to convince his audience of its existence. In doing so, Boyle turned to dramatic effect, demonstrating the functioning of his vacuum pump by using it to kill a small bird and so turning an invisible concept into an extraordinary spectacle.[39]

The concept of "engendering" aptly articulates the problem of turning imperceptible processes into objects of experimental-empirical inquiry.[40] Engendering processes in scientific practices make palpable the tension between experimental science's claim to universality and the fact that this is based on the performance of laborious procedures that can only be made and maintained in specially designed spaces; they are not to be found in "natural surroundings" and are not the subject of ordinary perception in everyday occurrences. Wolfgang Lefèvre's claim that our modern-day laboratory sciences tend to bring forth their own objects of inquiry instead of providing a context for researching phenomena that we encounter "out there" in nature can be formulated even more strongly in light of our historical considerations.

Engendering appears to be not a recent trend but a fundamental trait of experimental empiricism as it emerged in seventeenth-century Europe. Most important, this history shows us that engendering is intricately related to processes of aestheticization. If knowledge could no longer be considered to be the result of "immediate" ordinary sensory perception, the new scientists needed "to create a space of knowledge and demonstration in which science can anchor its authority."[41] Experimenters thus responded to the epistemological dilemma with the constitution of spaces of knowledge that allow for specific forms of experience.[42] The construction of such spaces posed an essentially *aesthetic* problem—understanding aesthetics here not in the constrained sense of a philosophy of art but as "a theory of culturally and historically embedded sensation and perception."[43]

Lefèvre distinguishes between "a strong and a weak sense in which research objects are engendered by the material means used in the research process."[44] Electricity and particle physics exemplify the strong sense, because the objects of these research fields are quite literally produced by the laboratory machines. However, life scientists also engender their objects of inquiry, modifying existing processes and putting them into the service of specific experimental systems. A good example of the strong sense of engendering in the life sciences is synthetic biology and bioengineering, an area of research that has attracted the attention of the public with headlines on "artificial cells," "tailor made bacteria," and "bio bricks."[45] Synthetic biology aims at designing and constructing new biological systems not found in nature, and to this end it applies engineering principles to biology.[46] In the last years, this line of research has raised not only questions about the possibility of new (engineered) life-forms but also existential and ethical questions about current definitions and conceptions of what life is.

The crisis of perception that marks the birth of what we today understand as modern scientific inquiry into nature has not diminished with time—on the contrary, we could say that it has become even more pressing in light of technoscientific developments in the twentieth and twenty-first century. The experimentalization of life and the molecularization of biology and the biomedical sciences, with their focus on life processes on subcellular levels, have pushed a significant part of research on living processes into imperceptible realms.[47] Most of the technologies used in molecular biology are directed at visualizing processes that take place in the depths of the living cell at scales far beyond our visual faculty. Contemporary laboratory practice is characterized by the prevalent use of precision micro measuring devices, including quotidian laboratory instruments like handheld pipettes and Eppendorf tubes, with which barely visible amounts of transparent liquids can be processed.[48]

The concept of engendering allows us to grasp the extent to which living processes and experimental technologies, or technologies for the observation of these processes, are entangled in contemporary biomedical and life sciences laboratories. The capacity to perform enormous and rapid replications of DNA sequences with polymerase chain reactions and the phenomenon of cellular life in vitro, which has developed from a curiosity into a ubiquitous laboratory technology over the course of the last hundred years, have transformed the experimental domain of life.[49] Not only do the processes that we generally encounter in today's life science laboratories lie beyond the range of what we can perceive with our "ordinary sensory apparatus" but the spatial settings of the laboratory have become increasingly sophisticated and complex. Phenomena have not just been alienated from their natural surroundings; they have become entirely dependent on these new spatial settings, not least because of their often total reliance on contamination control technologies. In the epilogue, I offer some reflections on the aesthetics of contamination-controlled spaces with a description of a cleanroom visit in a laboratory facility of the Department of Clinical Pharmacy and Toxicology at Leiden University Medical Center.

Experimental Empiricism versus Rationalist Epistemology in Seventeenth-Century Europe

The European scientific landscape of the seventeenth and eighteenth centuries is traditionally reconstructed in terms of a schematic opposition between English empiricism and French rationalism.[50] The American historian and sociologist Steven Shapin and the British historian Simon Schaffer saw the new experimental attitude toward knowledge production exemplified in the work of Robert Boyle. This English natural philosopher is closely associated with the institutionalization of experimental philosophy as a communal endeavor in the Royal Society of London for Improving Natural Knowledge, founded in 1660. Experimental philosophers affiliated with this institution understood the production of natural philosophical knowledge as a laborious process requiring attentive observation, well-devised experiments, and conscientious reporting of contrived events. They distinguished experiment-based forms of knowledge from traditional forms of *scientia* as they had been taught at European universities since the Middle Ages, and they considered the new science as superior to medieval knowledge systems that depended on textual exegesis. English natural philosophers have predominantly been linked to a practical and materially oriented approach toward knowledge making. With the establishment of the Royal Society, the experimental approach became institutionalized in England, and in modern historiography it became predominantly presented as a British accomplishment. Shapin has been influential in forging the *communis opinio* that the new science won a hard battle against a great rival model in Europe that was decisively shaped by the French rationalism of Cartesian origin.[51] Not surprisingly, perhaps, the English term *new science* came to be widely accepted

43

to denote the new experimental approach in seventeenth-century natural philosophy. René Descartes's epistemological masterpiece *Meditations on First Philosophy* (*Meditationes de prima philosophia*, 1641; here quoted from the first French authorized edition *Les méditations métaphysiques*, 1647; hereafter *Meditations*) came to be framed as a doctrine that was to be understood as the antagonist of an experimental approach.[52]

This stereotypical opposition between English empiricism and French rationalism resonates noticeably with a discourse about knowledge making that juxtaposes hand and mind. It is against the backdrop of such an antagonistic intellectual landscape that Shapin introduces the notion of "hands-on" into the debate. In "The Invisible Technician" (1989), he expounds on the effective use of rhetoric strategies portraying the new scientists in seventeenth-century England as "craftsmenlike practical doers."[53] We have seen a vivid example of a similar rhetoric in Galileo's previously cited experiment with balls rolling down an inclined plane, though it remains unclear in this earlier source who precisely crafted the experimental setup and prepared the materials used.

Shapin presents us with a body of rhetoric that opposed the new experimental science to the predominantly theory-laden work of learned scholars: "Unlike the 'barren scholasticism' of the universities, unlike the 'metaphysical' discourses of modern rationalists, the new experimental science was said to be hands-on practice."[54] In Shapin's account, the authenticity of scientific work became in the seventeenth century associated with practices that ought to evoke the laborious manual work of the experimenter instead of the lofty and bloodless realms of thought experiments and logical argumentations. Other historians have emphasized that the practice of experimental philosophy called for "skilled practitioners using instrumental technologies in the production of effects."[55] Shapin shows how new scientists attempted to frame their epistemic endeavors as a hands-on science without acknowledging, however, the role of invisible technicians in their experimental practices.[56] With his historical account of hands-on experiments he calls into question a portrait of modern science that is clearly meant to evoke associations with Cartesian epistemology: "The predominant biases in the Western academic world have traditionally portrayed science as a formal and wholly rational enterprise carried out by reflective individual thinkers."[57] In this seminal paper, Shapin takes issue with the idea that scientific knowing springs from the brains of rational individuals, even though he warns against too naive an understanding of the new scientists' hands-on attitude. That his critique has an anti-Cartesian flavor is not surprising.

44

Introductory curricula in European philosophy are still dominated today by readings of Descartes's *Meditations* as an unequivocal testimony to an exclusively mind-based rationalism that cannot easily be reconciled with the experimental empiricism of the new sciences. I will return in chapter 3 to the intriguing question of the conspicuous "invisibility" of laboratory technicians that Shapin raised here. But first I take a closer look at the apparent discrepancy between an emerging image of hands-on science and an exclusively mind-based epistemology associated with René Descartes.

The Wax Argument

In popular understanding, the French philosopher René Descartes has been identified with his famous "cogito," an image of a thinker who explored the possibilities of knowledge with his *ratio* alone. This image of a disembodied knower with a brain and no hands derives from Descartes's *Meditations*. The main protagonist of this text—a meditating first-person narrator—appears to renounce sensory empiricism completely in a famous passage that has become known as the "wax argument." The *Meditations*, which are generally considered to exemplify the Cartesian rationalist epistemology, played a key role in forging this persistent image of Descartes as a philosopher without hands.

Yet this image cannot be easily reconciled with a growing body of scholarship that portrays Descartes as an experimenting scientist.[58] How, then, can we make sense of this early modern philosopher and new scientist who appears to embody two opposing identities: the rationalist theorist and the hands-on experimenter? Let us take a closer look at Descartes's Second Meditation, which is often cited as a key moment in defining Cartesian philosophy as a rationalist project as it appears to convey a profound doubt about the validity of empirical knowledge (see figures 1.1–1.3 and "Note on Descartes's Texts and Their Translations," in this volume).[59] In the famous wax passage, the first-person narrator of the *Meditations* voices a deep mistrust of sense-based knowledge:

45

> Let us consider the most common things and what we think we understand most distinctly; that is, the bodies which we touch and see. I do not mean to talk about bodies in general, for these general notions are usually more confused, but of one in particular. Let us take, for example, this piece of wax, which has just been taken from the honeycomb: it has not yet quite lost the sweetness of the honey it contained, it retains some of the scent of the flowers from which

it was gathered; its color, shape and size are plain to see; it is hard, it is cold, [as] one touches it, and if you rap it, it makes a sound. In short, it has everything that can be known distinctly of a body. But, while I speak, look as it comes near to the fire: what rests from its taste fades away, the smell disappears, its color changes, its shape is lost, its size increases; it becomes liquid, it gets hot; one can hardly touch it, and if one raps it, it no longer makes a sound. Does the same wax remain after this transformation [*changement*]? It must be admitted that it does; no one denies it. So what is it that one knows of this piece of wax with such distinctness? Certainly, it cannot be anything of all that I noticed by means of the senses; for all the things that have been evident to the taste, or smell, or sight, or to the touch, or hearing have now altered, and yet the same wax remains. Perhaps it could be what I think now, that is, that the wax was not this sweetness of the honey, not this appealing fragrance of the flowers, not this whiteness, not this shape, not this sound, but only a body which shortly before presented itself to me in these forms, and which now exhibits different ones. But what, exactly speaking, is it that I am now imagining, after I perceived it in this manner? Let us consider it carefully, and, take away all the things which do not belong to the wax, and see what is left. Certainly, the only thing that remains is something extended, flexible and changeable.[60] (For original French, see figures 1.2 and 1.3.)

Indeed, this passage seems to confirm the idea of Descartes as an absolute rationalist; do we not witness here how a philosopher by mere introspection comes to the conclusion that we need to seriously doubt anything that we can perceive through our senses? Apparently, Descartes instructs us here to get our hands off any empirical objects if we seek true knowledge. After all, it is in the mental realm alone that we can grasp his concept of things as pure extension (*res extensae*). What we need to do is to concentrate on our mental capacities to acquire a clear and distinct idea of things. The meditating "I" figure demonstrates here that all we can know about bodies without any doubt is that they are extended in space and that we cannot derive this insight from our senses. When reading this passage, it seems obvious that we impute to Descartes an attitude toward knowledge making that runs counter to any empirical endeavors that operate by sight and touch. For Descartes, it seems, knowledge comes from cogitating, not from skilled hands and trained eyes that aptly make use of purposely designed apparatuses to contrive events

1.1–1.3 ⋆ Title page and "wax *experience*" on pages 26 and 27 of the first French edition of the *Meditations*, approved by Descartes. From Descartes, *Les méditations métaphysiques* (1647). Image source: gallica.bnf.fr/Bibliothèque Nationale de France.

For enlarged images of pages 26 and 27 (figs. 1.2 and 1.3), see the "Note on Descartes's Texts and Their Translations" in this volume.

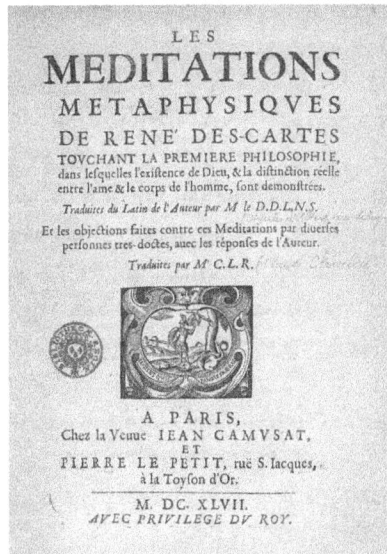

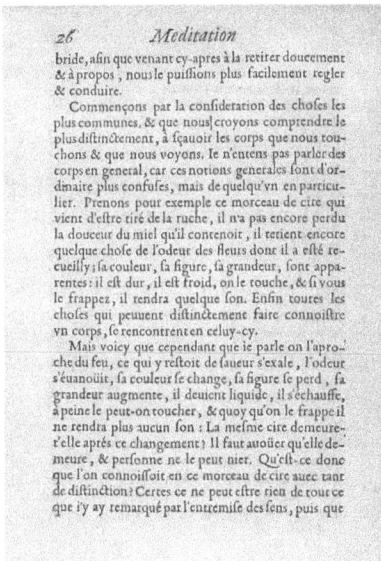

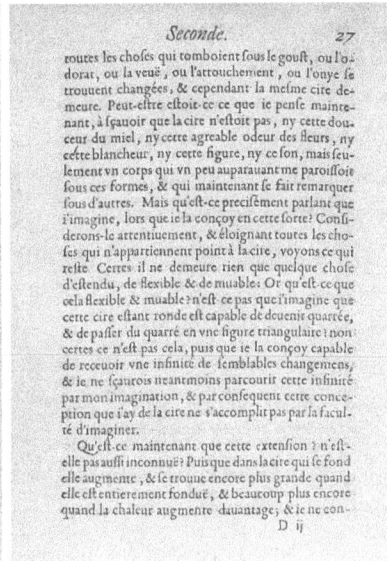

LES
MEDITATIONS
METAPHYSIQVES
DE RENE' DESCARTES
TOVCHANT LA PREMIERE PHILOSOPHIE,
dans lesquelles l'existence de Dieu, & la distinction réelle
entre l'ame & le corps de l'homme, sont demonstrées.

Traduites du Latin de l'Auteur par M le D.D.L.N.S.

Et les objections faites contre ces Meditations par diuerses
personnes tres-doctes, auec les réponses de l'Auteur.

Traduites par M C.L.R.

A PARIS,
Chez la Veuue IEAN CAMVSAT,
ET
PIERRE LE PETIT, ruë S. Iacques,
à la Toyson d'Or.

M. DC. XLVII.
AVEC PRIVILEGE DV ROY.

26   *Meditation*

bride, afin que venant cy-apres à la retirer doucement & à propos, nous le puissions plus facilement regler & conduire.

Commençons par la consideration des choses les plus communes, & que nous croyons comprendre le plus distinctement, à sçauoir les corps que nous touchons & que nous voyons. Ie n'entens pas parler des corps en general, car ces notions generales sont d'ordinaire plus confuses, mais de quelqu'vn en particulier. Prenons pour exemple ce morceau de cire qui vient d'estre tiré de la ruche, il n'a pas encore perdu la douceur du miel qu'il contenoit, il retient encore quelque chose de l'odeur des fleurs dont il a esté recueilly; sa couleur, sa figure, sa grandeur, sont apparentes: il est dur, il est froid, on le touche, & si vous le frappez, il rendra quelque son. Enfin toutes les choses qui peuuent distinctement faire connoistre vn corps, se rencontrent en celuy-cy.

Mais voicy que cependant que je parle on l'approche du feu, ce qui y restoit de saueur s'exale, l'odeur s'éuanoüit, sa couleur se change, sa figure se perd, sa grandeur augmente, il deuient liquide, il s'échauffe, à peine le peut-on toucher, & quoy qu'on le frappe il ne rendra plus aucun son: La mesme cire demeure-t-elle apres ce changement? Il faut aüoüer qu'elle demeure, & personne ne le peut nier. Qu'est-ce donc que l'on connoissoit en ce morceau de cire auec tant de distinction? Certes ce ne peut estre rien de tout ce que i'y ay remarqué par l'entremise des sens, puis que

Seconde.   27

toutes les choses qui tomboient sous le goust, ou l'odorat, ou la veuë, ou l'attouchement, ou l'ouye se trouuent changées, & cependant la mesme cire demeure. Peut-estre estoit-ce ce que ie pense maintenant, à sçauoir que la cire n'estoit pas, ny cette douceur du miel, ny cette agreable odeur des fleurs, ny cette blancheur, ny cette figure, ny ce son, mais seulement vn corps qui vn peu auparauant me paroissoit sous ces formes, & qui maintenant se fait remarquer sous d'autres. Mais qu'est-ce precisément parlant que i'imagine, lors que ie la conçoy en cette sorte? Considerons-le attentiuement, & éloignant toutes les choses qui n'appartiennent point à la cire, voyons ce qui reste. Certes il ne demeure rien que quelque chose d'estendu, de flexible & de muable: Or qu'est-ce que cela flexible & muable? n'est-ce pas que i'imagine que cette cire estant ronde est capable de deuenir quarrée, & de passer du quarré en vne figure triangulaire? non certes ce n'est pas cela, puis que ie la conçoy capable de receuoir vne infinité de semblables changemens, & ie ne sçaurois neantmoins parcourir cette infinité par mon imagination, & par consequent cette conception que i'ay de la cire ne s'accomplit pas par la faculté d'imaginer.

Qu'est-ce maintenant que cette extension? n'est-elle pas aussi inconnuë? Puisque dans la cire qui se fond elle augmente, & se trouue encore plus grande quand elle est entierement fondu, & beaucoup plus encore quand la chaleur augmente dauantage; & ie ne con-
D ij

47

for special observations. This popular image of a knower without hands is based on an overtly theory-laden interpretation of Descartes's philosophical writings, still commonly taught at Western universities.

I argue against such a reductionist reading of the famous wax passage in Descartes's Second Meditation. Instead of interpreting this passage as a straightforward argument against sense-based, empirical epistemologies,

I propose to read it as a testimony to the struggles of early modern natural philosophers to come to grips with an epistemological crisis. Can we understand historically *how* the Cartesian cogito took shape in the embodied mind of an early modern natural philosopher? I argue that we need to read Descartes's Second Meditation in the context of the epistemic practices of the new scientists and their obvious difficulties to write their experiential accounts into the existing conceptual frameworks of a sense-based Aristotelianism.[61] In this chapter I will portray René Descartes as a fervent experimenter who was actively engaged in the sciences of his day. I show, based on historical sources and studies, how he was manually forging a new theory of knowledge—with his mind *and* hands.

Recent scholarship in the history of science ascribes to Descartes an important role in the rise of the new science and groups him together with such leading figures as Nicolaus Copernicus, Johannes Kepler, Galileo Galilei, Francis Bacon, Marin Mersenne, William Harvey, Robert Boyle, Christiaan Huygens, and Isaac Newton.[62] While historians agree that the seventeenth century saw the rise of "a recognizable scientific community" marked out by "its commitment to a new kind of natural philosophy," it has by now become highly questionable that such a commitment was found only among natural philosophers of the Royal Academy.[63] Penelope Gouk argues, for example, that the work of Mersenne provided an important model for experimental investigations, emphasizing its impact on the developing knowledge cultures at the Royal Society.[64] Craftsmanship and artisanal knowledge were central to Mersenne's approach: working like an artisanal ethnographer avant la lettre, he collected information from professional musicians and instrument makers for his experimental studies in acoustics.[65] Mersenne, who was part of a wide network of natural philosophers, was in close contact with Descartes. He published Descartes's *Meditations* in its initial Latin edition of 1641 only a few years after publishing his own major work, *Universal Harmony*, which contained detailed descriptions of his own experimental observations. European natural philosophers on both sides of the English Channel experimented with different methods to gather knowledge of natural phenomena, including the accumulation of empirical data through experimental observation, and exchanges with craftsmen and skilled practitioners in order to supplement and revise an academic body of knowledge that was still, at that time, dominated by the text- and theory-laden Scholastic tradition. Hence, the idea that the experimental attitude can be seen as a trademark of an English tradition has been met with much criticism.[66] This more nuanced and differentiated perspective on French rationalism and Cartesian epistemology, on the one hand, and British empiricism and the experimental approach, on the other, provides the backdrop for my reading of Descartes.[67]

Weighing the Cartesian Legacy: A Page Count

To be sure, Descartes was a prolific writer; if we examine the French edition of virtually all his writings, the *Oeuvres de Descartes*, first published by Charles Adam and Paul Tannery between 1896 and 1901, we can see that Descartes produced during his lifetime the equivalent of about 8,800 pages in the printed modern edition (figure 1.4).[68]

Considering the total number of texts Descartes produced over the course of his life, it becomes more than obvious that he was an inquirer into natural phenomena who was actively involved in the emerging experimental sciences of his day. Descartes's written oeuvre of "scientific" correspondence and tentative essays on natural philosophical questions easily outnumbers his philosophical "classics": *Rules for the Direction of the Mind* (*Regulae ad directionem ingenii*, 1628, published posthumously in 1701), *Discourse on the Method* (*Discours de la méthode*, first published in 1637), and the first French authorized edition of the *Meditations* (*Les méditations métaphysiques*, 1647), this last with no more than 128 printed pages.[69] The philosophical works for which Descartes is famed today constitute only about 3 percent of his total writings: *Discourse on the Method* and the *Meditations* account for about eighty pages each in Adam and Tannery's modern edition, if we leave all the additional essays, responses, and objections aside. Together with the *Rules for the Direction of the Mind* at around 140 pages, these most famous works would easily fit into just *one* of the Adam and Tannery's eleven volumes that make up the complete works of Descartes.[70] It is, however, impossible to clearly separate metaphysical, ontological, and epistemological concerns in seventeenth-century natural philosophy. A vivid example hereof is the account earlier in this chapter of Descartes's contemporary Mersenne on the ontological and epistemological status of an inaudible tone.

What is perhaps more notable is the fact that almost half of Descartes's writings, namely, five of the eleven volumes, contain correspondence that he maintained with contemporaries. Roughly estimated, more than four thousand pages of the complete *Oeuvres* are devoted to letters that he wrote from the age of twenty-six, starting in April 1622, until his death in February 1650. His other writings, including his major works in natural

49

1.4 ★ René Descartes, *Oeuvres de Descartes*, 11 vols., new ed. (Paris: Vrin [1896–1901], 1996), edited by Charles Adam and Paul Tannery. On the right are the volumes containing the texts of the *Meditations* and the *Discourse on the Method* in Latin and French. On the left are the four volumes of correspondence and the volumes comprising Descartes's other treatises and miscellaneous notes.

philosophy concerning optics, meteorology, geometry, music, mathematics, physics, psychology, physiology, and anatomy, but also religious, theological, and epistemological questions, make up the other half of the volumes. What do these numbers tell us? For one, Descartes's vast correspondence clearly contradicts the image of a solitary thinker who meditated in total seclusion. Leafing through the volumes of letters, one encounters a palpable manifestation of the amount of time Descartes must have dedicated to discussions and conversations with other natural philosophers, often comparing and discussing experimental observations. A look at the sheer quantity of his complete writings firmly situates this allegedly reclusive thinker in the discussions and investigative inquiries of his time.

In Cartesian historiography we can observe a scholarly bias for Descartes's metaphysical classics and a tendency to ignore the rest of his writings. This has led Clarke to the following ironic conclusion: "Descartes is, in many ways, a victim of his own successes as a philosopher."[71] But the tide has turned: it is no longer a scarcely known historical detail, but a generally acknowledged fact, that Descartes was a "serious experimenter" whose experimental work is well documented in his treatises and letters.[72]

The textual evidence for Descartes's serious involvement in experimental natural philosophical inquiries of his day has prompted different interpretations. For Clarke it entails a radical reevaluation of Descartes's "scientific writings" at the cost of his philosophical reputation: he urges his readers to "interpret the extant writings of Descartes as the output of a practicing scientist who, somewhat unfortunately, wrote a few short and relatively unimportant philosophical essays."[73] Daniel Garber, in contrast, places emphasis on the intimate entanglement of philosophical and scientific interests in the "Cartesian programme" and attempts to explore how these two sides are mutually illuminating.[74] He also stresses the artificiality of distinguishing between Descartes "the scientist" and Descartes "the philosopher," echoing here an earlier criticism by Alexandre Koyré, who had convincingly argued that the insurmountable dividing wall that historians wished to erect between the Scholastic tradition and Descartes existed only in their imagination.[75] Descartes, who enjoyed an education at La Flèche, a Jesuit school with an excellent reputation, was not just familiar with but indeed well-versed in a tradition of Aristotelian and Scholastic natural philosophy as it was taught at the universities of his time.[76] The accounts of his student years create the picture of a brilliant and intellectually accomplished student who grew up in a bookish world: the young Descartes absorbed the Aristotelian doctrines that dominated inquiries into living phenomena, becoming the best pupil in his school.[77] The idea that Descartes and his fellow new scientists completely broke with Scholastic traditions has by now become much more nuanced. In a similar vein, we can ask whether a categorical distinction between Descartes the philosopher and Descartes the experimenter is not in the first place a *historiographical* problem that, on closer inspection, cannot be substantiated with historical evidence. This is especially the case if we keep in mind that we cannot project our current conceptions of what we call "philosophy" and what we call the "sciences" onto the seventeenth-century natural philosophical landscape. Instead of opposing the one to the other, historians have argued for a continuum approach or have discarded a distinction between philosophy and science altogether for the premodern period, emphasizing that both were part of the same domain of inquiry in early modern Europe.[78] Additionally, scholars have drawn attention to the anachronistic use of the term *scientist*, which, in fact, was only coined in the nineteenth century by William Whewell.[79] Penelope Gouk argues

51

this point nicely when she compares the long professional tradition of musicians to scientists, who "became recognised as a body only relatively recently."[80] Notably, Gouk uses the term *body* here in its double sense, referring to an acknowledged entity, such as a body of scholarship, and at the same time hinting at the long-understudied embodied dimensions of scientists' work in the history and philosophy of science.[81]

Descartes on *Experiences*

Of particular interest is, of course, what Descartes himself says about the place that writing and experimenting take up in his daily doings. In a much-cited letter, Descartes defends himself against the criticism that his views expressed in the *Principia philosophiae* had not been sufficiently confirmed by experience.[82] He writes:

> What I find most strange is the conclusion of the judgment you sent me, namely that what will prevent my principles from being accepted in School (*dans L'Escole*) is that they are not sufficiently confirmed by *experience,* and that I have not refuted the arguments of others. Notwithstanding, that I notably have proven with almost as many *experiences* as there are lines in my writing, and that I in general provided explanations in my Principles of all the phenomena of nature, I have explained, with the same means, all the *experiences* that can be made regarding the inanimate bodies—none of which, by contrast, have ever been well explained with the principles of ordinary philosophy (*la Philosophie vulgaire*)—I note with surprise that despite all this the followers [of this ordinary philosophy] continue to accuse me of a lack of *experiences.*[83]

Let us savor for a moment his remark: Descartes boasts here that his *experiences* almost match in number the lines of his writings. Given that the printed modern edition of his complete writings comprises approximately 8,800 pages (not lines!), we can indeed envision a mind-boggling number of handwritten lines—even if we exclude his correspondence from this total, and take into consideration that he likely wrote this letter in 1645, five years before his death. Obviously, Descartes displays a penchant for exaggeration—it is, after all, quite implausible that he could have conducted enough experiments in the course of his adult life to reach even a quarter of the lines that came down to us in the form of correspondence, essays, and

52

treatises. But even as an overstatement, this remark still illustrates that—in Descartes's own perception—he spent as much time on *experiences* as he did on writing.

It is important to note that Descartes's use of the Middle French term *experience* is varied and diverse. Desmond Clarke notes that he differentiates now and then between special experiences (*expériences particulières*) and everyday experiences (Lat., *experientia quotidiana*) or ordinary experience (*expérience vulgaire*).[84]

English translations of the above quoted letter illustrate this semantic ambiguity: the still authoritative English edition of Descartes's correspondence, published by John Cottingham et al. in 1991, translates the French term *experiences* here as "observations." My translation intentionally diverges from Cottingham et al., as "observations" in the sense of modern "scientific observations" is problematic and also more suggestive of an uninvolved observer than of a hands-on new scientist engaged in making phenomena knowledgeable by experience; others have preferred the translation "experiments" or simply left the original French word untranslated, as I did above.[85]

Throughout his writings Descartes uses the French term *experience* to refer to a range of perceptual modes or experiential events that scholars have circumscribed as "intellectual awareness, test, observation, experiment, phenomenon, and ordinary experience."[86] Yet scholars have also pointed out that he frequently employed the term in relation to what has been assumed to be experiment-based observations. Often it is clear from the context whether he is using the term *experience* in a sense that comes closer to what we understand today as a scientific experiment. Clarke makes a case for translating *experiences* as "experiments" in contexts that allow for special observations: "Those observations which were not common to all untutored observers of nature but which required skill and scientific knowledge to make are evidently experiments."[87] For him the context "makes it very clear in a great number of cases that '*expériénce*' does refer to experiments, in the precise sense of an artificially arranged or contrived observation which is designed to test some implications of one's scientific theory."[88] Today, historians are much more reluctant to translate early modern terms into modern scientific terminology, keeping in mind that our modern sense of scientific theories and experiments cannot be projected onto seventeenth-century natural philosophical inquiries. Nonetheless, it is important to note that Descartes reflects in the quoted passages on the fact that it becomes problematic to conflate the different meanings of experiences relating to experimental natural philosophical inquiries and those relating to commonplace or everyday acts of perception.

In *Discourse on the Method*, Descartes, writes, for example: "With regard to *experiences*, there are so many more necessary to advance our knowledge."[89] He then distinguishes between experiences that present themselves readily to our senses, and rare occurrences or experiences that have only been the subject of studies (and not of ordinary events); the latter, he explains, "depend always on such specific [*particulieres*] and small circumstances, that it is very difficult to notice them."[90] Further on, Descartes explains that if one desires to move from the most primal and ordinary effects and causes down to "those [natural effects and causes] that are more particular, it would require the use of several special experiences [*plusieurs experiences particulieres*]."[91] Note that here he employs the term *experiences particulieres* to demarcate special or specific forms of experiences demanding particular attention and circumstances from "ordinary" experiences, a distinction that resonates with Galileo's contrived occurrences.[92] However, scholars have found no systematic usage, with consistent linguistic indicators, of either experience or experiment in these two different senses.

In a much-cited quote he laments about the sheer number of *experiences* it would take to gain a better understanding of nature. It is worthwhile to pay attention to his precise choice of words: Descartes writes that it requires so many of them that "his hands and his income" would not suffice, not even if he could multiply his revenues by a thousand.[93] Descartes emphasizes here that "special" experiences require much more handiwork and costly investments than one natural philosopher alone could afford in a lifetime. Evidently, he knows that the new sciences call for a time-consuming amount of manual labor.

54

I find Descartes's ambivalent yet distinctive use of the term *experience* intriguing. It may point toward a phase of transition: a period in which it becomes necessary to develop a vocabulary that allows natural philosophers to describe more senses, or modes, of sensual perception than the rhetoric of the Scholastic tradition allowed for.

In an illuminating passage from an anonymous letter included in the preface to Descartes's *Passions of the Soul* (*Les passions de l'âme*, drafted in 1646 and published in Paris in 1649), the author clearly differentiates between two sorts of experiences. It has been suggested that the letter was written by Abbé Claude Picot (1601–1668), who was responsible for the work's distribution in France, or even by Descartes himself.[94] The author draws a distinction between experiences that are readily apparent to ordinary sensual perception and other experiences that come at a cost and require effort to discern: "These experiences are of two kinds: one of them is easy and only presupposes that we reflect on those things which are spontaneously presented to our senses. The other kind is more infrequent and more difficult, and cannot be had without some study and expense."[95] Clarke deduces hereof that it became critical to seventeenth-century natural philosophers to distinguish between "two kinds of experiential evidence."[96] Though it remains questionable whether we can understand seventeenth-century experiences in terms of a binary opposition, this discussion points us at an important concern. It shows us that aesthetic issues lie at the core of discussions about the use of the term *experience* in relation to seventeenth-century epistemic practices. With the rise of new sciences in early modern Europe, sensual perception became an issue in a new sense for natural philosophers. The move away from Aristotelian sense-based epistemology in doing observations with experiments called for new knowledge theories, concepts, and vocabularies that were not constrained to what could be understood as everyday experiences. The new sciences called for a language and conceptual framework that also work for experiential events produced with artificial setups and specialized apparatus.

Interestingly, in a later paper Clarke excludes any "particular" experiences from Cartesian epistemology. Citing passages from the *Rules* and the *Discourse*, Clarke argues that Descartes's favored concept of clear and distinct ideas that present themselves to us in a straightforward and indubitable manner leads him to privilege the reflection on everyday experience when he searches for a metaphysical foundation of the sciences.[97] Artificially contrived experiences, by contrast, appear to be more complex, complicated, and less reliable, and therefore of no value for Cartesian epistemology. For Desmond Clarke, "the epistemic foundation of Cartesian metaphysics is reflection on 'common sense' or on our everyday experience of the natural world."[98] In Clarke's analysis the concept of clear and distinct ideas becomes conflated with common sense or everyday experience. Such an interpretation is, in my view, rather problematic. After all, Descartes demonstrates over the course of several meditations that those clear and distinct perceptions to which he ascribes the highest epistemic value can only be attained methodologically with rigorous philosophical introspection.[99] The philosophical exertions he describes in his *Meditations* appear to be far from the ways in which early moderns have experienced everyday events. Descartes's recourse to intuitively given ideas that we can mentally grasp in a clear and

55

distinct manner can be better understood in the context of a philosophical operation which I discuss in more detail in the next chapter.

Ironically, we owe it, so to speak, to the "Cartesian spirit" to approach Descartes's meditations on how and what we can know as an exercise in pure reasoning with an entirely ahistorical stance. In the popular imagination this has led to reducing Descartes's rich philosophical thinking into the simplified motto "I think, therefore I exist."[100] More recently, historians and philosophers became interested in the philosopher behind the cogito, now trying "to understand Descartes as a living, breathing human being, who learns (and forgets) things, whose views develop and change over time."[101] We can take this as a plea for studying Descartes's thinking in the context of his daily doings and habits and for paying attention to processes of learning and (changing) ways of knowing. How, then, can we gain a better understanding of Cartesian epistemology in light of the historical fact that Descartes saw himself primarily as an experimental philosopher of natural processes?[102] What kind of embodied practices besides writing letters and drafting his treatises took up the bulk his time?

I want to find out whether—and how—this hands-on philosopher is inscribed into the *Meditations*. The *Meditations* has commonly been taken to be the text par excellence to embody an antagonistic attitude toward the new scientists' empirical attitude. What could a radically historicized reading of this text tell us?

Situating Descartes's work and thoughts within this context of changing practices of knowledge making and writing about knowledge allows us to see that the metaphysical and epistemological work of seventeenth-century natural philosophers is intimately entangled with their experimental inquiries. I argue that we cannot read the *Meditations* independently from Descartes's active experimental investigations into living processes. Instead, I propose in the following to read the famous wax passage of the Second Meditation in light of Descartes's experimental anatomical inquiries. Descartes, a dedicated hands-on practitioner in his quest for new knowledge, was part of the flourishing experimental scene in seventeenth-century anatomy. How did it *feel* to practice experimental philosophy? How did Descartes's *ways of doing* affect his *ways of knowing*?[103] What kind of hands-on experiences were involved in his practical search for knowledge, and how did these manipulations (from the Latin *manus* = hand) inform his theoretical writings?

My rereading of the famous wax argument in Descartes's Second Meditation will enable us to grasp the innovative thrust of the Cartesian epistemological program. It will point us at critical tools within the historicized, contextual, and aesthetic project of Cartesian epistemology to reflect on contemporary practices in the life sciences from a historically and philosophically informed hands-on perspective.[104] But before returning to the wax passage, I want to go back in time and take a closer look at some of the experimental practices Descartes engaged in before and, perhaps, while he was drafting the *Meditations on First Philosophy* in the 1630s. How can we picture Descartes as an experimentalist? What kind of experiments did he perform? The following section paints an impression of Descartes's practical doings during a prolonged stay in the Low Countries with the region's flourishing centers of anatomical research in Amsterdam and Leiden.

Vesalian Renaissance in the Low Countries: Hands-On Anatomy

When Descartes came to Amsterdam in 1629, he arrived in the midst of a Vesalian renaissance. In the beginning of the seventeenth century, new editions of Andreas Vesalius's (1514–1564) famous anatomical atlas *De humani corporis fabrica* were printed in the Low Countries with elaborate engravings.[105] At that time anatomy was not yet an established academic discipline with a theoretically and methodologically well-defined body of knowledge, as we know it today. Premodern and early modern anatomy involved explorative practices of the inner parts and processes in human and animal bodies at the time of a widely accepted religious-based human exceptionalism. Renaissance anatomists practiced human dissections not without risk; moreover, those who extrapolated findings from animal anatomies to the anatomy of human beings were prone to being accused of heresy. Descartes's keen interest in anatomy likely was incited by his new surroundings—or at least we have no evidence that Descartes had been interested in the study of anatomy or physiology prior to his arrival in Amsterdam.[106] The Dutch Republic, with its urbanized burgher society, nurtured no disdain for manual labor. Favored by many for their religious tolerance, openness of knowledge, and advanced crafts technologies, the Dutch cities provided a fertile ground for early modern anatomy, which flourished here under the motto "seeing with my own eyes."[107]

The revived interest in anatomical dissections is manifest in the famous Netherlandish painting *The Anatomy Lesson of Dr. Nicolaes Tulp* (1632) that portrays a well-known Dutch physician as a "new Vesalius"; Rembrandt painted this portrait in Amsterdam, the city where Descartes resided at that time, and where he continued to work on a treatise devoted

57

to the human body. The reanimation of anatomical traditions and their practical approach must have appealed to Descartes, who taught himself medicine and showed off a critical aversion to the book-learning paradigm of the Scholastic tradition. Descartes refrained, however, from publishing the *Treatise on Man* (*L'homme*), in which he sets out a mechanistic physiology of the human body, when he heard about the Inquisition's condemnation of Galileo Galilei's work.[108] In historiography, the *Treatise on Man* has long epitomized Cartesian mechanistic philosophy. Famously using a clock analogy as explanation, with this work Descartes set forth a theory of the body as an automaton.

We know that Descartes had been pondering physiological issues (including the physiology of animal motion), the workings of sensory perception, and problems in mathematical physics for years before he came to Amsterdam, but his interest in the study of human and animal bodies and physiological processes must have received a renewed vigor after he left France.[109] We can indeed assume that, before he arrived in the Low Countries, Descartes was first and foremost *thinking* about physiological questions and the working of sensual perception and visual cognition. He only began to complement his more theoretical inquiry into optics, the science of visual perception and the physiology of the eye, with practical studies when he started to instruct himself in anatomical experiments in the Netherlands.[110]

He likely started his anatomical work the same year he arrived in Amsterdam. In a letter to Marin Mersenne from that city, dated December 18, 1629, he remarks that he has too many other diversions to fully devote himself to finishing a treatise, adding that he had begun to study anatomy.[111] A year later, he makes a similar allusion, in another letter to Mersenne, dated April 15, 1630: "My work on it [a small treatise] is going very slowly, because I take much more pleasure in *instructing myself* than in putting into writing the little that I know. I am now studying chemistry and anatomy simultaneously; every day I learn something that I cannot find in any book. . . . Moreover I pass the time so contentedly in *instructing myself* that I never set myself to write my treatise except under duress."[112] The letter conveys that Descartes not only had studied the great anatomical works of his time but, apparently, also had begun to perform dissections. Descartes takes pride in claiming that the knowledge he thus acquired had not been written in any books.[113] His words testify to the high value he ascribed to knowledge gained from learning how to practice anatomy himself.[114] They characterize him as an outspoken autodidact in the study of anatomy, an important nuance that got lost in the English translation of Cottingham et al. The original French expression "ie prens beaoucoup plus de plaisir m'instruire moy-mesme," here translated verbatim, is phrased more elegantly, but less precisely, as "I take much more pleasure in acquiring knowledge."[115]

Writing to Mersenne on November 13, 1639, Descartes defends his interest in anatomy and his hands-on approach against the alleged prejudices of learned men. His favored pastime has apparently prompted some contemporaries to accuse him of seeking macabre diversions in the villages. He defends himself against such accusations by stating that there are more animals slaughtered in the cities than in the villages. He then recalls his daily visits to the butcher during his first winter in Amsterdam and comments on the practicalities concerning the supply of materials for his self-directed studies in anatomy. Finally, he explains that it is indeed no crime to be interested in anatomy: "During one winter in Amsterdam, I used to go nearly every day to a butcher's, to see him slaughter animals, and to have brought to my house the parts of the animals I wanted to anatomise at leisure."[116]

Descartes's memories speak to the imagination. We can picture him at the slaughterhouse gathering ox eyes and other organs and body parts from freshly slaughtered animals, then shouldering his prey in a sack and hurrying home at the end of the day through Amsterdam's cold and narrow alleys, the soggy mass still warm and resting against his back, perhaps even dripping with fresh blood when he makes his way back from the Kalverstraat, the Street of Calves, in the center of Amsterdam.[117] Maybe he had to smuggle these unsavory goods into his room, taking care that his landlord was not offended by what he was doing in the seclusion of his chamber.

This portrayal of Descartes as an autodidactic anatomist who immersed himself in the self-study of slaughtered body parts should not lead us, though, to envision him as a medical amateur. The surviving evidence of his extensive correspondence with learned physicians proves the contrary: Descartes discussed medical problems with famous physicians of his time, among them William Harvey (1578–1657), who had studied medicine in Cambridge and Padua.[118] Descartes and Harvey exchanged ideas on the movement of the heart, and the French philosopher was one of the first to promote Harvey's theory of the blood's circular course, though he did not agree with Harvey on all points.[119] Harvey's *De motu cordis et sanguinis* was published in 1628, a year before Descartes began to study anatomy. Descartes first read Harvey's book in 1632 and commented extensively on it in his own writings and letters; for example, in one of his letters he mentions in the same breath that he is dissecting animal heads and that he has seen Harvey's book *De motu cordis*, of which Mersenne had spoken to him.[120] Even though Descartes had not received any proper training in medicine, he actively participated in medical debates and evidently played an important role in early modern medical discourse.[121]

Imagining Descartes at work at the dissecting table, conducting detailed anatomical dissections, yields vivid images of his study of animate phenomena, but do these explorations situate him amid the pioneering work of the experimenting new scientists? In what sense can we actually relate the century-old discipline of anatomy to the rise of the experimental sciences? To answer these questions, we will need a better understanding of the branch of knowledge and set of practices that formed the discipline of anatomy in Descartes's time.

While we think of anatomy today as the study of body structures by dissection and observation, it would be wrong to limit the historical discipline of anatomy to a preeminently descriptive and observational field of study. On the contrary, the discipline of anatomy was in Descartes's time an explorative as well as an inquisitive practice centered on experiment.[122] Anatomical experimentation gained momentum beginning in the mid-sixteenth century, and the seventeenth and eighteenth centuries witnessed the development of unprecedented "artificial operations," including the compelling results that the famous anatomist Frederik Ruysch (1638–1731) achieved with wax injections.[123]

In his seminal studies on the disciplinary identity of physiology and anatomy before 1800, Andrew Cunningham puts the manual experiment firmly at the heart of seventeenth-century anatomy.[124] Seventeenth-century anatomists were "active experimentalists" engaged in an investigative and innovative discipline.[125] Cunningham warns us that we cannot project our current disciplinary conceptions of anatomy and physiology onto pre-1800 practices. Then, as now, anatomy and physiology were two distinct disciplines, but while physiology is understood today as an experimental science, and anatomy as a mainly descriptive study, this was not so in the seventeenth century. At that time, anatomy was characterized by a hands-on experimental attitude.[126] Physiology, in contrast, was known as an exclusively discursive practice that Cunningham describes as "a thinking and talking discipline—a discourse" and "not an investigative discipline, nor an empirical discipline, nor an experimental discipline."[127] It was understood as a purely intellectual activity, practiced in contemplation and writing. It was not until the nineteenth century that physiology was reinvented as a rigorous experimental science; before that time, as Cunningham emphasizes, physiologists were armed with pens rather than knives.[128] Anatomy and physiology as practiced and discursively engaged with before about 1800 thus provided divergent knowledge models for inquiries into the phenomena of life.[129]

Hence, we should understand Descartes's dispute with Harvey and his anatomical experiments concerned with the dynamic functions of the body (e.g., the motion of the heart) as part of Cartesian anatomy.[130] Harvey's breakthrough in the understanding of blood circulation was a milestone discovery in early modern anatomy, although retrospectively it is often understood as a physiological breakthrough.[131] The use of experiment in seventeenth-century investigations of life phenomena needs, therefore, to be understood today as an instance of the old *art of anatomy*.[132]

Building on Cunningham, we can observe a meaningful shift regarding knowledge fields and activities that previously had been understood as arts (*artes*) and that later, in the context of the experimentalization of natural inquiries, became part of early modern sciences.[133] The manual practice of anatomy, with experimental dissection as its main tool, was defined by Descartes's contemporaries as an act of the hand and mind. Francis Glisson, an English physician and contemporary of Descartes, discussing the epistemic value of "artificial dissections," describes anatomizing as an art that "implies not the *manual* dissection only, but in especial manner the *mental*."[134] Comparing the handiwork of anatomists to that of butchers, who also cut up bodies systematically, Cunningham emphasizes that in seventeenth-century discourse anatomy explicitly called for a combination of manual skills with mental activity.[135]

Cunningham's account of anatomy and physiology before 1800 describes a transitional period during which the manual early modern art of anatomy gained in importance relative to "old physiology," a previously more highly valued and strictly theoretical *scientia*. I argue that we need to situate Descartes's anatomical experimentations and physiological contemplations within the context of this transition. It was a process in which the writer's hand became more and more dependent on the experimenters' hands that could handle dissecting knives and probe with their fingers the inner parts of dead and living bodies.

Descartes's Anatomical *Experiences*

Many vivid accounts of anatomical dissections in Descartes's hand have survived from this Dutch period, which has been described as Europe's golden age of letter writing.[136] This flourishing epistolary culture offers us a glimpse into Descartes's anatomical doings. Of special interest to my argument are the ways in which Descartes conveyed his anatomical hands-on explorations to his correspondents. We see his mindful hands at work in his anatomical inquiries in the 1630s; indeed, he not only cuts and observes but also performs "artificial dissections," which implies not only description and contemplation but also experiment and manipulation. His correspondence conveys insights into seventeenth-century anatomical *experiences* (I reference here the Middle French term *experience*, sometimes used in the sense of an experiment in Descartes's writings).[137] A closer look at Descartes's precise wordings sustains the impression that this new scientist placed epistemic value on doing the manual work involved in these *experiences* himself.

On February 20, 1639, Descartes wrote to Mersenne: "In fact, I did not only take into consideration what Vesalius and the others wrote about Anatomy, but also many more remarkable things than the ones they described, [things] which I noted when doing myself a dissection of diverse animals. Over the last eleven years, such exercise has often taken up my time, and I think that there is no physician who has looked as closely as I have."[138] In their English rendition of this letter to Mersenne, Cottingham et al. have chosen—again—a more elegant but notably less concise translation. The engaged act of examining the inner body as closely as possible is translated into the detached idiom of an observing scientist: "I have spent much time on dissection during the last eleven years, and I doubt whether there is any doctor who has made such a detailed observation as I."[139] By comparison, the early modern French text is much more evocative, as my almost verbatim translation, provided earlier, shows. Readers of the French text can almost picture Descartes with his hands and nose in the inner parts of the bodies he dissected. English editors often choose to translate *experience* as "observations" when the French term is used in a context that we associate today with a scientific experiment. The word choice demands closer attention. Historians of science have shown, for example, that key epistemic concepts of the modern sciences, such as "objectivity" and "observation," are neither neutral nor universal, but they do have their own histories.[140] In modern times, "observation" has often been associated in the sciences with a positivistic ideal of an objective, neutral, and noninterventionist observer, an ideal that is also closely linked to a here disputed disembodied conception of knowledge. The probing explorations—feeling with one's own hands—and dexterous use of dissection instruments suggest an embodied practice that can perhaps better be described with the concept of "skilled vision" (see the introduction) than with the unattainable ideal of a hands-off observer. Certainly, the multisensory practices of early modern anatomists can hardly be reduced to an ocular rhetoric that prioritizes looking above sensing, touching, and feeling. The rise of (experimental) anatomy as an acknowledged knowledge-making practice depended on observations made by *handling* bodies and body parts and involved more senses than vision alone.

Notably, we do not encounter the rhetoric of a detached observer in Descartes's own anatomical reports. In contrast, many passages stand out for the vivid and palpable manner in which he reports on how his anatomical observations naturally depend on skilled instrument use and are intricately interwoven with manual explorations of the body. In contrast, it is striking that in his meticulous accounts of his manipulations, Descartes appears strikingly cool-blooded and distanced. Vivisections must have been gruesome tableaux and intense experiences on the level of all senses, about which he keeps silent. Descartes's reporting on his examinations of animals that he cut open alive thus appears in line with his mechanistic worldview that allowed him to conceptualize animal bodies as mere automatons. While we cannot project our current thinking about animals onto seventeenth-century practices and attitudes, we can wonder how mechanistic theories have played out in practice when we reimagine the sensory realities of vivisections during that period. Given the richness of the perceptual evidence and the important role eyes *and* fingers play in investigating how the living body works in these passages, it would be misleading, though, to reduce the anatomical *experience* to the sensory domain of vision alone. Similarly, medical historians have emphasized that the art of anatomy was a manual craft.[141]

Epistemics of Touch

Touch and tactile sensations were, for instance, of major epistemological significance for Descartes's theory of the motion of the heart. Descartes opens the second part of his unfinished work *Description of the Human Body* (*La description du corps humain*) with the following words: "It is beyond doubt that there is heat in the heart, for one can even feel it with one's hand when one opens up the body of a living animal."[142] This passage is perhaps outdated in light of our contemporary understanding of the actions of the heart—the obsolete theory of the raised temperature of the heart that lies at the core of Descartes's mechanistic explanation of heart movement and blood circulation had already been refuted by Harvey in the seventeenth century.[143] However, epistemologically, Descartes's writing on this matter remains highly relevant.[144] It tells us a lot about the vital role hands played in making knowledge claims in seventeenth-century anatomy, which was then a leading discipline for research on life phenomena.[145] Descartes understood himself to be an anatomist. In the early modern period this implies that he must have conceived of himself as an experimentalist and hands-on practitioner in a very specific sense.

Descartes ascribes an important role to the performance of hands-on explorations in the making of anatomical knowledge. He corroborates his mechanistic theory that refutes prevailing ideas about the movements of heart and blood with "palpable facts" that he obtained through experiment:

> I would like those who have never studied anatomy to take the trouble to look at the heart of some land animal, something reasonably large (for they are more or less similar to those of men), and having first cut off the end of the heart, to take note that there are two caverns or cavities inside, which are able to hold a lot of blood. If one then puts one's fingers in these cavities, towards the base of the heart (and from which it discharges its contents), to seek out the openings through which they receive the blood, what one will find there is that there are two very large ones in each: to wit, in the right ventricle, there is a large opening which leads the finger into the vena cava, and another which will lead it into the pulmonary artery.[146]

He urges his readers to repeat the *experience* and to feel out for themselves how the heart works. He literally gives instructions to put your finger inside the living animal to explore the most complex issues in anatomy.[147] There are yet more passages in the *Description of the Human Body* in which Descartes uses hands-on examinations to ground his theoretical account. In an attempt to prove his mechanistic theory of the movement of the heart, Descartes reports on another anatomical experiment, demonstrating once more the scrupulous finger work involved in anatomical knowledge-making practices. Descartes proceeds to confirm an observation made by Harvey, but gives a different explanation for the same effect. He reports on the shrinking movement of the heart: "He [Harvey] could have confirmed this by a very evident experiment (*une experience fort apparante*): namely, if one cuts the point of the heart of a living dog, and through the incision one puts one's finger into one of its ventricles, one will clearly feel (*on sentira manifestement*) that every time the heart shortens it presses the finger, and that it will stop pressing whenever it is elongated."[148]

While Descartes continues by explaining that the "*experiences* themselves give us often occasion to err," he does not discredit firsthand embodied experience as a vital part of new knowledge formation. Quite the opposite: he places great emphasis on detailed descriptions of his own hands-on *experiences* to argue his case. He further corroborates his findings with experimental evidence from the vivisection of a young rabbit: "If one cuts the point of the heart of a young rabbit that is still alive, the naked eye shows that its ventricles become a little larger and expel blood at the moment the heart hardens. . . . What makes this much less apparent in the heart of a dog or some other more vigorous animal than in a young rabbit is that the fibres take up more of the ventricles; they stiffen when the heart hardens and can press against a finger inserted into one of the ventricles."[149]

We can infer from these passages that Descartes promotes his standpoint and tries to refute rival theories with detailed descriptions of *experiences*, many of which he probably has performed himself. He argues his point, so to speak, with his hands up to his wrists inside the cut-open chest of a dog or a rabbit. Though it was not uncommon at that time to take recourse to the reports of great anatomists and to rely on observations of the ancients, we have seen from his writings that Descartes had indeed attended and conducted dissections himself.[150] He had, for example, undertaken dissections together with the Dutch physician Vopiscus Fortunatus Plempius (1601–1671) in Amsterdam before Plempius left the city to become a respected professor in medicine at the University of Louvain.[151] In a letter dated April 1, 1640, in which Descartes describes a dissection he had attended in Leiden, he refers, for example, to the anatomical skills he had attained through frequent animal dissections:

> Three years ago at Leiden, when I wanted to see it [the infamous conarium, better known as the "pineal body," that Descartes claimed to be the seat of the soul] in a woman who was being autopsied, I found it impossible to recognize it, even though I looked very thoroughly, and knew very well where it should be, being accustomed to find it without any difficulty in freshly killed animals. An old professor who was performing the autopsy, named Valcher, admitted to me that he had never been able to see it in any human body. I think this is because they usually spend some days looking at the intestines and other parts before opening the head.[152]

In a later letter from 1646 he recalls dissections he conducted for embryological research. Descartes describes how he had a cow slaughtered that had been impregnated earlier in order to examine its progeny. He writes that small unborn calves are particularly well suited for anatomical experimentation:

> But my curiosity had only grown; because at another time I had a cow killed, of which I knew that she was about to give birth soon, for the very reason to see the fruit [of her womb]. And when I understood afterward, that the butchers of this country often slaughter them when they find them pregnant, I have asked them to bring me more than a dozen of wombs comprising little calves, the ones as big as mice, the others like rats, & others like little dogs, in which I could observe [*observer*] much more things than in chickens, because their organs are bigger & better visible.[153]

When Descartes wrote that he wanted to begin with anatomy, he intended to teach himself by study *and* experiment the art of manually exploring the inner parts of dead and living bodies in order to investigate the phenomena of life.[154] This becomes manifest in the *Discourse on the Method*, in which Descartes provides a detailed lesson on the anatomy of the heart. He encourages his opponents not to reject his explanations on the movement of the heart without hands-on examination: "I would advise them that the movement I have just explained follows from the mere arrangement of the parts of the heart (which can be seen with the naked eye), from the heat in the heart (which can be felt with the fingers), and from the nature of the blood (which can be known through observation)."[155]

It is conspicuous that the experiential information is bracketed in the English translation in Cottingham et al.'s edition, quoted here, but not in the French original, where Descartes writes of looking with your eyes, feeling with your fingers, and knowing by *experience*: "Qu'on peut voir a l'œil dans le cœur, & de la chaleur qu'on y peut sentir avec les doigts, & de la nature du sang qu'on peut connoistre par experience."[156] Descartes employs the term *experience* for observations that can only be made by "intervening" in nature, because in order to see or feel how the blood runs through the body and how the veins and arteries work, he not only needed to open up the body but also needed to devise experiments that could make the phenomena of systole and diastole perceivable.[157] Descartes's anatomical inquiries included experimental interventions, on living and dead bodies. We should keep in mind that in seventeenth-century anatomy, vivisection and dissection were both part of anatomical practice.[158]

We can find ample evidence in Descartes's letters, his treatises—for example, in *Optics* (*La dioptrique*, 1637) and *Description of the Human Body*—and in his extensive (unpublished) notes that he conducted dissections and vivisections on various animals, including cattle, dogs, rabbits, and eels.[159] In light of Descartes's extensive correspondence with Plempius on animal dissections, it seems very likely that the anatomical accounts on the movement of the heart in the second part of the *Description of the Human Body* are based on vivisections Descartes had performed himself.[160] The dissection reports show that he was intensively engaged in the manual and increasingly experimental discipline of anatomy.

The Latin correspondence between the philosopher Descartes and the physician Plempius has been translated into English by Marjorie Grene, whose essay "The Heart and Blood: Descartes, Plemp, and Harvey" provides a detailed account of the experimental evidence that Descartes brings up for his own theory of the motion of the heart.[161] One of his letters dated to 1638 contains a passage in which he sets out to prove his own theory of the physiology of the heart. Here Descartes provides the physician with a meticulous description of a vivisection he conducted on a young rabbit: "Opening the thorax of a young rabbit and displacing the ribs so that the heart and the trunk of the aorta are exposed, I then tied the aorta with a thread at a certain distance from the heart, and separated it from everything adhering to it." He continues, "Then with a scalpel I made an incision between the heart and the ligature." The heart, Descartes writes to Plempius, "does dilate at the moment, as one can demonstrate by touch itself, since, held in the hand, it feels much harder in diastole than in systole." He closes his account with the remarkable conclusion: "Indeed not only the instantaneous stretching of the heart beat, but the whole fabric of the heart, its heat, and the very nature of blood: all these givens conspire so that we perceive by the senses nothing that seems to me more certain."[162]

In contrast to the widely accepted reception of Descartes as a "theoretical mechanist" and a pronounced opponent to sensory epistemology, Grene observes here that Descartes "seems positively to glory in the evidence of the senses, in what can be seen and touched."[163] Importantly, the evidence Descartes provides here is not simply given in ordinary perception; instead, it is the product of laborious hands-on study.

When Descartes performs a vivisection of a rabbit to investigate the processes of systole and diastole, he is clearly an anatomist at work, someone who contemplates the complex functioning of the living body while he performs an *artificial dissection*, a term coined by Andrew Cunningham to flesh out the experimental attitude of pre-1800 anatomists.[164] During the anatomical experiment with a living rabbit, Descartes examines these physiological processes by tying up the aorta using his hands to manipulate the natural course of the blood. Only by artificially blocking the bloodstream to and from the heart can he observe the workings of the blood. He differentiates here between what we can perceive "directly" with our eyes and hands, and what needs to be established through an *experience*, that is, an experiment that makes it possible to see or feel something we cannot perceive with our ordinary senses. Obviously, it is difficult to make a strict demarcation between ordinary and what might be called a weak form of "engendered" perception, since cutting open a dead or live body also appears to not belong to the domain of our ordinary perception. The transition between ordinary and enhanced modes of perception is fluid, or one could say that the two modes are intricately interwoven in the act of experimental observation, as all modes of observation are grounded in or built on ordinary perception.

The artificiality of contrived observations is sometimes countered rhetorically with an emphasis on direct experience (with one's own hands, with one's own eyes); most important, perhaps, are the minute descriptions that give credence to that which can only can be seen under specific conditions. Descartes's reports provide strikingly palpable descriptions of the manual work of dissecting living and dead bodies, demonstrating how not only observing but also knowing through touching is practiced as part of the experimental investigation of the phenomena of life. Anatomists such as Harvey and Descartes himself did not just engage in the observational practice of manual dissection; with more and more sophisticated experiments, they made the unseen processes of physiological speculation appear as sensory phenomena in the anatomical theater. Here, as Cunningham writes, life phenomena became tangible and visible to the skilled touch and vision of anatomists.[165]

So what does it mean that Descartes's epistemological masterpiece, the *Meditations*, had been drafted with hands that had been trained not only in writing with quill and ink but also in handling knives, tweezers, sponges, and other anatomical instruments?[166] It is significant for our understanding of Cartesian knowledge theory that Descartes gave shape to his new epistemology during a period when he supplemented his studies in geometric optics with extensive work in anatomy. The specifically manual and experimental nature of anatomy is manifest in his anatomical reports and deserves, in my view, special attention when one reads accounts of his later epistemological work, the *Meditations*. To understand the epistemic significance of

Descartes's anatomical engagement, we need to take a closer look at a prevailing knowledge model in early modern medical discourse.

A Fernelian Knowledge Model

The early modern conceptualization of anatomy as a hands-on investigative endeavor and physiology as an exclusively theoretical discipline is indebted to the writings of Jean Fernel (1497–1558), a sixteenth-century professor in philosophy and medicine. He was an erudite scholar well-versed in ancient medical scholarship who made a career in medicine and became court physician to King Henri II of France.[167] Fernel's medical magnus opus, which introduced the term *physiologia* to medicine, was first printed in Paris in 1554 and then quickly republished in other European cities as a comprehensive pocket-size medical manual.[168] The *Universa medicina*, as this work was titled in later editions, became a bestseller, with more than a hundred editions printed before the end of the seventeenth century.[169]

Fernel's medical work decisively shaped the conceptualization and understanding of anatomy from his time until well into the eighteenth century.[170] When Descartes, who had read Fernel, turned to the systematic study of life phenomena, his studies in medicine had surely been framed and guided by the Fernelian paradigm. Fernelian physiology was concerned with matters that go beyond the realm of ordinary perception, such as the body's smallest constituents, understanding what ultimately causes movement and change, and the explanation of complex bodily functions such as breathing, sensing, food uptake, and reproduction.[171] The Fernelian paradigm takes the discipline of physiology to be a proper science (*scientia*) that contemplates the phenomena of life in a purely theoretical manner, dealing in reasoning and searching for causes. Descartes's *Treatise on Man*, published posthumously in 1662, illustrates the persistence of the Fernelian paradigm in seventeenth-century natural philosophy. According to Cunningham, it is "precisely a discourse on the sub-divisible structures of the body, on the necessary mechanics of its workings beyond the realm of the senses, and on the grand functions of the body, and it is composed on the Fernelian model."[172] It deals with the smallest parts, invisible to the naked eye, speculating on how these induce movement and changes and how they bring about bodily functions.[173] Interestingly, this physiological treatise also draws on anatomical knowledge. In June 1632, Descartes writes to Mersenne that he has included all that he wanted concerning inanimate bodies in his treatise *The World* (*Le monde*) and that "it only remains for me

69

to add something concerning the nature of man," an inquisitive endeavor for which he had turned to anatomy.[174] At the end of the same year, Descartes reports that he had been instructing himself in anatomical experiments, writing to Mersenne: "I anatomize at the moment the heads of different animals, to explain of what imagination, memory etc. consists of."[175] We can therefore infer from Descartes's correspondence that he was actively engaged in anatomical investigations while also working in 1632 and 1633 on a treatise concerning the nature of man. Within the Fernelian knowledge model, physiological suppositions were the outcome of theoretical cogitation, but they depended on the experimental knowledge gained from anatomy.[176]

An anatomical investigation begins with the structure, relation, and connection of body parts. Cunningham describes it as an "integral three-stage investigation" that moves from structure further to explanations of actions, and then to the use and functions of body parts.[177] In "old anatomy" the theory of function is not the starting point but the end point: it is built on sensory evidence about structure, not the other way around.[178] Theory became grounded in the practice of manual investigators who understood research into life phenomena as a complex undertaking of the manual and the mental—the art of anatomy.

It is important to keep in mind that the same person could practice both disciplines.[179] Descartes, for instance, is a marked example of an investigator who switched between anatomical and physiological modes of inquiry.[180] His treatise built on recent anatomical studies, such as Harvey's descriptions of the circulation of the blood, and most likely also on anatomical experiments he performed himself.

Cunningham reminds us that anatomy was understood as the "senior discipline" and "the premier investigative, experimental discipline in the investigation of the phenomena of life."[181] Anatomists did not put to test the theories of physiologists. In fact, quite the reverse was true, as the theoretical discipline of physiology firmly built on the fruitful ground provided by anatomy. Thus before 1800 *theoretical physiology* was built on *investigative (experimental) anatomy*, or science built on art, as Cunningham put it.[182]

Breaking New Ground in Epistemology:
In One Hand a Knife, in the Other a Quill

We can imagine Fernel's model as a construction with anatomy at its base and physiology reaching out from the top of the building into the lofty realms of abstract thought. Whoever enters this building needs to pass on from hands-on anatomy to physiological cogitations, rising upward toward higher verities.[183] However, we must be careful to understand the building metaphor as a dynamic, not a static, model: a working space in which the investigators continuously need to move downward in order to head upward again. Cunningham thus shows how the *art* of manipulation is a requirement for the *scientia* of theoretical physiology: "For Fernel physiology was by its nature a discipline which took empirical information, sense evidence, only as its starting point, but which then soared aloft above such mere sense experience in order to reach truth by philosophising."[184] Despite anatomy's preeminent role in knowledge formation, it is the theoretical science of physiology that prevails in the Fernelian knowledge model. Within the Fernelian paradigm, the sensory and manual exploration of things can only provide a starting point in the process of knowledge formation; everything beyond that, and what really matters, takes place in the abstract realms of theory and speculation.[185] Nonetheless, it is this acknowledgment of the manual and the experimental as a necessary starting point in knowledge-making processes that provides a fruitful framework for my radical rereading of the Cartesian wax argument in the context of Descartes's anatomical practice.

I argue here that Fernel's take on medical knowledge formation offered Descartes a model in which he could fit his criticism of theory-laden and bloodless teachings of medicine in the Scholastic tradition, his newly discovered interest in instructing himself in the art of dissection, *and* his personal penchant for the theoretical sciences combined with a strong belief in the capacities of his mind. My analysis shows how in his epistemological writings Descartes addresses the challenges of an intimate intertwinement between the mental and the manual that he encountered in the pioneering practice of seventeenth-century life scientists, such as himself and Harvey.

Consequently, Descartes's anatomical investigations and his familiarity with the Fernelian model put the wax argument of the Second Meditation in a different light. My thesis is that the mental does not triumph over the manual in the famous wax meditation. Instead, I propose that the central passage of the Second Meditation is based on the Fernelian model of knowledge. Descartes does not dismiss the insights he gained previously from his hands-on explorations of a piece of wax. On the contrary, the moment that the piece of wax is probed by hand in the Second Meditation is central to the following meditations or cogitations: it provides the meditator with the necessary starting point for a cognitive quest for indubitable knowledge.

Hence, to read the wax passage in Descartes's *Second Meditation* as a categorical dismissal of sensory experience and hands-on knowledge is to misread the very architecture of Descartes's body of thought. The Fernelian model that takes the interdependence of manipulative and contemplative sciences as the beginning point of knowledge formation in the sciences provides, in my view, a much better framework for a reading of Descartes's epistemological cogitations. My thesis is that Descartes's new epistemology was grounded in the new sciences of his time and that Descartes extrapolated Fernel's model of manipulative and contemplative investigation of life phenomena to a general theory of knowing and knowledge making. But before exploring this further with a radical rereading of the so-called wax argument as a proper wax *experience*, I will discuss one more historical example of an experiment that Descartes performed. The evocative historical setting will set the stage for the philosophical discussion of the next chapter.

Setting the Stage: The Camera Obscura Experiment

In the same period in which Descartes instructed himself in anatomy, he maintained close contact with the famous polymath Constantijn Huygens (1596–1687), who assisted him with his dioptrical work for several years (from about 1635 to 1638) but without experimental successes, perhaps due to the mediocre work of the lens grinder that thwarted the collaborative efforts.[186] Nonetheless, with the help of Huygens, Descartes published *Optics*, one of the essays that followed the *Discourse*. Descartes's treatise on optics not only contained his "infamous physico-mathematical account of refraction" but also reported at length on his anatomical experiments, with extensive elaborations on the workings of the eye and human vision.[187] Descartes had already been engaged in optical studies in Paris before he went to the Low Countries. Seventeenth-century dioptrics, the study of light, vision, and refraction, can best be understood as a "mathematical blend of contemplation and manipulation."[188] Fokko Jan Dijksterhuis gives a detailed account of Descartes's activities that "were not just intellectual but equally immersed in material pursuits with which his physico-mathematical reflections were closely entwined."[189] In 1625, Descartes joined a Parisian scene that Dijksterhuis has described as being "heavily engaged with mirrors and lenses, trying to devise and realise the shapes that would focus rays perfectly."[190]

In 1627, shortly before Descartes left for Holland, he successfully collaborated with the mathematician Claude Mydorge and lensmaker Jean Ferrier, with whom he managed to produce a hyperbolic lens. Dijksterhuis notes that Descartes is often portrayed as "the theorist", who worked closely with a "draughtsman" and an "artificer," though he adds the success of this three-some must be ascribed to their collaborative effort that benefited from their individual skill sets.[191] Dijksterhuis emphasizes that Descartes's dioptric studies in Paris show that it is ill-founded to strictly separate manipulative from contemplative efforts in this collaborative work. This is even more so with respect to Descartes's pursuit of optical problems by means of ana-tomical studies of the eye. In this context, Dijksterhuis points to the con-siderable amount of experimental work that Descartes conducted in these years on a variety of eyes, together with the anatomist Plempius, much of which found its way into his chapter on the eye.[192] He emphasizes the dif-ficulty for a historian of mathematical sciences "to distinguish inquisitive from inventive activities" because "contemplation and manipulation are almost completely interwoven."[193] Descartes was hence already well famil-iar with experimental research before he arrived in Holland. I can imagine, however, that it must have been quite an experience for him to move out of Parisian salons and lensmaker workshops into anatomical theaters and slaughterhouses to spend his time in long sessions at dissection tables. The next historical example of an optical experiment that Descartes performed himself will set the stage for a radical rereading of the famous wax passage in Descartes's *Meditations* (see chapter 2).

In *Optics*, Descartes describes an experiment in which he wanted to find out how light is broken through the lenses in our eyes and how we can see things (figure 1.5).[194] To find out how our eyes work, Descartes inge-niously combined anatomical and optical experiments with the techniques of a camera obscura. In the fifth chapter, which is devoted to "the images in the eye," he gives a detailed description of an experiment that he probably performed himself. With this experiment he showed that the images we perceive from the things that surround us are cast on the back of our eyes upside down.[195] He provides detailed instruction on how we can observe the way in which images are formed on the retina. First, he explains that the mechanism of this phenomenon has already been well illustrated with an experiment simulating the effect of a pinhole camera in which the observer finds himself in a fully darkened room with only one light source—a small hole in the wall in which he inserted a glass lens; a white cloth is hung at a certain distance from this hole inside the darkened room. The objects on the other side of the wall, outside the darkened room, are projected through the pinhole on the cloth. Descartes modifies this experiment by suggesting that

73

his readers insert into the hole not a glass lens but "the eye of a newly dead person, or failing that, [the eye] of an ox or some other large animal."[196] We can assume that Descartes did indeed perform this experiment himself as he writes in detail about the difficulties involved in conducting it correctly in a letter to Mersenne from 1638:

> For the one who will cut the eye of an ox from the sort with which one can see the same as in the darkened room [*la chambre obscure*], as I have described it in the *Optics*, I assure you that I have made the *experience*, and that if it is done without much care and without any precautions, it is not likely that it will succeed; but I will tell you how to [do it]. I took the eye of an aged ox (one has to attend to this, because those of the young calves are not transparent), and having prepared half [of the inner skin] of an eggshell, which is fitted for the eye to be put inside and moved without changing its shape, I have cut around it with very sharp scissors and a little blunt at the tip the two skins [layers], *corneam* and *vueam*, without damaging the third one, *retinam*. And the size of the round piece that I cut out was only about as big as a coin (*sous*), and it had the optical nerve at its center. Then, when it was cut in this manner all around, without it being moved from its place, I only pulled at the optical nerve, and it came off with the *retinam*, which broke without damaging the vitreous humor at all, in such a good manner because it was covered by my eggshell, I could see behind it what I would like; because the eggshell was transparent enough to achieve this effect. And since then I have shown it to others in the same manner, even without an eggshell, with a paper behind it. It is true that the eye easily wrinkles a bit on the front, & because of that it renders less perfect images; but one can prevent that by pressing a little bit with the fingers at the sides, or also by taking the eye of a just freshly slaughtered ox and in keeping it always in water, until it is taken out for the test, even while one cuts the skins, until it has to be adjusted in the eggshell. That's it for your first letter.[197]

Descartes assures Mersenne that he performed the experiment himself, and he elaborates on the difficulties involved in performing it successfully. However, as Descartes explains, Mersenne need not worry, as he will tell him how to do it correctly; he then provides detailed instructions on how to prepare the ox eye for this experiment. A similar account of the full experimental setup can be found in the fifth chapter of *Optics*:

But you may become more certain of this if, taking the eye of a newly deceased man, or failing that, [the eye] of an ox or some other large animal, you carefully cut away at the back [of the eye] the three skin [layers] which enclose it [specified more accurately in the above letter to Mersenne], so as to expose a large part of the humor there [figure 1.5, M] without spilling any. Then, having covered [the exposed humor at the back of the eye] with some white body thin enough to let daylight pass through, as for example a piece of paper or an eggshell [figure 1.5, RST], you will put this [carefully prepared] eye in the hole of a specially made opening in the wall [figure 1.5, Z] so that the front [of the eyeball; figure 1.5, BCD] faces a place where there are various objects lit up by the sun [figure 1.5, VXY]; and its back [where you placed the white body; figure 1.5, RST] faces the inside of the room where you are standing [figure 1.5, P], and no light must enter [the room] except what comes through this eye, all of whose parts you know to be entirely transparent [figure 1.5, C to S].[198]

Descartes's description calls to mind the widely circulating recipe literature of his time, providing the reader with a quite detailed set of instructions to perform an experiment.[199] To modern ears, this style of writing resonates with the later genre of the scientific report and the protocol. We can gather from his remarks that it requires quite some dexterity to dissect the two layers of skin covering the eyeball without breaking open the third, the retina, and letting its inner fluids pour out. After the skin layers have meticulously been removed, the fiddly task involved inserting the eye firmly into the small opening at the correct angle in an almost dark, or only candlelit, chamber. His own amazement and satisfaction ring through when he describes the effect that can be observed if the experiment is conducted correctly: "For when this has been done, if you look at this white body [the white paper or eggshell covering the back of the eye; figure 1.5, RST], you will see there, not perhaps without admiration and pleasure, a picture [*peinture*] ingeniously representing all the objects in perspective that are located outside [figure 1.5, VXY]—at least if you do it in such a way that this eye retains its natural shape, in accordance with the distance of these objects."[200]

Descartes does not want his readers simply to observe the spectacle that presents itself to the eyes of the experimenter. He also instructs them to gently squeeze the carefully prepared eye to see how this manipulation affects the visual depiction of the objects projected onto the white paper or eggshell covering the back of the eyeball that has been inserted into the

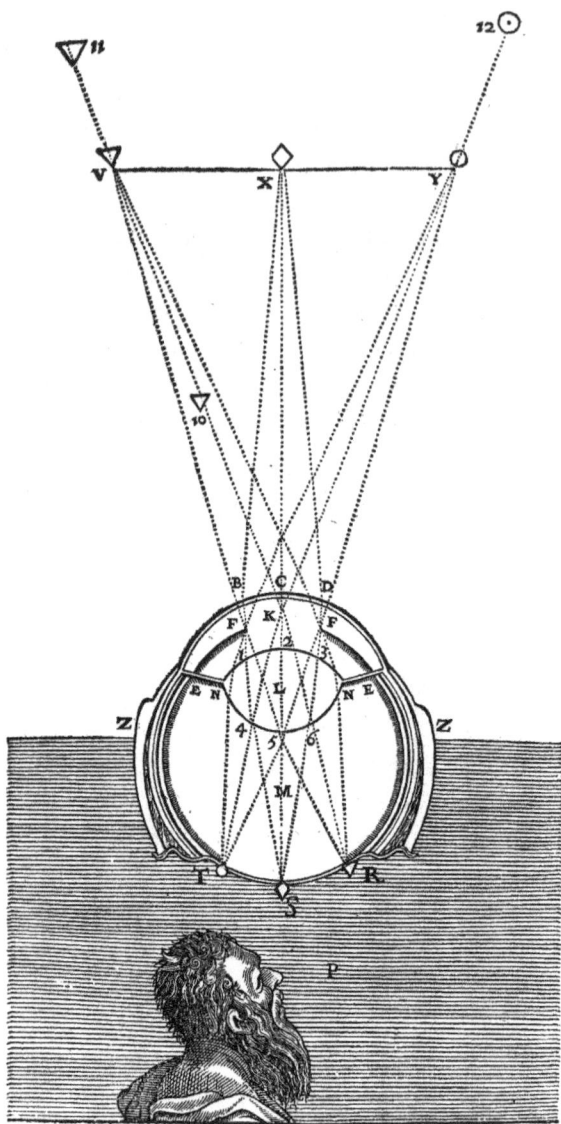

1.5 ★ Illustration accompanying Descartes's description of his ox-eye experiment in the Fifth Discourse of his *Optics* (*La dioptrique*): "Of the Images Formed in the Eye" ("Des images qui se forment sur le fonds de l'oeil"). Reproduced from Descartes, *Discours de la méthode . . . Plus la dioptrique* (1637), 36.

1.6 * Camera obscura, from Athanasius Kircher, *Ars magna lucis et umbrae* (1646).

small opening . As he explains, "For if you squeeze it more or less than is right, no matter how slightly, this picture [*peinture*] will become less distinct. And it should be noted that we should squeeze it a bit more, thereby making its shape a bit longer, when the objects are very near than when they are farther away."[201]

What struck me when I read about this experiment is the way in which Descartes combines techniques from different traditions (anatomical studies, the techniques of a pinhole camera) with the methods of a new experimental empiricism. Note how he places the experimenter into the camera obscura—just like the polymath and Jesuit scholar Athanasius Kircher pictures himself some ten years later in a famous drawing (figure 1.6)—turning this wondrous device into a "laboratory setting": a controlled environment with artificial conditions (no light except through a specially designed opening) in which it becomes possible for the experimenter to undergo particular artificially produced *experiences* that should give him insight into the workings of human visual perception. The ox eye experiment—an *experience* that is the result of a laborious and skillful manual operation—illustrates beautifully how an inquiry into our ordinary sense of sight involves a quite extraordinary experimental setup.

This experiment illustrates the difference between artificially contrived occurrences and the meaning of the word *experiment* in the Scholastic sense of mere everyday experience. We can see how an *experience* is created that is heavily dependent on manual modifications of "natural conditions." To observe and describe the phenomena that play a role in visual perception, Descartes first had to become a manually skilled hands-on practitioner trained in anatomy and well-versed in optical experimental setups. Furthermore, Descartes's *Optics* not only gives detailed explanations of the anatomy and physiology of our visual faculties but also, and perhaps most importantly, provides, as Dijksterhuis observed, "a means of perfecting vision by extending the capacities of the eye."[202] The ox-eye experiment makes the increasing "experimentalization" of anatomy palpable. At the same time, it illustrates the impossibility of reconciling the inherently artificial nature of the experiment with an Aristotelian sense-based epistemology. Descartes mobilizes here the potentialities of experimental research for investigations geared toward improvements of nature and enhancements of sensory organs rather than restricting himself to observations and descriptions of phenomena in the course of nature, a trend that has become characteristic of contemporary biomedical and life sciences research.

From Ox Eyes to Wax Balls

Let us imagine an ordinary day that Descartes spent at his lodgings anatomizing an ox eye at leisure.[203] Where is he standing, what is he holding in his hand, and what is doing? Descartes begins the Third Discourse of his treatise on the eye ("De l'oeil") in *Optics* with a schematic illustration of an eye's anatomical composition (figure 1.7). The image itself does not convey any difficulties or any part of the laborious process of anatomizing an eye, which Descartes refers to in his description: "If it were possible to cut the eye in half, without the liquids with which it is filled escaping, and without any of its parts changing their places, so that the plane of the section passed right through the middle of the pupil, it would appear such as it is represented in this diagram."[204]

To attain such a diagram of the eye, he writes, we would need to imagine the possibility of cutting an eye in half, without having all the fluids contained in the eyeball flowing out and without any of its parts moving about during the act of dissection. To arrive at the schematic illustration and the mathematical measurements it makes possible, Descartes needs to extend his anatomical experiments into abstract realms, into a virtual space in which he can do away with all the messiness that he faced when trying to dissect an animal eye. In imagining a dissection of the eye freed from all these "impurities," Descartes engages us in a thought experiment that results, one might say, in

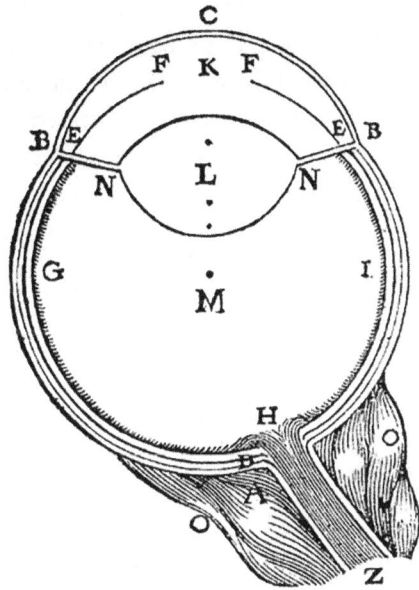

1.7 ★ Illustration of the eye's anatomy, indicating the thin layers of the eyeball and the optical nerve, in the Third Discourse of Descartes's *Optics* (*La dioptrique*), "Of the Eye" ("De l'oeil"). Reproduced from Descartes, *Discours de la méthode . . . Plus la dioptrique* (1637), 26.

the neat and clean scientific diagram. What precedes this thought experiment, though, is a real experiment. Imagine Descartes in the course of his anatomical examinations of the eye, which he so vividly described to Mersenne only three years before the publication of the *Meditations*.

As we have seen, Descartes reported to Mersenne at length on the practicalities of such a manual operation. We can infer from his detailed description of the inner parts of the eye that the explanatory drawing is the result of minute motor activity and laborious dissection work:

> ABCB is a rather hard and thick skin which constitutes something like a round vessel in which all the eye's interior parts are contained. DEF is another fine skin [layer] which is stretched in the manner of a tapestry inside of the former. ZH is the optic nerve, which is composed of a great number of small fibers whose extremities are extended throughout the space GHI, where, mingling with an infinity of small veins and arteries, they compose a sort of extremely tender and delicate flesh, which is like a third skin [layer] that covers the entire inside of the second. K, L, M are three kinds of very transparent glairs or humors which will fill the entire space contained within these skin [layers], and each of them has the shape which you see represented here.[205]

79

Indeed, we get the impression that all further theoretical explanations of the physiology of this sense organ are based on thorough hands-on work conducted at the dissection table and on optical experiments. We can envision Descartes standing at a dissection table, an array of instruments at hand, including different sorts of knives, sponges to mop up blood, saws, scissors, and pincers. We see him bending over the body of a dead cow, closely examining its head. We can imagine that he is engaged in the delicate operation of isolating an intact eyeball from the socket, carefully cutting off the optical nerve and separating the eye from the corpse and then struggling to handle this slippery sphere. With which techniques did he manage to dissect and investigate this perfectly spherical object while preventing it from turning into a difficult to handle, soggy mass of sticky tissue? We can envisage the difficulties of holding on firmly to this smooth, moist, and slimy eyeball without unwillingly tearing its fine layers of skin. How did he fasten the ox eye to the table to "anatomize" it at length? How did he go about placing the tip of his dissection knife steadily on its surface, ready to make the first cut? And how did he proceed when the inner fluids streamed out too quickly? We can see, as it were, how his hands explored different ways of handling different body parts, learning which instruments are best to use to cut, pull out, remove, or perforate different parts of bodily tissues.

All of this makes me wonder: How might his anatomizing of ox eyes have fueled his thinking about wax balls? Associating the eyeball in the hand of the anatomist with the piece of wax in the meditator's hand elicits a comparison that is highly evocative, yet the images invoked by the historical material are so tangible and suggestive that, to my mind, the similarities cannot be ignored. Keeping the historical context of seventeenth-century anatomical research practices, the Fernelian paradigm, and this account of the slippery ox eyeball in mind, I invite you now to a radical rereading of Descartes's wax meditation.

Descartes's Manual Meditations

Meditation is something you do.

Steve Hagen, *Meditation Now or Never*

hands-on involving or offering active participation rather than theory: hands-on practice to gain experience.

epistemology | noun Philosophy. ORIGIN mid 19th cent.: from Greek epistēmē "knowledge," from epistasthai "know, know how to do."

New Oxford American Dictionary, 2nd edition

Writing *Scientiae*: Literary Innovations
at the Intersection of Spiritual Exercises
and Hands-On Experiments

Shortly after the first Latin edition of René Descartes's *Meditations on First Philosophy* (*Meditationes de prima philosophia,* 1641) appeared in print, Descartes asserted in a much-cited letter to Princess Elisabeth of Bohemia from June 28, 1643, that his pursuit of knowledge had never been restricted to thinking alone: "And I can say with truth that the principal rule I have always observed in my studies, and that which I believe has served me the most in acquiring some bit of knowledge, is that I never spend more than a few hours each day in thoughts which occupy the imagination, and very few hours a year in those which occupy the understanding alone, and that I spend the rest of my time *giving the senses free rein* and resting the mind."[1] Descartes states here that he prefers to restrict purely theoretical cogitations to a minimum, but what is perhaps even more important is that his remarks imply that his thinking about what knowledge is and how one can acquire knowledge has substantially been shaped by experiences that cannot be located in the mind alone. What are the philosophical implications of the claim that we cannot separate Descartes's investigative doings from his epistemological thinking?

In this book I reassess the constitutive role hands-on engagements played in Descartes's practices of doing science and writing epistemology. I argue that the Cartesian epistemological project is not the result of pure thinking but was decisively shaped by a hands-on attitude toward knowledge production.

How did Descartes write about hands-on experimental observations? What were the literary tools of the early seventeenth century that paved the way for the literary technologies of the emerging modern natural and medical sciences? The close reading in this chapter investigates the rhetorical forms that René Descartes deploys in his famous *Meditations* (*Meditationes de prima philosophia,* 1641; here quoted from the authorized French edition, *Les méditations métaphysiques,* 1647). Descartes's literary innovations exemplify new tools for writing about and reflecting on knowledge making in early seventeenth century. His famous discussion of a malleable piece of wax shows how he experimented with literary forms of the Christian tradition (Ignatian versus Augustinian forms of meditation) against the background of his own experiences as an experimental anatomist. In contrast to most conventional readings of the "wax argument," and in line with my historical reconstruction of Descartes as a hands-on practitioner, I argue here that Descartes's literary experiments exemplify a broader shift toward experimentation in writing *scientiae*. Thinking and writing about knowledge making was, it seems, not a contemplative business constrained to mental activity but an active engagement that took place while he was feeling out the inner parts of living and dead bodies and experimenting with manipulative interventions. How can we understand the literary innovations in Descartes's *Meditations* and in his reports on anatomical investigations in the context of the rising need for new writing technologies that came about with the emergent knowledge practices of the "new experimentalists"?

My textual analysis shows how spiritual literary forms are transformed in the (bloody) hands of Descartes into means for shifting from the "meditator" to the "experimenter."[2] More generally, I suggest that seventeenth-century new scientists, of whom Descartes was one, developed important "precursors" to modern literary tools when engaging with complex mixtures of literary forms and hands-on knowledge practices in anatomy and the natural sciences. Descartes's textual experiments point, for example, to literary innovations, such as the "procès-verbal de l'expérience," that gained importance in the seventeenth century, and possibly anticipated modern genres of scientific reports and protocols.[3]

In this chapter I provide a close reading of a key passage in Descartes's *Meditations*. My aim is to engage my audience in a radical rereading of the wax argument as a "wax *experience*." The next chapter takes a closer look at the methodological and philosophical implications of this reassessment of Cartesian epistemology from a hands-on perspective.

Descartes's Meditation on Wax:
A Close Reading

The Second Meditation in Descartes's *Meditations* contains perhaps one of the most widely cited passages from his oeuvre: the so-called wax argument, here quoted from the 1647 French edition (see figures 1.1–1.3).[4] In this chapter I scrutinize this text fragment for traces of the intimate engagement between Descartes's two ascribed roles: the role of modern epistemologist, or rationalist thinker, and the role of new scientist and experimental hands-on practitioner. Whereas Descartes's famous meditation on wax is generally taken to epitomize the rationalist stance of a philosopher who prefers the mind to the hand as an instrument in knowledge making,

I reread this passage in light of the hands-on attitude that was constitutive of the new sciences of his day. How can we rethink the birth of modern epistemology from this perspective? The following close reading of the famous wax passage explores and exposes an intricate intertwinement of hands-on attitudes and hands-off theorizing from which a new model of the knower emerged.

The section of the Second Meditation in which the protagonist scrutinizes a wax nugget forms the heart of what is often referred to as the wax argument.[5] It is commonly argued that the main aim of the Cartesian wax argument is to convince Descartes's contemporaries that knowledge is not—as they were used to thinking—first and foremost obtained through the senses, but that to find true knowledge, it is necessary to turn away from the senses and turn to reason. Descartes had to anticipate that his audience would not easily let go of the deeply ingrained beliefs founded on the epistemological primacy of the senses. He had to find a way to convince his readers to embark with him on a journey to an unknown destination that would go against the prevailing dogma that "nothing is in the intellect that was not first in the senses."[6] With the *Meditations*, Descartes set out a path that was to transform those who followed it through to the end. His *Meditations* engage the reader in a transformative experience that reverses the relation between the senses and the intellect, thus replacing an Aristotelian sense-based conception of knowledge with a new epistemology grounded in the Cartesian cogito. Hence, a common and widely shared interpretation of the wax argument goes roughly like this: Descartes is a rationalist thinker who lets the meditating "I," the protagonist of the *Meditations*, voice the new epistemology that certain knowledge comes from rigorous thinking and not from the hands or the eyes. As such, for the Cartesian knower, the foundation of knowledge resides in the mind and not in the senses. Yet perhaps this text passage is not as straightforward as it might seem today to our modern eyes.

In this chapter I argue that such a reading ignores the subtleties of the epistemological dilemma faced by the seventeenth century's new scientists and unquestioningly projects a modern thinking in mind-body and rationalism-empiricism dichotomies onto this rich historical source text. My rereading proposes that Descartes's *Meditations* were designed not only to win over a reluctant audience but also to comment and reflect on the shifting meaning of concepts such as *experience* and sensual or ordinary perception in light of the experiment as a powerful new methodological device for natural philosophical inquiry. I suggest that the wax passage not only records the triumph of reason over the senses but also documents how practitioners struggled to get a grip on the new ways of doing empirical observations and explores possible conceptual and epistemological frameworks for these doings.

To begin, it is significant that this passage is often referred to as the wax *argument* in the Cartesian canon and scholarship. The term implies that Descartes chose to convince his readers by discursive argument, but is this indeed the case? Let us take a closer look at the original source. How does Descartes write about the wax? Does this passage have the structure of a logical argument? When we look at the original text, I would say this is not what first comes to mind. Descartes carefully introduces the wax passage with the following remarks: "But I see what it is: my mind enjoys wandering off and will not yet submit to being restrained within the bounds of truth. Very well then; just this once let us give it a completely free rein, so that after a while, when it is time to tighten the reins, it may more readily submit to being curved."[7]

What happens here? We encounter a first-person narrator who is often conflated with the author himself. We should keep in mind, however, that Descartes constructed here a literary figure: the meditator. With this term, scholars have emphasized that the voice constructed in the text does not refer to an uninvolved raconteur but rather invokes an engaged and embodied subject that undergoes a meditative experience.[8] I will push this argument even further, foregrounding the rich meaning of "experience" in the wax passage: I propose to picture the performer of this wax *experience* as a meditator turning into an experimenter, who reports on firsthand observations to *make* an epistemic argument by *hand* and mind.

This first-person narrator of the *Meditations* is not further identified. In today's terms, we could imagine this meditating "I" as he/she/they, though traditionally scholars have gendered the meditator as a male figure. In my radical rereading I will refer to the meditator as she/her and turn her into an experimenter the moment she takes up the piece of wax. I chose to do so for two reasons: gendering the meditator as she/her functions for me as a Brechtian alienation effect; it presents the text of the wax argument in an unfamiliar way, disrupting a conventional reading experience and preventing us from simply conflating the first-person narrator with the historical figure of the author, Descartes. Second, it emphasizes that the meditator invites us to identify ourselves with this literary figure, to imagine ourselves in the main protagonist's place performing a wax *experience*. Being a woman,

85

I found it rather odd to report my reading experience in the third person of a male reader. Therefore, I chose to use female pronouns to highlight that here I present *my* reading of the wax passage, the reading of a twenty-first-century female scholar.

In this introductory paragraph, the first-person narrator tells us about her inner thoughts. The detailed description of her mental state creates the impression that we enter, as it were, into her mind. There is no explicit reference to a narratee in the text, but we can ask questions about the implied reader. Am I, the reader, asked here to follow a stringently structured argument? The meditating "I" appears to invite me to an imaginative and associative exploration rather than expecting me to follow an exercise in logical reasoning. The meditator calls on me, the reader, to enter into a specific state of mind, to let my thoughts wander off, and to refrain from calling on my intellectual capacities to follow a rigorous line of argument. When she observes her own thoughts wandering off, she decides to give them free rein: "But I see what it is: my mind enjoys wandering off and cannot yet contain itself within the bounds of truth. So, let us loosen its reins once more, so that, removing them hereafter gently and at the right moment, we can then regulate and direct it [the mind] more easily."[9]

Instead of disciplining her mind, she allows for a playful moment, not in order to stop her meditations or quest for knowledge but perhaps, rather, to see what happens if she extends her exploration beyond her discursive capacities. She makes no preparations to step out of the meditation, but it seems as if she allows herself a well-defined time to experiment with an impulse that she has in the course of this mindful moment.

The introductory remarks read as an invitation to join the meditator in her mental preparations. Picturing the narrator as a meditator affects the narratee: the meditator shares an experience with us instead of relating a narrative from a distant position. The implied reader is not addressed as an opponent who has to be convinced; instead, I get the impression that the meditator primes herself—and us, the readers—for an imminent experience. So, what happens after the meditator has let her thoughts wander off?

> Let us consider the most common things and what we think we understand most distinctly; that is, the bodies which we touch and see. I do not mean to talk about bodies in general, for these general notions are usually more confused, but of one in particular. Let us take, for example, this piece of wax, which has just been taken from the honeycomb: it has not yet quite lost the sweetness of the honey

it contained, it retains some of the scent of the flowers from which it was gathered; its color, shape and size are plain to see; it is hard, it is cold, [as] one touches it, and if you rap it, it makes a sound. In short, it has everything that can be known distinctly of a body. But, while I speak, look as it comes near to the fire: what rests from its taste fades away, the smell disappears, its color changes, its shape is lost, its size increases; it becomes liquid, it gets hot; one can hardly touch it, and if one raps it, it no longer makes a sound.[10] (See figure 1.2 for the original French text.)

With the appearance of the wax, there is a conspicuous change in the idiom. Descartes switches the authorial voice from the contemplative mode "Let us consider" (*Commençons par la consideration*) to the active mode of "Let us take" (*Prenons*). The introductory paragraph is dominated by words that pertain to mental activities, such as "consider" (*consideration*) and "think to understand" (*croire comprendre*). In the second paragraph, the wax is not merely contemplated but the target of a range of actions: taking it up, testing its texture and temperature, rapping it, holding it near the fire. The wax is described not only with state-of-being verbs ("it is hard," "it is cold") but also with action verbs to emphasize a process rather than an object: it *has not quite lost* the taste of the honey; it *retains* the smell of the flowers; it *loses* its taste, smell, shape, and sonic qualities; it *changes* its color; it *increases* in size; it *becomes* hot and liquid.

Of course, it is a well-conceived choice to bring a piece of wax on stage for a meditation on the ephemeral, fleeting, and deceitful character of sensual impressions. Yet I want to draw attention to the fact that Descartes chose to write about the wax ball not in a contemplative but rather in a performative mode that pictures the meditator *in action*. The object is not rendered as an inert entity but transforms in the manipulating hands of the meditator. Descartes's choice of words is important: Why is the first-person narrator not simply calling on our imagination to *think* about a piece of wax? Indeed, the meditator does not ask us to *imagine* a piece of wax; instead, she seems to give instructions. She wants the readers to *do* something when she utters the words: "Let us take, for example, this piece of wax" (*Prenons pour example ce morceau de cire*). When I read this inviting request, I do not picture a motionless figure sitting in front of a fireplace, as in a cartoon, with a thought balloon displaying a piece of wax. Instead, what comes to mind is someone in action, someone who extends her hand, picks up an object, raises it to her nose to scrutinize it more closely, and then holds it near the fire. We see her weighing the wax in her hand while she tells us about its origin; we can imagine her probing the soft substance while inspecting the wax and reporting her observations. We might even feel prompted by her inviting gesture to take up a piece of wax and to experience her observations with our own hands and eyes. The investigation of the wax is pictured as a tangible act and not as contemplative imagining. Playing around with the piece of wax in your hand, feeling its soft, smooth texture against your skin, maybe even noticing imprints of your fingertips in the malleable material and sensing how it absorbs the warmth of your body is quite different an experience from placing a chunk of wax in the middle of a table, moving around it, and describing it as an immobile object of spectatorship. These lines of text are evocative of a palpable experience, a thorough examination of a malleable substance that comes closer to manipulation than mere observation.

What I witness here are no longer the inner states of a meditating person in front of a fireplace. The "meditating I" that gives an account of an inner thought world is replaced by an "acting I." Examination is performed here as a *mani*pulation; the object is not subjected to the gaze of a contemplator but is kneaded in the hands of an investigator. With the switch of the authorial voice, the text takes the form of a recipe or a how-to text.[11] We can read this passage like a manual, intended for a reader who will repeat the hands-on exercise to assess the reported observations. This reading allows for an alternative interpretation of the wax passage: not as wax argument, but as a wax *experience* (for a discussion of the term *experience* in the context of Descartes's anatomical hands-on experimentation, see chapter 1). I argue here that the transformative power of the meditative texts is mobilized in this fragment in an unexpected sense: it is not just the piece of wax that is transformed and undergoes different phases in the course of this passage; we also witness here a transformation of the narrator, who turns from a meditator into an experimenter. In my reading, the experimenter does not simply put forward an argument; instead, she employs experiment as a methodological device to explore and demonstrate a problem.

Turning the Meditator into an Experimenter

What happens if we read the wax passage as if we are called on to conduct not a thinking but a hands-on exercise, as if we are invited to perform an experiment?

Approaching the wax passage as a manual exercise opens up different perspectives: it allows us to read Descartes's epistemological considerations against the background of his diverse modes of inquiry into life and nature. From this hands-on perspective we can relate Descartes's writings to the daily activities that he was engaged in at the same period that he probably conceived of and perhaps even drafted parts of the *Meditations*. Considering the wax passage not only as a cognitive exercise but as an invitation to *perform* an investigation, we can integrate Descartes's stimulation of independent thinking (using one's own mind) with his call for performing practical explorations in natural philosophy (with one's own hands and eyes). Earlier examples of this rhetoric can be found in other writings. In *Discourse on the Method* (*Discours de la méthode*, 1637) and *Description of the Human Body* (*La description du corps humain,* published posthumously, 1664), for example, Descartes calls on his readers to perform an *experience* in order to follow his further explanations.[12] I invite you to read the original passage again and to substitute an experimenter for the meditator. What happens?

> Let us consider the most common things and what we think we understand most distinctly; that is, the bodies which we touch and see. I do not mean to talk about bodies in general, for these general notions are usually more confused, but of one in particular. Let us take, for example, this piece of wax, which has just been taken from the honeycomb: it has not yet quite lost the sweetness of the honey it contained, it retains some of the scent of the flowers from which it was gathered; its color, shape and size are plain to see; it is hard, it is cold, [as] one touches it, and if you rap it, it makes a sound. In short, it has everything that can be known distinctly of a body. But, while I speak, look as it comes near to the fire: what rests from its taste fades away, the smell disappears, its color changes, its shape is lost, its size increases; it becomes liquid, it gets hot; one can hardly touch it, and if one raps it, it no longer makes a sound.[13] (See figure 1.2 for the original French text.)

89

The experimenter brings a chunk of wax from the honeycomb into her workshop for further inspection—just as Descartes collected ox eyes, or other body parts of cattle, at the slaughterhouse for the purpose of anatomical investigation back at his lodgings.[14] First the piece of wax is detached from its natural surroundings—we see that investigating it and grasping its nature means to displace and decontextualize it. This isolation from its "natural habitat" and the process of its examination disassociated from any of its

Table 2.1 The Sensory Data of the Wax Experiment

	Wax before being heated	Wax after being heated
Felt temperature	cold	hot
Material condition/texture	hard	liquid
Smell	flowery	none
Handling	easy	difficult
Sonic capacity	sounds when rapped	none

(daily) use contexts turn this thing into an object of "scientific" inquiry. The experimenter gives detailed instructions for how to gather information by carefully investigating the piece of wax with all our senses: tasting it, smelling it, describing its outer appearance (what is conspicuous is the brevity of the visual description in comparison to the other senses) and how it feels: "It is hard, it is cold," and one can touch it. Another instruction then follows: "Rap it with your knuckle" and register what happens. Following this, an experiment is set up: What happens if we hold the piece of wax near fire? In short, the object is observed under different (extreme) conditions—rapped and held close to fire. Next, a new set of observations is made and compared with what already had been established about this object: after the wax is exposed to the heat, "the residual taste is eliminated, the smell goes away, the colour changes, the shape is lost, the size increases; it becomes liquid and hot." It cannot be handled anymore without difficulty because now "it becomes liquid and hot." We can hardly touch it anymore, and it has lost its reverberant sonic quality. We could imagine here a table in which the experimenter notes the changes that have taken place (table 2.1).[15]

After the experiment is terminated, we are to sit back, as it were, look at our notes, and see what they tell us. The detailed report of the manipulative moment formalizes the encounter when hand and wax meet. Reading the wax passage as a how-to text leads me to think about modern forms of observational recordings and how sensory data can be placed in tables to allow us to deduce distinct phases of a material. It makes me aware of how tables create a distance between the experimenter and the object of inquiry: one moment the experimenter still feels the sensation of the smooth texture in the palm of her hand; the following moment this intimate encounter is translated into a tabular presentation, exposing the object of inquiry, while the investigator disappears behind the grid. What the tabular presentation leaves out are the subtleties of the encounter as the wax took on the temperature of the hand that held and probed it, or the flowery smell and fatty feeling that remain on the hands after the wax is dropped into the fire.

We can read the wax passage as part of early modern knowledge cultures familiar with recipe texts and in the context of the emerging genre of the experimental essay.[16] Toward the end of the seventeenth century, genre conventions for textual accounts of experimental observations became institutionalized, for example, with the establishment the Royal Society of Science, anticipating the rise of scientific textual genres such as experimental reports and protocols that became increasingly formalized and standardized in the experimental sciences beginning in the nineteenth century.[17] A modern protocol is a set of instructions that allows for specific experiences and simultaneously provides a template to translate them into sets of "observations" or "data."[18] On the one hand, one could say, a protocol orchestrates an intimate material encounter and ensures the "code of procedure or behavior" that frames and structures this event. On the other hand, a protocol text prepares for a withdrawal, detachment, or act of disengagement the moment this encounter is reported in the form of "observations." Reading the wax passage as a recipe or how-to text and—from an anachronistic vantage point—as an experimental protocol brings to the fore some interesting issues regarding distance/closeness, detachment/engagement, and universality/immediateness that demanded attention when early modern knowledge making became intricately intertwined with experimental hands-on practices. I will return to anachronism as a fruitful research method in chapter 3 on methodology.

In the wax passage, the meditator extends her hand not only to the piece of wax but also to the reader, inviting us to get a feel for the methods of the new sciences. The manipulation of the wax creates an awareness for the problem of what it means to leave behind "ordinary perception" and to turn from observations that take place in the course of nature to artificially contrived experiences that take place in a particular time and space. It is important to pay special attention to the delicate and meticulous way in which Descartes lets the experimenter handle the piece of wax: how she arrives at her insights while gently kneading and squeezing. The manipulation of

91

the piece of wax is staged here as an attentive hands-on engagement, calling attention to the fact that skillful handling becomes crucial to knowledge-making processes. As William Gilbert (1544–1603), a physician and distinguished man of science, asserts in the preface to his groundbreaking book on lodestones, *On the Magnet* (*De magnete*, 1600), written in Latin for an international audience: "To the candid reader, studious of magnetick philosophy: . . . Who so desireth to make trial of the same experiments, let him handle the substances, not negligently and carelessly, but prudently, deftly, and in the proper way; nor let him (when a thing doth not succeed) ignorantly denounce our discoveries: for nothing hath been set down in these books which hath not been explored and many times performed and repeated amongst us."[19] Many scholars before me have argued that Descartes deliberately opens our minds to the whole richness of sensual impressions in the wax passage, yet insisted that Descartes uses this rhetorical move only to demonstrate what we will be required to discard in the following step. In the generally accepted reading of the wax argument, the close examination of the wax only serves to show in abundant detail that sensual experience has no epistemological value at all.

My point is that this passage is not merely geared toward loss, deception, denigration, and discarding. The key moment of the Second Meditation cannot be reduced to a simple rejection of ordinary sensual perception. I take issue with reductionist readings of the wax examination as a straightforward epistemological disqualification of *experience*.[20] Instead, I propose that the experimenter here makes the constitutive role of hands-on experiences for a new epistemology palpable. In my radical rereading, the meditator arrives at an epistemological insight with the help of the experimenter and *through* hands-on experience.

Turning the figure of the meditator into a hands-on experimenter has strategic advantages. What other figure could better embody a new model of the knower grounded in the conviction that truth finding does not depend on religious revelations or classical learning?[21] Descartes's radical conception of the cogito rejected the authority of the book and the learned scholars as authorities of the intellect. Such an attitude chimes remarkably well with the practical approach of experimental natural philosophers and their outspoken do-it-yourself attitude, so to speak. The Cartesian call for a liberation of the intellect and the experimenter's mistrust of passed-on dogmata and inherited knowledge resonate with key aspects of artisanal epistemologies. Pamela Smith has convincingly argued for a reevaluation of hands-on knowledge in understanding the rise of the experimental sciences.[22] While

92

we can discern a dominant trend in historiography of philosophy and history of science that has long obliterated or ignored the relevance of the "manual," there have also been sustained efforts to acknowledge the hands-on attitude and material literacy of artisans as part of a broader history of knowledge formation in seventeenth-century Europe. What can we learn from historical scholarship that redescribes the Scientific Revolution as a result of many hands rather than a few minds?

Edgar Zilsel's (1891–1944) work on the influence of artisanal ways of knowing, which became known as the "Zilsel thesis," ascribes a decisive role in the emerging new sciences to early modern expert makers, in particular in ancient and developing fields of craftsmanship and practical mathematics, including architecture, engineering, shipbuilding and navigational arts, surveying, and manufacture of instruments and measuring tools.[23] Such a reevaluation of the role of artisans in early modern knowledge production casts the Scientific Revolution in a different light. The argument has been significantly broadened in recent years, claiming a central role for a great diversity of knowledgeable makers in professional and domestic contexts in the transformation of natural philosophy into an experimental inquisitive endeavor. Into this historical and historiographical debate, Pamela Smith weighed in with the "body of the artisan," arguing that learned knowers appropriated the hands-on attitude of expert makers and built on their practice-based know-how and material literacy to reform natural philosophy into an actively embodied way of doing natural history and philosophy.[24] The increasing institutionalization of an emerging "new philosophy" in academies and the incorporation of hands-on teaching into universities' curricula raised the status of bodily acquired knowledge. Smith draws our attention to practices of "gaining knowledge neither through reading nor through writing but rather through a process of experience and labor."[25] The subversive thrust of such experimental approaches appears appealing for envisioning a new model of the knower freed from the bookish weight of ancient authorities.

While a persistent ancient scheme of knowledge had firmly located manual operations and bodily doings at the bottom of the epistemological hierarchy and outside the realm of *scientia*, this became increasingly problematic for seventeenth-century philosophers, who went out of the library and into the slaughterhouse, armed with knives and dissection instruments. For early modern medicine—being an art and a philosophy—any strict distinctions between practice- and theory-based disciplines did not fit well.[26] According to Harold Cook, the seventeenth-century revolution in anatomical knowledge arose not so much from new concepts or theories as from epistemic practices that relied heavily on skilled handiwork and material-technical innovations; new modes of thinking and theorizing living, healthy, and sick bodies came with the development of "new investigative techniques" and insights resulting "from difficult investigative labors using new materials."[27] These developments, together with an increasing acceptance of anatomical experiments in making knowledge claims, prompted natural philosophers to rethink the body's role in knowledge making. Rereading Descartes against this background, we can imagine that the figure of the hands-on experimenter can serve as a powerful ally to forge a new epistemological program.

In my reading of the meditation on wax, the experimenter comes to the meditator's assistance. Perhaps to win Descartes's fellow new scientists for his epistemological program? Perhaps the long-established cognitive constructions of Descartes's time proved to be too persistent to be swept away by mental effort alone? The voice of the experimenter speaks to the imagination of an audience that Descartes was intimately acquainted with, and that was already well accustomed to hands-on experimentation. In the eyes of his contemporaries, Descartes likely embodied both—the idea of a new science and a new epistemology. Descartes intended to reach an audience beyond the universities and chose to publish not only in Latin but also in the vernacular:

> And if I am writing in French, my native language, rather than Latin, the language of my teachers, it is because I expect that those who use only their natural reason in all its purity will be better judges of my opinions than those who give credence only to writings of the ancients. As to those who combine good sense with application— the only judges I wish to have—I am sure they will not be so partial to Latin that they will refuse to listen to my arguments because I expound them in the vernacular.[28]

Descartes wrote his *Discourse on the Method (Discours de la méthode pour bien conduire sa raison et chercher la verité dans les sciences; Plus la dioptrique, les meteores et la géometrie, qui sont des essais de cette méthode*, 1637), containing detailed descriptions of the anatomy of the eye, in French, and, as noted above, a French translation of his Latin *Meditations* was published with his approval in 1647. Marjorie Grene observed that the new scientists Harvey and Descartes wanted their books "not to be read simply, but to be worked through as exercises."[29] Grene saw here a correspondence between Harvey's anatomical writings and Descartes's *Meditations*, while maintaining a clear distinction between Descartes's epistemological writings and his anatomical practice. But why? Descartes might just as well have considered to put his anatomical *experiences*, performed under the motto "seeing with one's own eyes and feeling with his own hands," in service of his epistemological program. Not only Harvey but also Descartes encouraged his audience to practice anatomy for themselves. Where Grene sees a break, I argue that the famous meditation on wax reveals a continuum between Descartes's hands-on work in pursuing knowledge and his theoretical considerations about knowledge making. We can observe in the epistemological reflections of this radical thinker an intimate entanglement between hands-on experimentation and innovative writing practices drawing on a diverse array of textual traditions.

Turning a Wax Ball into an "Epistemic Thing"

Ordinary perception is not simply given to our senses to be rejected; Descartes needs to make the concept of "ordinary perception" perceivable for his readership, and to do so he makes use of a contrived experience. It is only by means of a staged "experiment" that the piece of wax becomes this exemplary, ephemeral, shapeless, deceptive object that embodies and signifies the idea of the deceitfulness of the senses. The disqualification of ordinary perception is made palpable in a carefully crafted mise-en-scène. We, the readers, need to get a sense of what it means to question the epistemic value of ordinary sensory perceptions. We first need to grasp the idea of deceitful ordinary perception in order to let it go. And though we let the object go, we gain the method. I argue that the wax *experience* is key to Cartesian epistemology as it invites us to *feel* how perception becomes a matter of manipulation, how it gains in complexity, and how this complexity demands new epistemological vistas and vocabularies. What I show here is that we cannot strictly disassociate the practitioner's doings from the epistemologist's thinking.

    Let us take a closer look at the choice of a specific material object, a piece of wax, above other possible examples. Wax is not only illusive and ephemeral but also a distinctively malleable, plastic, manipulable, and workable material. Descartes had to make the concept of the deceitfulness of the senses into something palpable that we, his readers, can experience here and now. Historians have shown how early modern natural philosophers had to cope with increasingly mathematical explanations for natural phenomena, and that experiments offered ways to make this new science "palpable and convincing."[30] The Cartesian experimenter engages us in a wax *experience* to prepare us for a new epistemology, similarly to the way in which Robert Boyle later made a spectacle out of empty space to give weight to the concept of a "vacuum."[31] I suggest here a reading of the wax passage that comes close to what Florian Nelle has called a "transformation of a metaphysical moment into an aesthetic artefact."[32] In the experimenter's hands the wax ball is transformed, so to speak, into a shapeless, ephemeral blob and an aesthetic artifact that at the same time becomes this ultimate "epistemic thing." Considering that the *Meditations* present a meticulously constructed story, we need to pay scrupulous attention to every sentence and every image that work evokes, to understand its argumentative force. On close reading, I noticed, for example, that the meditator begins her exploration not from vision but from taste. Is it a coincidence that she begins with a reference to the most intimate sensual impression possible? Taste and smell are invasive senses; they are intrusive and contaminate any ideas of clearly designated subject-object positions. In the core section of the wax passage, Descartes's rhetoric foregrounds moments of immediacy and intimacy. The wax *experience* can be read as a well-orchestrated, intimate encounter.

The *Meditations* as a Manual for Minds and Hands

Other scholars have called attention to the literary technologies of this philosophical text and pointed out that Descartes did not write an argumentative treatise but a meditation.[33] They have stressed the importance of the literary form that Descartes chose to make his epistemological program public.[34] Instead of presenting the reader with an argument up front, Descartes makes an effort to tutor his readers in a way of thinking that must appear to them counterintuitive: the meditational form has been considered as a methodological device that Descartes deliberately employed to train his readers to turn away from the senses in search of genuine knowledge.[35] Descartes's choice for this literary form might have been motivated by the challenge of convincing a hostile audience that was committed to an opposing epistemology.[36] Does the literary form, then, bear only an arbitrary relationship to the epistemological content of the text? Taking the meditational form as a mere methodological device would imply that we can conceive of Descartes's new epistemology as something that exists outside of the textual framework. Can we ignore the stylistic choices and go straight to the "message"? This would mean that we can reduce the wax *experience* to a wax *argument*, stripping the text of all its layers of meaning and summarizing its supposed meaning as an argument that reveals its core message. This "thinning" of the text would allow us to shorten the meditational exercise into an argument without losing its sense.[37] This raises questions about the (dis)entanglement of aesthetic and epistemological issues: Is the literary form merely chosen to "transport" an argument, or is it an integral part of the Cartesian epistemology?

In the following, I explore the rhetorical devices employed in this text. Building on the work of others, I show how these devices operate also on an affective and not purely cognitive level to create an experiential rather than a discursive reading. I argue that Descartes had good reasons to compose his text in the form of a meditation. Others have drawn attention to the performative thrust of the *Meditations* and laid emphasis on the fact that the text takes the form of *mental* exercises.[38] Extending this argument, I propose here that the meditation on wax gains compelling force from spiritual practices that call for bodily and material engagements.

Amélie Oksenberg Rorty has pointed out that this text does not present its readers with a sequence of meticulously composed acceptable or rejectable theses; instead, it invites the reader to follow the narrator and engage in "meditational exercises."[39] Indeed, as Gary Hatfield has argued, it matters that Descartes's *Meditations* take the form of "guidebooks."[40] The literary form not only is a deliberate choice to convince skeptical contemporaries but is somehow inherent in the epistemological argument that unfolds before our eyes. We can understand Descartes's literary choices from within the historical and cultural context of Descartes's education.[41] Hatfield traces the writing style of the *Meditations* back to literary forms that Descartes was well-versed in through his education in the Roman Catholic tradition.[42]   As an attendee of the Jesuit school and the Parisian oratory, Descartes was certainly familiar with both the Ignatian and the Augustinian tradition.[43] As Hatfield convincingly shows, Descartes was well-acquainted with spiritual writings that take the form of meditative manuals to reveal spiritual truth rather than presenting logical arguments.[44] Hatfield makes clear that reading "the *Meditations* as meditations" shows how this text is constructed to guide its readers through particular experiences.[45] He stresses that they function as "a set of instructions" and must be understood as "cognitive exercises."[46] Indeed, the *Meditations* are not composed as a preconceived argument but as an exploration: Descartes invites the reader to embark on a journey with a yet unknown destination.

The literary form of a meditational manual also has an effect on today's readers who have no familiarity with the spiritual tradition from which Descartes drew. Rorty has shown how the style in which the *Meditations* are written urges us to let go of a contemplative stance vis-à-vis the text and to engage in a transformative experience.[47] We cannot sidestep the "tedious path" set out by the *Meditations* and get straight to the core of Descartes's thinking. Such a rationalist, reductionist, and ahistorical reading of the *Meditations* fixes all the attention on the outcome, while preventing us from paying careful attention to the rich reading *experience* constructed in the text. Hatfield has identified this as the remarkable "experiential thrust" of the *Meditations*.[48]

I draw for the following exposition on Hatfield's essay "The Senses and the Fleshless Eye: The *Meditations* as Cognitive Exercises."[49] Meditations persuade not only by means of argument but also by evoking an immediate experience. They carry their force in at least two ways—as "considerations" that appeal to our understanding and as "exemplifications." Traditionally, Descartes's *Meditations* have readily been related to the literary form of considerations that appeal to the faculty of reason, but Hatfield provides us with a thorough analysis of Descartes's use of the spiritual form of exemplification. He identifies two sources on which Descartes must have drawn for his revised and reenvisioned use of exemplification—the Augustinian tradition and the Ignatian tradition. In these traditions, spiritual exercises characteristically bring the reader into the appropriate mental state with carefully arranged settings and preparatory exercises (51). The Second Meditation and in particular the wax passage is marked out as one of the "key points in the text" where "exemplification in intuition is at work" (52). Hatfield lays great emphasis on the performative thrust of the *Meditations* and foregrounds Descartes's indebtedness to the Augustinian tradition for which "exemplification involved illumination of the intellect and will by divine light, yielding direct apprehension of eternal truth"(49–50). Within the Augustinian tradition, imagination and memory are only invoked as a foil against which Augustine (354–430 CE) could discover that God cannot be imagined in the same way as other physical things that can be evoked as pictures in our memories (51). Instead of calling on a heightened imagination of sensory experiences, Augustine achieved spiritual illumination through an incisive use of skeptical doubt to clear out former opinions and "through contemplating the immateriality of his own mind" (52). Hatfield demonstrates that Augustine's meditative mode displays striking parallels to Descartes's later *Meditations*: "By turning away from the

98

senses and discovering his own intellect as an invisible, immaterial, and yet mutable power, he [Augustine] was led to see with the fleshless eye of the mind the invisible, immaterial, immutable deity" (52). By contrast, in the Ignatian tradition, imagination is invoked in an explicitly sensual sense: Hatfield writes that "Ignatius invites one to see, feel, hear, taste, and smell the horrors of hell" and argues that the epistemic value of Ignatian meditations remains quite limited and cannot reach the illuminating heights of Augustinian exemplifications (51). Hatfield arrives at the conclusion that Descartes's *Meditations* mark a radical break "with the mainstream Ignatian tradition" of his own time (51). Drawing primarily on immaterial and intangible spiritual practices deriving from Augustine, Hatfield locates the "experiential thrust" of Descartes's "cognitive exercises" first and foremost in the mental realm. In this reading the wax *experience* is transformed into a genuinely mental affair.

However, Hatfield's claim that Descartes radically broke with Ignatian forms of meditational practices is not as convincing as it may seem at first glance. Among scholars it has, for instance, been a long and hotly debated question whether Augustine or Ignatius had more impact on shaping the Cartesian meditations.[50] Drawing on Hatfield's own illuminating descriptions of both spiritual traditions, it is indeed striking to see how much the wax passage in the Second Meditation resonates with the sensory thrust of the Ignatian tradition in which sensible objects and palpable experiences play a central role. Descartes makes a strong appeal to our imagination and lets us feel, smell, taste, and hear how the wax transforms, as it were, in our hands. I find it hence more convincing to argue that Descartes deliberately evokes a range of literary forms, spiritual and secular, in the Second Meditation. This suggests that his recourse to Christian meditative texts may have served more ends than Hatfield acknowledges: my hypothesis is that Descartes mobilized an Ignatian appeal to imagination in an unorthodox way, allowing him to bring his own quest for a "new epistemology" in conversation with the new scientists' quest for literary forms that can convey new forms of experience-based knowledge.

Bernard Williams argued that Descartes set this text forth as "an encouragement and a guide to readers who will think philosophically themselves," and, correspondingly, the thoughts "are presented as though they were being conducted at the very moment at which you read them."[51] Literary technologies that evoke a sense of immediacy and turn readers into eyewitnesses to unfolding events can, of course, be found not just in spiritual writing traditions. As Descartes seems to readily permit himself to revise and reenvision the meditative literature of his time in service of his own epistemological project, we can imagine that his innovative (re)use of literary technologies was not confined to spiritual writing traditions. Here I will take Hatfield's emphasis on the "experiential thrust" a step further.[52] I will examine the meditation on wax more closely, not only in the context of Descartes's school education and spiritual upbringing but also in the context of his later manual experimental inquiries.

Stylistic similarities between the first-person narrators in meditational texts and the rising genres of the experimental essay and observational report provided Descartes with a new rhetorical model that could be applied to a different goal than spiritual illumination, namely, to actively engage the reader in (thought) experiment. Reading the *Meditations*, as Hatfield suggests, as "sets of instructions," "exercises," or "guidebooks" resonates not only with meditational manuals but also with literary forms stemming from the long tradition of how-to texts and recipe literature, and with literary technologies in seventeenth-century experimental accounts. Imagining that the wax passage unfolds in a space of hands-on action and experimentation resonates with a reading of the wax passage as a set of instructions, a recipe, or a *"proces-verbal de l'expérience."*[53] Such a reading situates Descartes's epistemic reflections within the extremely popular genres of early-modern recipe compilations, books of secrets, craft handbooks, and household manuals. Much of the surviving how-to literature and technical writings from the sixteenth and seventeenth centuries testify to early moderns' keen interest in documenting and sharing diverse practical knowledge, including recipes that have been transmitted in writing over centuries, but also know-how that has been attained through (further) trials and experimentations.[54]

The Cartesian "I" of the wax *experience* evokes not only Augustinian and Ignatian meditators but also first-person descriptions of experimental trials and contrived observations, abundant in early modern recipe collections and in the correspondences exchanged among members of the Republic of Letters. While scholarly readings of Descartes's meditation on wax have traditionally contextualized and interpreted them as cognitive exercises, in my radical rereading I argue that this emphasis on *mental* meditations diverts our attention away from the remarkable fact that the wax argument is cast in the form of a *manual* exercise and an investigative *material* exploration. A rereading of the wax meditation in light of Descartes's deep involvement in the experimental practices of his time brings other epistemic meanings to the fore: the close examination of the wax ball not only evokes an intellectual engagement but also chimes with aspects of an experimenter's skilled bodywork and a writer's hand that could convey the epistemological thrust of firsthand experiences and observations. Descartes had trained himself in specific techniques, such as the probing and thorough examination of body parts and the exploration of the phenomena of life by means of the anatomical experiment, like the dissection and close examination of an ox eyeball—in one hand the knife, in the other the quill, setting forth his arguments, supported by detailed reports of his anatomical observations. We can imagine here a continuum between Descartes's extensive experimental investigations of the phenomena of life and the way in which he engages materials in his epistemological considerations. My reading of the Second Meditation suggests that Descartes knowingly employed the experiential thrust of spiritual texts, while adapting them to his own epistemological needs. Pushing the argument about Descartes's deliberate use of literary forms with an experiential thrust further, I propose here that the wax passage evokes a firsthand wax *experience* that speaks to the imagination of mindful meditators and hands-on experimenters alike.

On Spiritual Retreats: Seventeenth-Century Workshops and Virtual Laboratories

Descartes stages the meditation on wax as a wax *experience* in a way that resonates with the experimental method of the new scientists. The text creates the impression that we can grasp the object of inquiry immediately, as if no special efforts have to be made. In reality, however, there has already been a whole process going on to make the wax "plain to see," and Descartes himself refers to it. Someone had to go out and pick the wax from the honeycomb. A chunk of proper size must be collected (and this always involves aesthetic choices such as the following: How big? The homogeneous colored one or the marbled one? A softer or more brittle chunk?). Then the wax must be brought into the intimate and solitary, isolated sphere of the meditator's spiritual retreat, thereby turning it into an experimenter's (virtual) laboratory.

This reading resonates with other studies that have drawn attention to similarities between the spiritual tradition and emerging experimental practices. Picturing the protagonist of the *Meditations* as a meditator *and* as an experimenter aligns with Steven Shapin's study on the rhetoric of solitude in seventeenth-century England. "Solitude," Shapin writes, "is a state that symbolically expresses direct engagement with the sources of knowledge—divine and transcendent or natural and empirical. At the same time, solitude publicly expresses disengagement from society, identified as a set of conventions and concerns which act to corrupt knowledge."[55] Shapin and others point to similarities between spiritual and experimental scenes of knowledge, and his main argument chimes remarkably well with my re-reading of the wax passage. For it is not only the meditator who seeks the reclusion of spiritual retreat to "'withdraw the mind away from the senses' in order to perceive the primary truths of metaphysics."[56] Experimenters also take refuge in their laboratories in order to contemplate on matters in isolated spaces where "fundamental truths could be apprehended more clearly and securely than in the natural world."[57] Contemplating "the place of knowledge" in seventeenth-century endeavors to make the invisible visible, Adi Ophir and Shapin point to the delineation of physically and culturally "segregated spaces, distinct from the space of everyday objects."[58]

As we have seen in chapter 1, these "special scenes of knowledge" became sites of artificially contrived experiences that call for new laboratory epistemologies: "The truths established in these scenes do not hold precisely in the world of everyday experience. They were not meant to do so," James McAllister argues, "they were intended to hold in an abstract network of laboratories, both concrete and virtual. This network," he goes on, "spanned the world, but did not exhaust it: outside the walls of the laboratory, nature continued its ordinary course."[59] With the rise of laboratory sciences, an epistemic discourse emerges that clearly distinguishes between ordinary perceptions of natural occurrences and insights into natural phenomena derived from site-specific experimental manipulations.

Shapin and McAllister call attention to rhetorical similarities in spiritual and empirical stories of enlightenment that are located in special spaces of isolation. Expanding on these observations, I suggest that the stage in the Second Meditation can readily be transformed from a (spiritual) refuge into a (virtual) laboratory. What has become known as the "Cartesian ideal of the solitary investigator" does not stand in opposition to the experimental thrust of the new scientists in the early modern period.[60] The character of the solitary investigator in the *Meditations* can be embodied by both a meditator and an experimenter. Descartes might have drawn on personal experience when carefully constructing this literary figure.

Descartes left Paris to work in the Low Countries in relative seclusion in late 1628 or early 1629, and while there he apparently changed residences several times in the course of the following decade to ensure his privacy.[61] As we have seen in chapter 1, Descartes recalls in a letter to Mersenne that he visited the slaughterhouse at the places where he had stayed in order to see the butcher slaughter animals, and to have brought to his house the parts of the animals that he wanted to "anatomise at leisure."[62] We get the impression that Descartes spent the years prior to the publication of the *Meditations* withdrawn from the lively Parisian salon culture, devoting his time to writing and experimenting, but not in intellectual seclusion. He supposedly reserved one day a week for his extensive correspondence that testifies to lively discussions with his peers.[63] We can imagine him switching continuously between his writing and his dissecting table or even writing and experimenting simultaneously as he writes in his letter to Plempius on February 15, 1638, in which he reports on the vivisection of a young rabbit: "For this is refuted by a most certain experiment, which I am not displeased to have seen several times in the past and to see again today while I am writing to you."[64] The letter includes detailed first-person descriptions of the experiment in the following passage.

For Descartes, hands-on science was apparently not yet an inherently communal endeavor. He expressed, for example, little trust in experiments not performed by himself. In a letter dated to 1643, he writes with regard to calculations for the construction of a water jet: "And, because I do not put my trust in *experiences* which I have not done myself, I would have made a pipe of twelve feet for this jet; but with the few hands I have, & the artisans that make so poorly what one asks them to do, I have not been able to learn anything else, except for [followed by detailed calculations on the diameter and length of a bent water pipe]."[65]

For Descartes and his contemporary new scientists, the repetition of experiments was of crucial importance. Initially, this meant that the investigator himself ought to repeat an experiment several times to establish a scientific fact. In response to Steven Shapin and Simon Schaffer's influential study on the early history of scientific experimentation, scholars have argued that individualistic epistemologies only gave way to social conceptions of knowledge and scientific inquiry later in the seventeenth century.[66] Daniel Garber argues, for example, that it was only with the establishment of the Royal Society ten years after Descartes's death in 1660 that the constitution of experimental facts became a communal business.[67] In Descartes's time, Garber writes, scientific research was not yet associated with a "cooperative conception of the creation of new knowledge" or with the idea that "many Heads and many Hands" are involved in making knowledge.[68] According to Garber, the belief in a "new sense of reproducibility" grounded in "the claim that experiments must be repeated by a variety of hands" or "performed

103

by many hands" was not yet applicable to early modern experimenters.[69] He claims that early forms of observational accounts evolved into modern genres of scientific protocols and reports with the rise of "new conventions for reporting the outcomes of experiments" grounded in the idea that "a given experimental result must be capable of being reproduced by different hands and eyes."[70] For Descartes, Garber concludes, experimental science was still a solitary activity for which "he could not usefully employ other hands than his own."[71] Steven Shapin has noted that the idea of the pursuit of knowledge as a solitary individual endeavor has prevailed in modern accounts, even in the age of Big Science.[72] Garber's emphasis on "hands," and in particular on diverse and many hands, in making scientific knowledge is conspicuous but is not traditionally associated with Cartesian epistemology. This is partially due to modern translations of Cartesian writings. Well known, for example, are Descartes's own lamentations that he will never be able to conduct all the necessary experiments. Many scholars have quoted the following passage to illustrate that Descartes ascribed an important role to experiment in the pursuit of knowledge: "But I see also that they are of such a kind and so numerous that neither my dexterity nor my income [*ny mes mains, ny mon revenu*] (were it even a thousand times greater than it is) could suffice for all of them. And so the advances I make in the knowledge of nature will depend henceforth on the opportunities I get to make more or fewer of these observations."[73]

What usually goes unnoticed is that in this quote Descartes alludes to his hands (*mes mains*) when he expounds on the need for more experiments. The editors of the English edition created a slightly distorted meaning in translating Descartes's literal remark that he "has not enough hands for the job" as "dexterity." The standard English translation by Cottingham et al. thus contributes to a trend in the interpretive canon for removing references to *hands-on knowledge* from Cartesian writing. Moreover, the image of a solitary investigator should be put in context with historical evidence that documents Descartes's various collaborative efforts.[74] In addition, others have emphasized the strong link between Cartesian experimental practices and the lively salon culture of seventeenth-century France.[75]

Bodily Experience and a New Epistemology

The wax *experience* marks a vital step in the thought experiment of a hyperbolic doubt that the meditator performs in the *Meditations*. The Descartes scholar Myles Burnyeat reminds us that the radical doubt experiment is not performed with an example of *someone's* body, but that the meditators *"own hands and body* take over the centre of the stage."[76] Descartes thus introduces with the *Meditations* a radical methodological innovation—*one's own experience* is constituted, as any other, as an object of description.[77] Moreover, Burnyeat claims that Descartes is here the first to "put subjective knowledge at the center of epistemology."[78] We can extend Burnyeat's argument in another direction: in the Second Meditation an experimenter takes over the center of the stage with a piece of wax in hand. We witness a performance in which the experimenter's own body is actively involved in the pursuit of knowledge. The wax *experience* then invites reflection on how one's own body is turned into a research tool and what kind of epistemological problems arise when bodily labor and experience migrate to the center of a new philosophy. One of the challenges is to think through how these moments of embodied intimacy can be reported in a way that supersedes the individual and anecdotic and make subjective experience into an objective knowledge claim. The experimenter's wax *experience* addresses the problem of the formalization of intimate and (mediated) encounters that could more aptly be described as contaminating events than as rational observations of a distanced investigator.

As Hatfield reminds us: "Descartes stands at the beginning of modern philosophical investigation into the relation between the knower and the known. Even if we no longer are taken with his foundationalist enterprise, his epistemic individualism, or his account of cognitive faculties, the problematic of characterizing the contribution of the knower (or community of knowers) to the very constitution (or fabrication) of knowledge remains."[79] I have turned to Descartes's *Meditations* to investigate whether we can find in his epistemological explorations clues for a more dynamic conception of the knower's involvement. What puzzles me is this general question concerning embodied modes of knowing and the manual fabrication of knowledge. I propose to read Descartes's meditation of a wax ball not as a straightforward disqualification of sensory experience but as a probing investigation of aesthetic dilemmas that arise when exploring new modes and techniques of perception, designed to make (invisible) phenomena (indirectly) perceivable. If new scientists' experimental practices introduced new ways of making and knowing that necessitated a more complex vocabulary of differentiation between artificial and ordinary modes of perception, we have good reason to assume that Descartes, a keen thinker and experimenter, would have meditated on the epistemic challenges and potentials they pose. In my reading the famous wax *experience* hints at the deep implication of

105

DESCARTES'S MANUAL MEDITATIONS

one's own body in experimental inquiry as it emerged in the early modern period from artisanal practices.[80]

At the same time, Pamela Smith observes that seventeenth-century new scientists became increasingly "unsettled by the involvement of the body in cognition and they sought to control the bodily dimension of empiricism at the same time that they began to distance themselves from artisans and practitioners."[81] We can read Descartes's Second Meditation also as a response to a legacy of a sense-based Aristotelian epistemology that consists of two increasingly conflicting positions: on the one hand, the epistemological priority that Aristotle ascribed to the senses and, on the other hand, the disdain toward manual labor and exclusion of any bodily work from domains that pertain to the indubitable knowledge of *scientia*. While attention to the body had a long tradition in ancient philosophy with the cultivation of healthy bodies and bodily care as an aim and precondition of the learned elite, this ancient ideal provided no guidance in dealing with the bodily issues that forced themselves on hands-on investigators with an unprecedented and unfamiliar vigor in their daily experimental work. The problem that Descartes and his fellow new scientists needed to address was how to preserve an ideal and pure conception of *scientia* in the face of their own messy hands-on investigations. The epistemological reevaluation of praxis and techne that informed their experimental attitude endangered, so to speak, the well-defined domain of certain knowledge: *scientia*.

As the new scientists turned science into an active manipulative endeavor dependent on mental and manual operations, they blurred the boundaries of ancient traditions in which manual labor was excluded from the domain of *scientia*. The productive hands-on attitude of the experimenter offered new vistas. These had remained unexplored in the Scholastic tradition, with its focus on book learning that Descartes increasingly experienced as paralyzing. However, the rising importance of bodily know-how also posed a threat to an idealistic conception of *scientia* as pure contemplation.

Natural philosophers wanted to preserve a hierarchical system of knowledge indebted to an ancient division between *scientia* (philosophy) and *ars* (the practical arts) and to protect their knowledge-making endeavors from less highly regarded practices that involved manual labor. Steven Shapin argues that the idea that the new scientists revolutionized natural philosophy by becoming hands-on practitioners needs to be carefully scrutinized. He shows that Robert Boyle, working in the late seventeenth century, did not identify his scientific work with the handiwork involved but was happy to leave the manual work to his paid assistants. "Despite the clamor of seventeenth-century English scientific rhetoric commending a hands-on approach," Shapin emphasizes, "natural philosophy was still overwhelmingly a gentlemanly activity, and the traditional contempt that genteel and polite society maintained for manual labor was pervasive and deeply rooted."[82] The new scientists likely had professional, societal, and political interests in eliminating hands-on notions from disciplinary discourses when conceptualizing their practices as new forms of natural philosophy and scientific inquiries. However, I contend that even though the handiwork was delegated to technicians or remained in the hands of craftsmen in collaborative endeavors, we can hold that for the new scientist a hands-on attitude—that is, the pursuit of knowledge by one's own or others' hands—emerges as an alternative knowledge model that competes with the epistemic endeavor of learned scholars who make knowledge through exegesis of authoritative texts and disputation.[83] More important, Shapin seems to ignore that delegating hands-on work often necessitates that these tasks be at least partially mastered in order to be delegated in the first place. As we have seen earlier, Descartes complained that he did not trust anyone else to do his handiwork for him.[84] It is precisely this hands-on attitude that we can discern in Descartes's anatomical and optical experiments.

Descartes, who so to speak conceived of his new epistemology with both hands up to his wrists in an animal corpse, wants to counter a skepticism that calls into question the value of knowledge gained with new empirical methods. Yet linking the sciences of nature to the contingencies of hands-on investigations is precarious. This would imply that the ideal domain of pure knowledge would become prone to contaminating influences that could endanger its superior status. For seventeenth-century philosophers it must have appeared objectionable to commit themselves to a model that grounds *scientia* in hands-on practices.

In sum, as it became plain that the body weighed in with cognition, the embodied aspects of empirical research became an epistemic problem.[85] Historical explanations thus reveal a long-standing and persistent trend in early modern modes of knowledge formation in natural philosophy to obliterate manual labor and hands-on notions.[86] However, what merits more attention is *how* the new scientists, as Pamela Smith put it, attempted "to rework the artisanal bodily engagement with nature into a new disembodied epistemology."[87]

How to turn knowledge gained from hands-on explorations into conceptions of *scientia*, associated with the hands-off domains of *theoria* and *contemplatio*? In the previous chapter, I suggested that the influential work

Old Physiology = *scientia*; contemplative, speculative, mental, writing, thinking; "discipline of the pen"	Hands-off thinking practice; draws on results from Old Anatomy	Hands-off theory of knowledge
Old Anatomy = *ars*; investigative, manual, active, manipulative, empirical, experimental, instrumental, invasive, descriptive, doing; "discipline of the sword"	Hands-on manual practice; first senior discipline; provides speculative *scientia* with "data" and tangible insights	Hands-on moment Hands-on exploration Hands-on experience

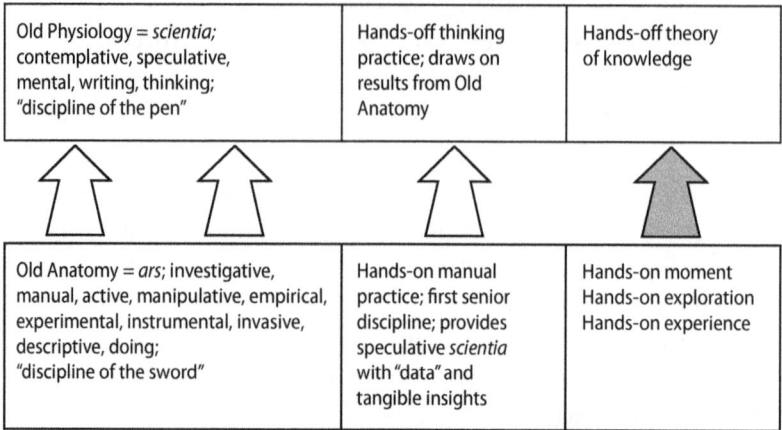

2.1 ★ This schematic illustrates key features of a Fernelian knowledge scheme (*left column*) and shows how one could translate them into hands-on and hands-off categories (*center column*). Reading from left to right, one can trace an evolution from a disciplinary theory of knowledge making to a universal theory of knowledge (*right column*). Image courtesy of Boo Chapple.

of the French physician Jean Fernel could have provided a contemporary epistemic framework that might have offered a way out of this dilemma. The Fernelian paradigm could have allowed the new scientists to reform and expand on the ancient and Scholastic concept of *scientia*. It provided a knowledge scheme that distinguished between practice- and theory-based domains of knowledge, and at the same time linked the hands-on domain of anatomy to the hands-off domain of physiology. Within the Fernelian paradigm, the interdependent relationship between the hands-on discipline of old anatomy and the purely speculative, hands-off endeavor of old physiology is defined interdependently (see figure 2.1).[88] Grounding medical theory in anatomical practice, the Fernelian scheme takes its starting point from an active hands-on conception of natural philosophy.[89] This knowledge model attributes a constitutive role to hands-on knowledge in attaining theoretical knowledge. Practices of speculation and theoretical reflection are firmly grounded in practices that consist of active (manual) investigation (see chapter 1).

Figure 2.1 illustrates how hands-on practice can be constitutively linked to theoretical domains of knowing. This schematic presentation envisions how a Fernelian disciplinary knowledge scheme could be extended into a universal theory of knowledge. A close reading of the wax passage against the historical background of Descartes's experimental explorations illuminates the ways in which Cartesian epistemology is shaped by a hands-on/hands-off dynamic. The Cartesian wax *experience* can be understood not as a purely mental exercise but as a hands-on experiment that takes its point of departure in the material explorative practices that Descartes engaged in, focusing here on anatomical experiments he performed in the 1630s. I shifted the focus from the outcome of the epistemological trajectory and its pervasive legacy—the static conception of a Cartesian mind-body dualism—to the meditative *process*. Into this process Descartes incorporated a hands-on moment that marks a vital step in Descartes's attempt to forge a new epistemology.

The model of knowledge that emerges from my reading explicitly distinguishes between active investigative manipulation and theoretical contemplation, but it also acknowledges and promotes an intrinsic relationship between practices of manipulation and practices of contemplation. My rereading opens the meditation on wax to a more dynamic understanding of knowledge-formation practices for which hands-on moments and processes of aestheticization play a vital role. I argue that this shift in perspective from the meditator to the experimenter allows us to identify the hands-on moment as a decisive step in the epistemological trajectory.

However, an appropriation of a Fernelian disciplinary knowledge scheme for a universal theory of knowledge also reveals an attempt to reserve a purely contemplative domain as the space of certain knowledge. The Fernelian scheme provided no radical break with ancient knowledge traditions: it allowed for a conceptualization of a knowledge domain—associated with *theoria* or *contemplatio*—that still complied with an ancient ideal and the Scholastic concept of *scientia*. The Fernelian solution for the integration of practice-based knowledge into the theorization of medical knowledge domains can also be read as an attempt to reserve a purely contemplative domain as the space of certain knowledge that is *anchored in* but *distinct from* the investigative domain of hands-on research. In this sense, the Fernelian model remains indebted to conceptions of *scientia* associated with the textual authorities of university curricula in contrast to "unorthodox" ways of knowing that were perceived as questionable, including alchemy, natural magic, and mechanical and practical disciplines.[90] Hence, this "new model" testifies at the same time to a conservative attitude toward an ideal conception of *scientia* as a purely theoretical domain. Descartes's seventeenth-century model of the new knower still reflects a persistent

109

ancient Christian tradition that cultivated a disregard of the body in any truth- or knowledge-seeking endeavor.

Returning now to the scene of action and cogitation in the Second Meditation, it is important to note that the mental space that we have entered into at the beginning of the *Meditations* is not the same as before. After the wax experiment it has substantially changed: first it was a space of deception, dreams, and diversion, and now it has become a "space of certainty."[91] How does this transformation come about? How does the manual exercise, with its explicit sensual richness, contribute to the possibility of constituting a space that allows for indubitable knowledge? I offer a reading of the wax passage in which a hands-on moment serves as a point of departure or a stepping stone for a thinking that establishes a space of certain knowledge in the mental realm. I take here the perspective of an embodied knowledge *maker* and propose that it is the experimenters' *handling* of the wax that enables a leap toward the mental realm of innate ideas.

The Cartesian epistemological project is not the result of pure thinking but was shaped decisively by a hands-on attitude toward knowledge production. Revising our interpretation of Descartes as the thinker in light of the historical figure of the anatomical experimenter thus raises questions about the role of hands-on engagements in making and theorizing knowledge. What are the philosophical implications of the claim that we cannot separate Descartes's investigative doings from his epistemological thinking?

My overarching argument is that Cartesian epistemology emerges from a hands-on/hands-off dynamic and embraces a model of knowledge that, on the one hand, explicitly distinguishes between active, investigative manipulation and theoretical contemplation but, on the other hand, acknowledges and promotes an intrinsic relationship between practices of manipulation and practices of contemplation. My experimental reading of the wax passage diverges from the dominant rationalistic reading and explores how Descartes himself prompts us to reflect on the involvement of the knower in knowledge production and the role of sensual perception and bodily involvement in knowledge making. My analysis of the wax *experience* leads to a dynamic conception of epistemology as a movement that alternates between hands-on and hands-off moments. It resonates with recent scholarship that highlights the role of hands-on work in understanding the rise of experimental sciences in early modern Europe and in shaping modern conceptions of scientific knowledge. I demonstrated how an emerging new model of the knower is grounded in a hands-on attitude characteristic of the experimental investigative practices of the new scientists.

I have argued that even if the path set out in Descartes's *Meditations* culminates in purely mental categories (*res extensa* and *res cogitans*) and a categorical distinction between body and mind, the wax passage provides a key moment, inviting us to think about epistemological issues in terms of a vital interplay between the mental and the manual. I also draw attention to the question of why hands-on notions have been mostly excluded from modern conceptions of scientific knowledge and from epistemology, the branch of philosophy that has traditionally been concerned with a unified theory of knowledge.

In what ways did the inception of a disembodied epistemology relate to pressing bodily issues and concerns that early modern philosophers faced? Indeed, what needs further philosophical inquiry is *how* the new disembodied epistemology took shape in the philosopher's hands and mind. What kind of philosophical constructions can we "dig up" that could have become so persistent and dominant in Western thinking that we still struggle today to articulate the diverse hands-on aspects of knowledge making? How can we make sense of the philosopher's desire to erase the experimenter's body and hands from early modern sites of knowledge making?

In the next chapter I reflect on methodologies. First, I will discuss the philosophical tools that helped me unearth a constitutive hands-on moment in Cartesian epistemology. Then I describe how a new disembodied epistemology was forged with philosophical operations that have obliterated the constitutive hands-on moment in modern epistemological thinking.

Making Modern Epistemology

It is to forget that from the tool kit to the hand that uses them, the tools themselves are being formed, that is to say, they appear less as entities than as plastic
forms in perpetual transformation. Let us think rather of malleable tools of wax
that take on a different form, signification, and use value in each hand and on
each material to be worked.

Georges Didi-Huberman, "Before the Image, before Time"

On Methodology and the Cartesian
Unterschiebung

We have seen that Descartes's *Meditations* are not the brainchild of a
thinker without hands but sprouted from the dexterous hands of a writer

and experimenter, anatomizing living and dead organisms, making knowledge by *experience*.[1] Why, then, is Cartesian epistemology readily associated with a Western conception of science and epistemology "as a rather atemporal, disembodied and theory-driven practice"?[2] How can we understand, not only historically but also philosophically, what it means that the birth of a modern "disembodied" epistemology is intimately entangled with the messy and bloody business of early modern knowledge making?

Apparently, it is not just a historiographical coincidence that we have forgotten to think about Descartes as an experimenting anatomist. Instead, I suggest that this misconception confronts us with a *philosophical* problem. The question is: How could the powerful image of Descartes as a thinker without hands so effectively obliterate the historical material that portrays this same philosopher as a fervent hands-on practitioner? And how could we retrieve the hands-on moment in Cartesian epistemology to reassess its epistemological significance? In what follows, I provide an explanation for the historiographical gap in our modern reception of Cartesian epistemology and its influential impact on Western thinking. I argue that this omission can be explained as the outcome of a *philosophical* operation. I do this in two steps. First, I give a methodological explanation for my reassessment of Descartes's epistemological project: To recapture the historical figure of Descartes as a hands-on practitioner, I used historical methods, but to recover the epistemological significance of the hands-on moment in Descartes's *Meditations*, I employ a philosophical *tool*, the *epoché*. Second, I argue that the elimination of hands-on notions from the Cartesian epistemological project is the result of a philosophical operation which can be described with the concept of *Unterschiebung*, developed by Edmund Husserl (1859–1938) in his masterwork *The Crisis of European Sciences and Transcendental Phenomenology* (*Die Krisis der europäischen Wissenschaften und die transzendentale Phänomenologie*, first published in German in 1954).[3]

In the first step, I explain how a philosophical instrument, known as *epoché*, made it possible to "unhinge"—as it were—the hands-on moment from the meditational flow and to reassess it in the context of historical evidence that testifies to Descartes's experimental investigations (the letters and treatises in which he reports on his anatomical experiments and the contemporary theories of anatomy and physiology that are discussed in the previous chapters). The *epoché* illuminates the epistemological significance of the hands-on experience and recaptures it as a vital constituent of a modern epistemological project. Then I draw on the phenomenological and methodological reflections of Jenny Slatman (born 1968) and Georges

Didi-Huberman (born 1953) to reassess my reading of Descartes's Second Meditation while I was conducting ethnographic research in a life science laboratory. I reflect on how the hands-on moment revealed itself to me in its epistemic significance and discuss how the concept of a displaced resemblance and an endorsement of anachronism as a heuristic method foster a rereading that opens up unthought dimensions in Descartes's meditation on wax (discussed in chapters 1 and 2).

In a second step, I introduce the philosophical notion of *Unterschiebung*, which I borrow from the German philosopher and phenomenologist Edmund Husserl. I show how this Husserlian concept can help explain why modern readers of the *Meditations* tend to remain oblivious to the epistemological significance of the hands-on experience. The technical term *Unterschiebung* allows me to explicate that this forgetting is no historical contingency but can be understood as the result of a philosophical operation.

Epoché

An *epoché*, in the sense of a tool to practice philosophical thinking, can already be found in ancient Greek philosophy. The method was used by the Pyrrhonian skeptics, who tried to solve the problem of having to choose between too many mutually contradictory philosophical systems. Instead of taking sides, the adherents of ancient Pyrrhonism cultivated the act of pausing and a philosophical attitude that remained in a state of withdrawal from involvement.[4] This philosophical attitude was later reassessed, in a slightly altered sense, as an abstention of judgment about the truth, or untruthfulness, of an idea.[5] Subsequent philosophers, René Descartes among them, have since then made use of the method of *epoché* and further adapted it to their needs. We can learn much from Edmund Husserl's reading of Descartes, first published as *Cartesian Meditations*.[6] Husserl provides therein a concise methodological analysis, describing how Descartes takes recourse to the *epoché* of the ancients when he lets the meditator withhold his assent from any former beliefs in the course of his meditations in order to explore the possibility of an indubitable foundation of knowledge. Descartes leads his readers in the Second Meditation toward the radical idea of an all-encompassing doubt. At the end of the First Meditation we encounter a "malicious genius no less cunning and deceiving as powerful who has employed all his might to deceive me."[7] In the beginning of the Second Meditation this malicious deceiver becomes so powerful that the meditator turns from a suspension of judgment with regard to certain things or beliefs to the method of hyperbolic doubt. The meditator now imagines anything that allows for the slightest doubt to be absolutely false.[8] Husserl describes how Descartes develops here the method of universal doubt *from* an attitude of suspension of belief that can be traced back to the ancients. Descartes's motivation for this radicalization is epistemological: the meditator pursues the universal doubt to discover whether there remains any certainty if one lets go of everything one thought one knew for sure. In his analysis Husserl does not seem to refute or embrace Cartesian epistemology. Husserl's interest lies not in the *outcome* of Descartes's meditation but in Descartes's ways of *doing philosophy*. Husserl dissects Descartes's text in search of the philosophical tools that this great early modern philosopher used to rethink the epistemological beliefs of his own day. In his analysis, Husserl refrains from carrying out the all-encompassing doubt up to its culmination in a universal negation—the moment when the meditator calls on us, the readers, to pretend for a time that all our beliefs and former opinions are utterly false and imaginary. What Husserl pursues instead is a *methodological* interest: he *retraces* the train of thought of the Cartesian meditator to single out this moment of suspension, that is, the moment when the meditator is about to "bracket" all former beliefs.

Already in an earlier text, Husserl had elaborated on the *epoché* in Descartes's *Meditations*. In his first book, *Ideas Pertaining to a Pure Phenomenology and to a Phenomenological Philosophy* (1913), Husserl asks his readers to refrain from engaging in the Cartesian thought experiment of a universal doubt.[9] What Husserl wants to do instead is to explicate the method that Descartes employs for his thought experiment. He wants to isolate the methodological moment from the whole of philosophical analysis that Descartes enters on in the *Meditations on First Philosophy*.

What makes Husserl's reflection so intriguing is that he, in fact, preaches what he practices: he uses an *epoché* to carve out the very *same* methodological tool from the larger whole of Descartes's philosophical reflection. Husserl demonstrates how we can put the whole epistemological project that Descartes unfolds in the course of six meditations, including the declared objective of this epistemological reflection—"the disclosure of an indubitable sphere of absolute being"—"on hold" in order to arrive at *a methodological insight*.[10]

Husserl is not especially interested here in the philosophical insights that can be attained with the Cartesian method of hyperbolic doubt. He wants to explore the potentialities of a philosophical thought process that departs from an act of suspension, rather than from an act of refutation or negation. The meditator does not simply reject the world we perceive; the meditator puts our beliefs in the truthfulness of this world on hold. What Husserl wants to single out is this phenomenon of bracketing or "switching off" and to explore the potentialities of this powerful philosophical tool. He concludes that this phenomenon of bracketing is not intrinsically bound up with the method of the radical doubt. Rather, the moment of bracketing can be "unhinged" and understood as a methodological step within Descartes's thought experiment of the universal negation. He shows how we can home in on this methodological moment and separate it from the end point of Descartes's philosophical meditation.

It goes beyond the scope of this book to enlarge on how Husserl develops the method of a phenomenological *epoché* in his *Cartesian Meditations*, but a short elaboration will be helpful.[11] The phenomenological *epoché* is a method that allows us to distance ourselves from a straightforward involvement with things and the world. Husserl distinguishes between a "natural attitude" and the "phenomenological attitude." The latter allows us to step out, as it were, of the natural attitude and take a reflective standpoint from which we can scrutinize, in a philosophical analysis, the objects and intentions that we do not question in the natural attitude. Crucially, this is an act of suspension or, better, an act of bracketing, which means that we do not negate or doubt how we engage with things and the world in the natural attitude, but only that we enclose our straightforward engagement with the world in brackets: Husserl describes this suspension of the natural attitude also as "switching it off" (*ausschalten*), but it is precisely the term *bracketing* (*einklammern*) that best illustrates what he wants to achieve with the *epoché*: when we put something in brackets, it is still there; we still have it before, or with, us, but in the act of bracketing we "put it *out of action*" (*ausser Aktion setzen*) and do not make use of it. Husserl explains that this does not mean the natural attitude is rendered innocuous in its entirety; instead, our attitude undergoes a modification the moment we bracket our natural attitude because we do not get rid of it; we do not revert into its negation but instead retain it in such a way that we gain a modified attitude that appears simultaneously with the bracketing of the natural attitude. For Husserl, the natural attitude is inherent in the natural sciences (*positiven Wissenschaften*) because it implies an attitude in which the world is given to me not only in an unquestioned natural-practical sense as an "always already there" (*im voraus seiende Welt*) but also as something that is for the natural sciences potentially graspable and ascertainable: "a universal ground of being [that allows] for knowledge progress in experience and thinking" ("einen universalen Seinsboden für eine in Erfahrung und Denken fortschreitende Erkenntnis").[12]

We will now take a closer look at the philosophical operation that Husserl has explicated *and* performed in his own *Cartesian Meditations*. Husserl shows us how we can retrieve the philosophical method of the *epoché* from Descartes's text, without conceding to the denouement of Descartes's *Meditations* that results in an epistemological primacy of a "disembodied" mind. Husserl thereby capitalizes on the methods of a great philosopher without consenting to the outcome of Descartes's epistemological project.

I follow Husserl's example by zooming in on the wax passage and singling out a methodological moment—the hands-on moment—while I refrain, in my reading, from carrying out the full Cartesian thought experiment that culminates in an epistemological devaluation of sensory experience and an attempt to rethink bodies as purely abstract extended things (*res extensae*). I put, so to speak, the readily assumed objective of this thought experiment—an epistemological devaluation of the senses—on hold. I borrowed the method of the *epoché* to return to the moment within the *Meditations on First Philosophy* when the meditator takes up a piece of wax and to examine this hands-on moment independently from its further elaboration into a radical hands-off contemplation.[13] I put, as it were, all the rest of the text into brackets in order to articulate a meaning of the wax *experience* that had been muted in modern readings of Descartes's wax argument—an epistemic meaning that had been buried by the reception history of the *Meditations* as the masterpiece of a rationalist thinker. In my philosophical education, I had, for example, been trained to read the hands-on moment immediately together with its radical negation. Therefore, it took some effort to carve out the wax examination in a sense that "withholds" its epistemic dismissal: I needed to mobilize the *philosophical method* of the *epoché* to reexamine *how* this hands-on experience allows for unfolding the idea of a hands-off epistemology in this text.

My short elaboration on Husserl's reflections shows that the *epoché* is not an arbitrarily applied tool; it is not extrinsic to the content of the *Meditation on First Philosophy*. When we employ an *epoché* to arrive at a radical rereading of the wax passage in the Second Meditation, we do not operate from the outside with the brute force of a crowbar; instead, we tune in to the very method of Descartes's own philosophical text.

Singling out the hands-on moment with an *epoché* allowed me to reread the wax *experience* in the context of Descartes's anatomical experiments, as shown in the previous chapters. My recontextualization of the wax *experience* is not only an attempt to situate Descartes's thinking historically. It is also the outcome of a philosophical operation with its own philosophical implications. My attempt to historicize Descartes's meditation on wax is not a backward movement or an attempt to reveal a "premodern" moment in his new epistemology. The method of bracketing does not imply the possibility of a return to an "innocent beginning point" in Descartes's thinking.[14] I want to show how the *epoché* allows the wax passage to appear to us in a modified sense. I employed the *epoché* as a method not only to read the wax passage in terms of its rationalist end point—an epistemological devaluation of the senses—but also to scrutinize the texture of this passage for an epistemic richness that may contain more leads than those that have been actualized in Descartes's own philosophical project.

Generative Repetition

The Dutch philosopher Jenny Slatman has coined the French term *une répétition génératrice* for an active engagement with philosophical texts. She describes a *generative repetition* as a way of retracing a line of thought such that it generates new potentialities.[15] It is a way of understanding philosophical texts as fecund sources that derive their status as iconic texts from repetitive and generative readings. This implies that we can engage with philosophical texts in a direct way without, however, ignoring their historicity, or constraining their meaning to one "transhistorical truth." Slatman builds here on methodological reflections by Martin Heidegger (1889–1976) and Maurice Merleau-Ponty (1908–1961). She describes how Heidegger has reframed the status of philosophical texts in a manner that allows us to understand them as potentialities for philosophical thinking rather than documentation of great thoughts. Heidegger introduces the concept of "the unthought" to describe this generative potential of philosophical texts, referring to the quality of the texts of great thinkers in terms of *Das Ungedachte*, that which is not thought (out) yet. Merleau-Ponty quotes Heidegger in his essay "The Philosopher and His Shadow" when he reflects on "the unthought-of element" in a philosopher's work "that opens out on something else": "'When we are considering a man's thought,' Heidegger says in effect, 'the greater the work accomplished (and greatness is in no way equivalent to the extent and number of writings) the richer the unthought-of element in that work. That is, the richer is that which, through this work and through it alone, comes toward us as never yet thought of.'"[16] Slatman's *répétition génératrice* is a further elaboration on this concept of "the unthought." The method of the generative repetition explicates how philosophical texts invite us to engage in a thought process that allows us to think *with* the texts but also allows us to think *beyond* them.[17] Considering that Descartes's *Meditations* are set up as an *exercise* to instruct the reader in the art of philosophizing, they form a particular case in point for such a *répétition génératrice*.

How, then, are we to read a "long-seller" like the *Meditations* that has belonged to the canon taught at Western universities for generations? Can we choose to read philosophical classics solely as historical documents of their own time, simply ignoring the compelling power they may exert on readers today? Or should we focus first and foremost on their argumentative power as if the meaning of these texts unfolds itself in a timeless realm, outside of all historical contingencies? The very idea that extant source texts can prompt such questions points at a textual overdetermination or semantic exuberance in philosophical writing. Perhaps we need to acknowledge that an interpretation can be sought neither solely in their past nor in our present. Instead of choosing one or the other, I am intrigued by the compelling temporality of philosophical classics. They prompt us to search for methodologies that allow for dynamic engagements with the abundance of meaning they offer us.

The Heuristics of Anachronism

PHILOSOPHY LIBRARY, UNIVERSITY OF AMSTERDAM

I sat in the library with a copy of Descartes's Meditations on First Philosophy *in my hands and tried to sort out my thoughts on what I had experienced during my fieldwork in life science research and teaching laboratories. While leafing through the pages, I wondered how I could connect these experiences with more theoretical reflections of the kind I was familiar with from my readings and writings in philosophy. At some point the wax passage caught my eye. I remembered reading the* Meditations *during my undergraduate studies. But in my memory the short fragment in which the meditator takes a ball of wax in his hand did not stand out in my reading experience other than in paving the way to the supreme insight of the* cogito, ergo sum. *What I mostly remember is how taken aback I was by the breathtaking rigor of philosophical thinking. But today this passage stood out to me in an unexpected manner. It appeared to bear a striking resemblance to a totally different text form, one I had just recently become acquainted with during my internship in a molecular biology laboratory. I had a vague feeling of recognition that—against all odds—linked this iconic philosophical fragment with this other utterly prosaic*

119

and mundane textual genre. When I returned to the philosophy library after months of working in life science instruction and research laboratories, the wax passage somehow reminded me of the plainly written sets of instructions, the experimental protocols that I had used and read daily when I interned with the biologists. At once, things in my head became "displaced and more complex" when I thought of protocols while rereading the famous wax argument of the Second Meditation.[18]

In the previous chapters, I historicized the so-called wax argument, reading it in the context of anatomical experimentations Descartes and his fellow new scientists performed in the seventeenth century. In this chapter, I explain how a reading experience that links this historical source text to contemporary laboratory observations made such a historicization possible in the first place. I argue that an anachronistic reading of the wax passage as a scientific protocol sheds new light on the epistemic significance of this text. I discuss anachronism as a heuristic method and the philosophical tools that allowed for a radical rereading of the wax argument as a hands-on wax *experience.* Then I reflect on the philosophical implications of this intervention with the Husserlian concept of *Unterschiebung.* My philosophical analysis reveals a double movement in Descartes's new epistemology that first discloses the manual as a point of departure and then brings to light a second movement with which manual experimentation had been erased from epistemological discourse. In the following chapters, I show how my analysis allows us to rethink epistemologies for laboratory life sciences from a hands-on perspective not in opposition to but from *within* Descartes's *Meditations.*

For the disputed use of anachronism, we can draw an analogy between the primary objects of study in visual art history and philosophy: images and texts. An anachronism is defined as an act of attributing a custom, event, object, or concept to a period to which it does not belong.[19] In historical studies an anachronism is considered to be a blunder, distinguishing the ignorant from the historical scholar who knows what is genuine to a particular period in time. The French philosopher and art historian Georges Didi-Huberman describes the rejection of anachronism as the golden rule of historians: what is to be avoided above all is "'to project,' as they say, our own realities—our concepts, our tastes, our values—on the realities of the past, the objects of our historical inquiry."[20] Anachronism is seen as a historian's ultimate faux pas. In opposition to this adage, Didi-Huberman embraces the inevitable "necessity of anachronism" in art historical studies.[21]

He argues that the very material presence of historical artworks in our own time confronts us, the viewers, with an anachronism. We encounter the works of Fra Angelico, Leonardo da Vinci, Rembrandt, and others "displaced," as it were, from the period that brought them forth. Didi-Huberman suggests that the compelling presence of historical artworks can grip our attention in a peculiar way, despite the many centuries that lie between their making and our encounter with them.[22]

It seems to be genuine to surviving artworks that they confront us with an exuberance of meaning that can be delegated neither solely to their past nor to their current presence in our own times. It is this overdetermination of the object of inquiry that the historian has to account for in their interpretation. In a first approximation, then, anachronism would be the temporal way of expressing this very complexity and overdetermination of images.[23] Didi-Huberman thus argues for an approach that can account for the paradoxically temporal structure of a cultural artifact from the past: embracing the necessity of anachronism challenges us to come up with meaningful interpretations that do not recoil from the work's compelling presence. An aesthetic experience of an artwork cannot be reduced to a historical understanding of a painting as a cultural relic of its time; art historians also have to account for the artwork's vivid material presence and the way we can be deeply moved at the sight of a painting that is hundreds of years old. As a consequence, Didi-Huberman argues that art history should be written as "a history of objects that are temporally impure, complex, overdetermined."[24] He also reminds us that not all the images that survive into our time become objects of art history, nor do all philosophical texts transmitted from the past become objects of philosophical inquiry. Given how much of our cultural past has, in fact, sunk into oblivion, we can wonder to what extent historic artifacts owe it to their exuberant presence that they become the subject matter of scholarly investigation in the first place.

Didi-Huberman's reflections on the historicity of artworks result in a positive reevaluation of anachronism as a heuristic tool for their study. I would like to propose that we encounter a similar abundance of meaning, when reading Descartes's *Meditations*, as when we are facing an artwork of the past. Regarding the peculiar historical status of philosophical texts, there is much we can learn from this discussion.

The main aim of Didi-Huberman's criticism is directed against the unquestioned assumption that art historians can disclose the meaning of artworks by interpreting them in the context of textual sources of their own time. He argues that much of art historical historiography is motivated by an unobtainable ideal: namely, the art historian's striving for an interpretation that achieves full congruence between object and text through recourse to sources of the period.[25] But how can we rightly define an authoritative source of the period? Are we referring to the same year, decennium, or century? Moreover, Didi-Huberman questions the very idea that contemporaries who happened to take up a pen should be considered as the only voices that matter when interpreting an artwork. In his view, art historians' striving for "temporal concordance" is futile: we will not overcome the time gap between now and then by attempting to become a ventriloquist, voicing the written words of those who lived in the past.[26] He reminds us that "contemporaries often fail to understand one another any better than individuals who are separated in time: all of the contemporaneities are marked by anachronism. There is no temporal concordance."[27] What Didi-Huberman argues for instead is a positive reevaluation of anachronism as a heuristic *method*. What can we learn from this reevaluation of anachronism when facing philosophical texts, handwritten and printed in the past?

In contrast to art historians, philosophers tend to turn philosophical texts into ahistorical subjects of inquiry by attacking philosophical questions directly while ignoring their historical origins. Philosophical interest here outweighs historical interpretation. That philosophers think it legitimate to distinguish between "philosophical interest," on the one hand, and "historical sophistication," on the other, exposes a general stance in philosophical scholarship that is in stark contrast with the "historian's ideal," criticized by Didi, "to interpret the past using the categories of the past."[28] What counts in one discipline as an ultimate faux pas appears to strike the right tone in another: whereas in art history students are encouraged to turn to the promised authority of "sources of the period" when facing their objects of inquiry, in Western philosophy departments, it appears to be an accepted practice to divorce primary sources from their historical contexts and to restrict oneself in their interpretation to the expressive force of philosophical ideas.

Philosophers today, it seems, have fewer reservations about engaging with philosophical texts without asking about any time gaps, or any procedures of demarcation that have produced the physical source text and philosophical subject of inquiry. How past philosophers have thought about ways in which knowledge can be obtained, what we can know, and what valid or reliable knowledge is, is often taught at Western universities from a transhistorical standpoint. By and large, knowledge theories from the ancients to today are compared in a seemingly seamless universe of printed or digitized pages from modern pocket editions and translations. I cannot remember, for

example, that reading Descartes's *Meditations* (in modern English translation) as a philosophy student and trying to follow the reasoning of the wax
argument raised many historical questions in the seminar room. It is not
uncommon, when teaching philosophy, to approach epistemological questions about "what, how, and how much we can know" with a universalist
outlook that is in search of an answer that ought to be valid everywhere at
every time, and for every reasonable human being.

But, as Daniel Garber rightly asked, to what extent can we "do philosophy historically"?[29] To pose this question goes against the grain of teaching
traditions in Western universities where philosophical classics are read in
core curriculum classes as part of an abstracted development or evolution
of ideas. Notably, Descartes's own writings have been fundamental in defining modern epistemology as a universalist project.[30] Yet the call for more
historical awareness in interpreting key epistemic texts and concepts, such
as "observation" and "objectivity," as historically and culturally situated phenomena has gained prominence.[31] In contrast to branches of philosophy that
aim to keep epistemological questions free from historical contingencies,
we see an increasing interest in the history and philosophy of science, and
notably in history of knowledge as an emergent discipline in its own right,
to understand the generative processes by which phenomena have become
objects of scientific inquiry and philosophical reflection, and in redefining
epistemology as an intrinsically historical, culturally and socially situated,
embodied, and embedded endeavor.[32] My radical rereading of Descartes's
meditation on wax chimes with these developments.

A Displaced Resemblance

When I read the wax passage of Descartes's Second Meditation at the library,
I experienced a moment of "inappropriate" recognition: the passage in this
seventeenth-century text stood out as being evocative of the experimental
protocols I had been using in a twenty-first-century life science lab. I include this autoethnographic snapshot story to introduce the phenomenon
of a *displaced resemblance*, a term Didi-Huberman adopted to describe an
experience of a strange similarity that cuts across different temporalities.[33]
A displaced resemblance strikes a discordant note in the logic realm of
chronological order. Didi-Huberman adopted the term to refer to experiencing a sense of similarity that is in a very literal sense misplaced and
mistimed. He uses the notion of the displaced resemblance to describe an
incident that is, in his view, not by definition a sign of malpractice but, in
contrast, may turn out to be of special heuristic value. He argues that a

displaced resemblance in itself does not hold enough interest for any prolonged inquiry, but that it can lead to "the emergence of a new object to
see."[34] A displaced resemblance refers to an act of recognition that is out of
place—that is, so to speak, inappropriate for the situation in which it occurs.
Something springs to mind because it appears reminiscent of something
that we encountered somewhere else, yet this appearance of similarity is
at once striking and improper, like thinking of Jackson Pollock's famous
"drip" paintings from the first half of the twentieth century while viewing
late medieval frescoes by the Italian painter Fra Angelico.[35] The recognition
surfaces unanticipated and produces a surprised feeling of a remarkable resemblance between two things that are *perceived* in accordance at once, but
the logic that connects them seems to lie beyond one's comprehension. For
Didi-Huberman this is precisely what links a displaced resemblance to the
"paradoxical fecundity of anachronism" and reveals its heuristic value.[36]

Didi-Huberman was not the first to use the concept of displaced resemblance as a heuristic. He borrowed it from another discipline, ethnoarchaeology, where it figured prominently in methodological debates concerning
the use of ethnographic material for the interpretation of archaeological
data.[37] For archaeologists, the use of ethnographic observations for the interpretation of archaeological findings had become "an accepted device for
the formulation of arguments" by the 1980s.[38] But the use of ethnographic
analogies connecting contemporary observations to surviving traces of ancient societies has also generated considerable debate. In a paper on Aboriginal cultural practices published in 1989, the archaeologists Richard Robins
and David Trigger introduced the term *displaced resemblance* to distinguish
it from the idea of a "direct resemblance."[39] The latter refers to ethnographic
analogies that are more commonly accepted for studies of the more recent
past where a continuity of artifacts and cultural context could be claimed.
Such claims about "proper" ethnographic analogies are based on assumptions about what might be an appropriate spatial and temporal distance
between archaeological finds and the heuristics used to interpret them. In
this debate about the validity of ethnographic analogies, Robins and Trigger
shift the focus from an assumed truth value of "appropriate" analogies to
the heuristic value of working with analogies in a broader sense, not limited
to temporal or spatial proximity. They stress the knowledge-generating potentiality of a displaced resemblance as a method that brings the present
in conversation with findings from the past. Drawing on earlier scholarship in archaeology, Robins and Trigger emphasize the usefulness of (ethnographic) analogy for reviewing accepted assumptions and formulating

124

hypotheses.[40] Citing the influential writings of the American archaeologist Lewis Binford (1931–2011), they flesh out how analogies can work as heuristics: "Analogy should serve to provoke new questions about order in the archaeological record and should serve to prompt more searching investigations rather than being viewed as a means for offering 'interpretations' which then serve as 'data.'"[41]

Even though the term *anachronism* is not mentioned once in Robins and Trigger's article, the thrust of their argument is similar to Didi-Huberman's reevaluation of anachronistic methods. Both the ethnoarchaeologists and Didi-Huberman dispute the possibility that archaeological data can be reconstructed or the meaning of artworks disclosed by a strictly synchronic approach. The methodological conviction that one can best rely exclusively on sources from the same period for the interpretation of ancient artifacts derives from the rarely questioned idea that historical findings and extant objects can best be understood in terms of the period in which they were made. The primacy of a chronological logic and this dictate of temporal or spatial proximity has long dominated the methodological debates in these fields. By contrast, the heuristics of a displaced resemblance can be mobilized to argue for an approach that allows for a nonlinear dynamic.

The term *heuristic* derives from the ancient Greek εὑρίσκειν (to find). The anachronism of a displaced resemblance initiates a process of discovery: a new object of scholarly inquiry comes into view. The concept helped me to understand this strange reading experience as a heuristic, as an invitation to engage in a generative repetition of the philosopher's line of thought. This *répétition génératrice* started out from a rereading of Descartes's wax argument, in which I refrained from reading the Second Meditation as a path that would lead me straight to the cogito. Instead, I singled out the wax passage to explore other pathways that may be activated from the rich texture of this text and Descartes's *Ungedachtem* (the *unthought* dimensions that this passage opens up). An anachronism functions as an opening that creates a fissure, an exploratory vein. In the present case an anachronistic reading of the wax passage as an experimental protocol contaminates, as it were, the ahistorical knowledge claim inherent in the new epistemology.

Traditionally, the wax argument is presented as a thought experiment that we can repeat in the same way in order to arrive at the same insight through reasoning alone.[42] The conceptualization of the wax *experience* as an exclusively mental operation has dominated its reading and reception history in today's philosophy textbooks. Such an ahistorical and mind-biased reading obscures the richness and epistemological significance of the Cartesian wax

125

experience. I focus in my reading on the moment *before* the meditator turns the gaze inward, bracketing the reconceptualization of the wax passage as a purely mental effort. From the experimenters' perspective, we can reassess how this wax *experience* palpably stages a *manual investigation* as a necessary step in the meditator's attempt to "dig out" the foundations of indubitable knowledge. Perhaps it is no coincidence that we encounter this figure of speech conjuring up bodily exertions and manual labor in *The Stanford Encyclopedia of Philosophy* entry titled "Descartes' Epistemology."[43] Descartes's innate ideas (*res extensa* and *res cogitans*) can only appear in full clarity and distinctness to the mind's eye *after* the manual meditation is turned into a mental operation.[44] Digging out the wax *experience*—obliterated by its reconceptualization as a "purely mental scrutiny"—was no easy task.[45] It called for a range of methods, including the here discussed *generative repetition*, *epoché*, and displaced resemblance.

My reading hence departs from a potentiality *within* Descartes's writing to open up vistas that we usually do not associate with Cartesian epistemology, namely, a constitutive role for hands-on attitudes in knowledge making. From this point of view, I scrutinize modern conceptions of science and epistemology to call attention to the conspicuous deficiency in hands-on notions and descriptions of manual engagements in theorizing knowledge formation in technoscientific contexts. Theory-biased conceptions of science have in the past dominated academic discourses in the disciplines of philosophy and history of science. In other domains, for example, in engineering and science institutions such as the Massachusetts Institute of Technology, it is, by contrast, common to acknowledge the interplay between hands-on making and scientific thinking as in that school's institutional motto *mens et manus*, or "mind and hand."[46] This is also true for the much younger discipline of science studies or science and technology studies (STS) that describes processes of knowledge making in experimental sciences also in terms of manual "tinkering."[47]

So why should we continue to tell the story about the rise of experimental sciences and modern epistemology in the West in terms of "a *conceptual* revolution, a fundamental reordering of our *ways of thinking* about the natural," and not in terms of natural philosophers' new ways of engaging with living and dead matter?[48] Why, then, should epistemology be understood primarily as a theoretical project focusing on mental operations, instead of connecting embodied practices to cognitive and discursive processes?

Crafting a new space for thinking and knowledge making, a new philosophy and a new epistemology, not only was an effort of the mind but also involved the work of the hands. When anatomical reasoning wanted to free itself from the corset of Scholastic textual traditions and authoritative voices from the past, it did so by opening bodies *and* books, handling knifes *and* pens. Yet, we are confronted with a conspicuous lack of theoretical awareness and vocabulary with which to talk about bodily activities, in particular the embodied practices that played a role in the study of life phenomena—not only in historical studies but also with regard to contemporary laboratory practices that we encounter in the life sciences today. To reflect on their epistemological significance calls for a better understanding of knowledge making as an embodied activity.

To study the life sciences of the past and the present, we need epistemologies that can help us make sense of the heterogeneity of experimental knowledge-making practices and the interdependence of hands-off contemplation and hands-on manipulations. More scholars have drawn attention to our lack of "elaborate vocabularies for talking about the bodily processes of knowledge making."[49] This inarticulacy has become only more conspicuous in light of the abundant and indiscriminate use of the term *tacit knowledge* for all kinds of nonpropositional and embodied ways of knowing. Investigating how hands-on knowledge making shapes cognition resonates also with a paradigm shift in the cognitive sciences that brings understudied embodied, embedded, extended, and enactive dimensions of cognition into prominence.[50]

In the following I argue that the conspicuous absence of hands-on notions in Western modes of thinking is not merely a historical coincidence. The aim of this project is to reassess the Cartesian program in light of current efforts to historicize and pluralize current epistemologies of scientific knowledge making. What kind of *epistemological* explanations could be given for the persistent tendency to talk about sciences predominantly as mind-driven rather than understanding the experimental sciences as specific forms of embodied work? I will address this question with a reexamination of two key concepts in Descartes's *Meditations*: *res cogitans* and *res extensa*—a conceptual dualism also known as the Cartesian mind-body dichotomy.

Bodily Experience and Extended Things: The Disembodiment of Epistemology

To develop a modern conception of knowing as a mental activity was a process that cannot be reduced to rationalistic thinking; neither can mental activity be explained—as is popular today—as "nothing but brain activity."[51]

Phenomenologists, like the French philosopher Maurice Merleau-Ponty, situate our thinking in our doing. Handiness, then, must not be understood as limited to manual skills but refers to our active bodily engagement with the world. Because this holds also for Descartes, we need to situate his thinking about how to obtain certain knowledge in the context of the embodied knowledge practices that he was engaged in and that—among other things—required dexterity and specific manual skills. We should be aware that the new space of the Cartesian knower was not a vacuum or a bloodless realm but a busy workspace filled with lenses, telescopes, air pumps, knives, tweezers, mortars, and other instruments that allowed for new forms of bodily encounters with animate and inanimate materials. New ways of thinking about how to obtain certain knowledge did not sprout directly from the brain of Descartes's meditating alter ego—these ways of thinking that paved the way for modern epistemology involved a changing attitude toward extracting knowledge from nature through bodily work.[52]

Descartes's distinction between two substances, the *res cogitans* (an indubitable thinking "thing" that we discover through the meditating "I" figure) and the *res extensa* (a theoretical concept that denotes all existing bodies as things extended in space), paved the way for conceptualizing knowledge formation in the sciences as a predominantly mental affair. Traditionally, the notion of *res extensa*, deriving from geometric concepts, has readily been associated with Descartes's interest in the abstract science of mathematics. But can we also understand the need for a *res extensa* concept from the viewpoint of a hands-on experimenter? That is, not only from the frame of reference of a mathematician who sees in the certainty of analytical geometry a model for philosophy but also from the standpoint of an anatomical practitioner who was preoccupied with the messy manual business of dissecting living and dead bodies? Bodies and bodily issues were matters of practical concern that Descartes apparently dealt with on a daily basis during his investigations of human and animal bodies. Indeed, Descartes was involved in an exemplary bodily business, anatomical experimentation, at the time he developed the epistemic notion of bodies as things extended in length, width, and depth. How can we understand the need for the *res extensa* concept from the perspective of Descartes as an experimenting anatomist?

To answer this question, I turn to the work of the Dutch philosopher Jenny Slatman. In *Our Strange Body* (2014), Slatman describes Descartes's reconceptualization of bodies—including implicitly my own body—as a mere extension in space as a process of alienation:

By conceiving the body as an extended substance, he [Descartes] underlines the notion of the body is not essentially different from all other physical things in this world. They all have a size and take up space: they are "extended." It does not matter whether it involves living or inanimate things, or things that can have a soul or cannot have one. In essence, all are the same, namely: extended. In this logic, the body becomes something that is no longer typical of human beings or any other living beings. In this way, in a sense, our body is taken away from us. The body I call my body comprises nothing that is my own.[53]

The moment that my own body becomes a mere extended thing among all other extended things, I can feel alienated from it. This body is to me no longer the very body that I call my own. I can perceive it now as a mere thing, not essentially different from any other things. Descartes's mode of thinking allows us to conflate different body experiences into one, leveling out any differences between the body I own and other bodies, living and nonliving things, and equating human and nonhuman bodies, alive or dead. The *res extensa* concept thus assimilates all bodies to the same operative, logico-mathematical notion. According to Slatman, this process of alienation does not just point toward a sense of loss. When we experience our own body as something that is estranged from us, we become aware that we can alternately experience our bodies as our own *and* as something other. Slatman's phenomenological reading shows that we can think about bodies as extended things not only in relation to a mathematical reordering of our thinking about nature but also in relation to bodily experiences. That being so, the *res extensa* notion is not necessarily at odds with an embodied perspective—the Cartesian concept can also be derived *from* a bodily experience.

Touching Hands and Strange Bodies

Departing from a hands-on experience of one's own body, the German philosopher Edmund Husserl has discussed the phenomenon of two hands touching and being touched. With this famous example, Husserl described how we can experience our own hands in two different ways. In her reading of this famous Husserlian example, Slatman discusses how the *Leib*-experience cannot be separated from the *Körper*-experience.[54]

When I touch my left hand with my right hand, I can experience my left hand as a thing that lies in front of me on my desk. I can palpate my hand

and register certain qualities; my hand feels, for example, smooth and warm. I can experience my hand in this mode just as I experience other things or bodies in my surroundings: the stiff skin of my leather-bound notebook, the finely grained surface of my wooden desk, the furry ear of my cat. When I palpate my hand in this manner, I can experience this tangible "bodily thing" in front of me as *Körper*, a German term for "body" used by Husserl. Husserl's description in this example clarifies how I can differentiate between this sensory perception of my hand and the sensation that I have when I shift the focus from the touching hand to the touched hand. Husserl uses another German term, *Leib*, when he describes the bodily experience of my hand being touched. Germanic languages, like German and Dutch, have two terms for "body," *Körper* and *Leib*, which allow Husserl and Slatman to distinguish between different bodily experiences: perceiving my left hand as tangible thing (*Körper*) and sensing how my left hand is touched by my right hand.[55] In the latter sense I experience my body as *Leib*: I am aware that my left hand is not just a passive tangible object that can be palpated, but that it also has the sensation of being touched. When I sense that I am being touched, I feel that my left hand is not just a thing among other things, but that it belongs to *me*. I experience my hand in this second sense as my body—in Husserl's terminology, as *Leib*. Husserl thus explains two ways in which my body can appear to me: my body as a thing among other things (*Körper*) and my body as being my own (*Leib*).

Slatman emphasizes that we surely can make sense of *Körper* without talking about *Leib*, but not vice versa. With the term *Körper* we can denote something that can solely be grasped as a thing. It is, however, impossible to make sense of *Leib* without an understanding of *Körper*: "The hand felt as *Leib*, must also be *Körper*."[56] My *Leib* experience is inscribed into a bodily experience and can thus never be separated from the "thing-like" experience of my body. More precisely, "the body's being a thing, its *Körperlichkeit*" is a precondition for the possibility to experience my own body also as *Leib*: "The *Leib*-experience," Slatman explains, "is an experience of *me-ness* that is based on corporal experience. It is a form of self-consciousness that becomes apparent in and on the body."[57] It is peculiar to my bodily experience that I can alternate between feeling how my right hand palpates my left hand as a *Körper* and sensing how my right hand touches a part of my body (*Leib*). The embodied experience of my body in which it appears to me as *mine*, and not just as a thing, is impossible without the possibility of experiencing my body in its mere corporeality (*Körperlichkeit*). I can have a *Leib* experience *only because* my body also can appear to me as a thing. In other words: "The *Leib* presupposes and confirms the *Körper*."[58]

Leiberfahrung thus means that I can feel or sense that I have a body, while *Körpererfahrung* means that my body can appear to me as a thing. According to Slatman, this "complex relationship between *Leib* and *Körper* can be interpreted as the relationship between what is our own and what is other or strange."[59] One body experience relates to a sense of ownership—I experience my body as *mine*. The other body experience relates to a sense of alienation or estrangement—I experience my body in its otherness. This analysis shows how I can derive from a bodily experience that other bodies are essentially not different from mine, namely, the sensation that I can perceive my body as if it is a foreign object that I can examine, palpate, and observe as something alien or external, and that this "thing" is essentially not different from any other bodies in my surroundings, animate or inanimate.[60] Note that an experience of alienation is here not *opposed to* but *anchored in* bodily experience. Slatman demonstrates how we can start out from a hands-on experience to arrive at a thinglike body concept reminiscent of Descartes's notion of *res extensa*. In the context of anatomical practices, a conceptual distinction between *Körper* and *Leib* has its heuristic advantages. A sense of alienation or estrangement is also manifest in Descartes's recourse to geometry. Philosophically, it becomes possible to think of bodies in terms of coordinates. The Cartesian meditator can abstract from corporeal experiences and engage in a mode of thinking that situates all bodies within geometry's sterile, bloodless, and fully transparent space.

In the context of Descartes's anatomical investigations, the heuristic advantages of conceiving of my body as a thing that can be examined, prepared, and dissected become apparent. Pertinent questions raised in early modern anatomical practices can be addressed with a reconceptualization of bodies as extended things: How is it possible to turn living and dead, animate and inanimate bodies, including my own, into objects that are essentially the same and can be approached with the same methods? How can dead and living bodies become things that can be known with experimental hands-on methods? Thinking about bodies as extended things seemingly enables us to understand bodies in such a way that we can gain access to their entire epistemic potential. This new way of thinking potentially allows me to make all bodies, including my own, into objects of experimental inquiry that can yield extrapolative insights.

That I can conceive of my hand as a thing that can be examined and scrutinized like other things makes it possible to understand my body as being essentially the same as the bodies that I am cutting open and dissecting. I can now relate the insights of my anatomical research to the inner workings of my own body. I can experience my body as if it were a dead thing. It is this alienating perspective that becomes manifest in a mechanistic worldview and Cartesian materialism. A mechanistic philosophy sees, as it were, the potential of a "dead body perspective" to facilitate a better understanding of the workings of living bodies. The assimilation of all bodies into a logical-geometric notion as *res extensae* brings them under the instrumental control of the experimenter. Understanding all bodies, including my own, as *res extensae* makes all bodies fully accessible to the explorative grasp of the hands-on investigator. A homogenizing body concept allows the experimenter to extrapolate anatomical insights gained from animal bodies to human bodies and from dissections of dead bodies to intact living bodies.

We can see this process of alienation as a way of objectifying the world: to objectify all bodies, including our own, means to make them available for the hands-on exploration of the new scientists. To locate the soul outside of the body means, in practice, that we can fully explore bodies with the new experimental means. The soulless body is "disenchanted," as the Canadian philosopher Charles Taylor writes; it is understood as a "mere mechanism . . . devoid of any spiritual essence or expressive dimension."[61] As such, it puts up no resistance to the experimental approach: there is no intrinsic difference between experimenting with a lump of iron or a living rabbit, a chunk of wax or an eyeball. Animate and inanimate bodies, human and animal bodies all become available for the experiments of the hands-on practitioner.[62] Taylor links the objectification of one's own body and the world to the position of an uninvolved external observer in *Sources of the Self: The Making of the Modern Identity* (1989): "We have to objectify the world, including our own bodies, and that means to come to see them mechanistically and functionally, in the same way that an uninvolved external observer would."[63]

In Taylor's account an alienated body perspective is inscribed into the figure of a hands-off observer. But Taylor notes that in the process of objectifying our embodied experiences, the body is not neglected but becomes "an inescapable object of attention" to the mind.[64] An alienating perspective and its radicalization into the logical-mathematical concept of the *res extensa* work, however, not only for an "uninvolved external observer" but also for an actively and intimately engaged anatomist: when all bodies are made interchangeable and transparent, they become accessible to the hands-on experimenter and can be fully grasped and understood by the mind. The idea that the mind can grasp the (inner) workings of bodies as a mechanism is linked to the idea that the world can be understood as a domain of possible instrumental control.[65] The Cartesian distinction between mind and matter makes matter fully accessible to the understanding of the mind. With this modern conception, certain and indubitable knowledge is located in the human mind (and not, like Plato's ideas, in a metaphysical realm). The hierarchical order of knowledge is restructured into an internal immaterial thinking process and an external material world. Building on Taylor's explication of this epistemological restructuring, I argue that Descartes's wax meditation shows how these distinct domains are intrinsically linked in the figure of the experimenter and the meditator. My reading of the wax *experience* demonstrates how the insights yielded by the hands-on practice of the former lead to the domain of pure thinking of the latter. I return to Slatman to explicate how we can start out from a hands-on experience to trace the genesis of *res extensa* as a philosophical concept.

The possibility that I can conceive of my body as a soulless thing, or *Körper*, is turned into an absolute in Cartesian epistemology. So when Descartes gives primacy to *Körper* at the cost of *Leib*, we have to keep in mind that *Körper* takes priority over any other way of talking about and experiencing bodies when we practice philosophy. The meditator performs this operation on an analytical level with a philosophical instrument: the methodological doubt. Hence, the categorical distinction that Descartes draws between body and soul is not a distinction that chimes with our daily experience; it is a *philosophical* distinction.[66] Descartes postulates a separate domain of pure thinking in which he locates the concept of *res extensa*. Hence, *philosophically* it becomes possible to think of bodies in terms of extended substances that are situated in geometry's sterile, bloodless, and fully transparent space, even though we do not experience bodies solely as things in our everyday experience. My perception of other bodies is never fully separated from the complex perceptual experience of my body because I experience myself as a situated body that can perceive and not just as a thing that can be perceived from the viewpoint of an external observer.[67] Building on Taylor and Slatman, I argue that we have lost sight of the epistemic interrelation between hands-on experience and hands-off thinking in scientific practice.

In the *Meditations*, Descartes presents the conceptualization of bodies as *res extensae* as a first philosophy, that is, as an idea we can bring to light when we philosophize, but which de facto precedes the more ambivalent and

133

complex body experiences that we encounter in daily life. Cartesian episte-mology thus gives primacy to logico-mathematical notions that allow us, on an analytical level, to ignore our bodies' opacity and ambiguity. Descartes's postulation of first philosophy, literally in the title of the first Latin edition of his *Meditationes de prima philosophia* (1641), is in fact a foundational-ist philosophical enterprise with which the intrinsic interrelation between embodied engagement and cognition becomes obliterated. I argue that the hands-on experience plays a constitutive role in the making of modern epis-temology, but that we have lost sight of its epistemic significance. To expli-cate the deliberate and systematic obliteration of bodily knowledge making in Cartesian epistemology, I turn to Edmund Husserl's concept of *Unter-schiebung*, which expounds an act of inversion as surreptitious substitution.

Unterschiebung: Forging a First Philosophy and a New Model of the Knower

The powerful image of a thinker without hands who declares the primacy of a thinking that is fully transparent to itself overshadows and dominates the laborious processes that lie behind this carefully constructed image. I argue here that the philosopher's work of clarification is deliberately concealed and that we can explicate this concealment with the Husserlian concept of *Unterschiebung*. Edmund Husserl introduces the term *Unterschiebung* in *The Crisis of European Sciences and Transcendental Phenomenology*, writ-ten between 1934 and 1937 and published posthumously in 1954.[68] *Unter-schiebung* has been translated into English as "substitution," but I choose to stick to the German term, which has stronger connotations of the labor or exertions implicated in an act of inversion.

134   Husserl calls attention to a crisis in the European sciences almost three centuries after the publication of Descartes's *Meditations*. He offers a critique of a tendency in Western philosophy to use geometry as a methodological model for metaphysics. Husserl, himself a mathematician, starts to question early on in his philosophical writings the presumed epistemological primacy of geometry. He turns against an idealization and universalization of geom-etry that he ascribes to the work of Galileo Galilei and the new scientists. In his late work Husserl develops a critique of the mathematical conception of space that paved the way for a "Mathematisierung der Natur" as a core characteristic of the emerging new sciences. He finds this "mathematiza-tion" later epistemologically undergirded in Descartes's *Meditations* and in

the metaphysical work of the Enlightenment philosopher Immanuel Kant (1724–1804). Husserl criticizes in particular Kant's attempt to ground epistemology in geometric concepts that were understood to be in a foundational relationship to our perceptual and cognitive faculties.[69] He was concerned with the way in which theoretical constructs take precedence over everyday experiences, but his critique is not aimed at the knowledge-pursuing project of the modern sciences per se. Husserl's phenomenology calls attention to the problematic way in which modern sciences and modern epistemology claim a foundational primacy for theoretical constructs. To illuminate this point he discusses philosophers' preference for geometric concepts of space to explain spatial experience: "the surreptitious substitution (*Unterschiebung*) of the mathematically substructured world of idealities for the only real world, the one that is actually given through perception, that is ever experienced and experienceable—our everyday life-world."[70] He argues that the foundational relationship between our everyday lifeworld and our scientific explanations of the world have been turned upside down with an act of *Unterschiebung*. He describes this operation in the *Crisis* as an act of thought (*Denkakt*) with great implications. It is a philosophical notion with which I explain an operation that takes place on an analytical level. In the following I use the Husserlian concept of *Unterschiebung* to explicate the move in Descartes's *Meditations* from a hands-on pursuit of knowledge to the inception of a hands-off epistemology as first philosophy.

In the Second Meditation, Descartes performs a thought experiment that departs from a hands-on moment, the vivid description of the wax *experience*. The experiment results in the reconceptualization of bodies as *res extensae*, finally culminating in the cogito as an apodictic truth. However, instead of conceiving of the cogito as an end point of a movement of thought, Descartes postulates the outcome of his experiment as a first truth or first philosophy. The radical doubt experiment that he performs in his *Meditations* is modern in the sense that it postulates the possibility that we can return to an ahistorical beginning point in thinking. What makes it modern is precisely this retrospective negation of the historicity of his own inquisitive endeavor. Retrospectively, Descartes separates his hands-off epistemology from its hands-on inception. On the analytical level, his modern epistemology disassociates the link between thinking and doing.

We have lost sight of the epistemic significance of the *manual* in the making of modern epistemology. Husserl's concept of *Unterschiebung* allows us to explicate that this forgetting is no historical contingency but can be

135

understood as the result of a philosophical operation. I employ the concept of *Unterschiebung* here as a *technical tool* to expose the deliberate obliteration of hands-on experiences in Cartesian epistemology.

We can read the wax passage in the *Meditations* as a response to the pressing issue of the involvement of the body in cognition, which here is addressed with a philosophical operation and which we can describe as an act inversion (i.e., *Unterschiebung*). The postulation of an act of thinking (the cogito) as a first truth, the radicalization of the ancient ideal of *theoria* into the apodictically given certainty of the cogito, obstructs the constitutive role of a hands-on moment in the Cartesian new model of the knower. In my reading, the operation of *Unterschiebung* allows the meditating philosopher, as it were, to tame the bodily issues the new scientists encountered in their daily practice. It is a radical philosophical solution to ambivalences that arise with the increasing importance of hands-on engagements and the accompanying shift in the epistemological status of the body. This development goes hand in hand with the increasing objectification of bodies from a modern perspective that grasps matter "mechanistically and functionally as a domain of possible means."[71] The objectification is geared toward transparency; bodies become coordinates in geometry's seductively transparent space, where they are transformed into objects that are made fully available to the mind. In popular thinking, it seems that the philosophical operation of *Unterschiebung* has been so successful that Cartesian thinking is commonly equated with a hands-off stance toward knowledge making: the powerful image of a thinker who explores the possibilities of knowledge with his *ratio* alone. Awareness of *Unterschiebung* makes explicit that this purification can only be attained *in retrospect*.[72] We have lost sight of the laborious process—a *manual* meditation—that enabled Descartes to construct a "disembodied epistemology." This dominating image of the rationalist philosopher has overshadowed the historical figure of Descartes as an anatomical experimenter who first and foremost explored life and the living from a hands-on perspective.

The philosopher's work of clarification is deliberately concealed. Through an act of *Unterschiebung*, Descartes can declare a philosophical insight as a precondition, a first principle that must be thought of as preceding all investigative endeavors or attempts at knowing. The apodictic certainty of the cogito must function as a foundation for the new sciences, guaranteeing the intelligibility of the world and the objectivity of knowledge. With the postulation of this foundation, Descartes carries out a reversal at the conceptual level. I describe this philosophical operation schematically as an act

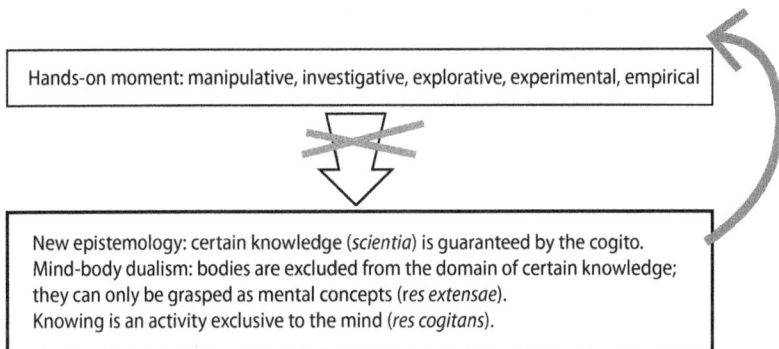

| Hands-on moment: manipulative, investigative, explorative, experimental, empirical |

| New epistemology: certain knowledge (*scientia*) is guaranteed by the cogito. Mind-body dualism: bodies are excluded from the domain of certain knowledge; they can only be grasped as mental concepts (*res extensae*). Knowing is an activity exclusive to the mind (*res cogitans*). |

3.1 * This schematic illustrates how the end point of an epistemological project is substituted for its starting point and turned into a philosophical foundation. With this inversion (up arrow), a theoretical superconstruction emerging from a hands-on genesis is postulated as that which comes first (Descartes's *prima philosophia*), thus obliterating its own making process and origin in manual practice (down arrow). Image courtesy of Boo Chapple.

of "substituting" a philosophical insight and declaring it as a precondition for experimental knowledge making (illustrated in figure 3.1).

My reading aims to reveal this inquisitive movement of thought so that we can better address problems we face in discussions on the interplay of mind and hand today. What has been forgotten is that Cartesian thinking departs from a hands-on *experience*. With a close and radical rereading of Descartes's *Meditations*, we can call into question the presumed disembodiment that has cut epistemology off from its productive source in hands-on science. To reconstruct how this *Unterschiebung* operates helps us to gain a deeper understanding of the conspicuous obliteration of the hands-on perspective in modern discourses about scientific knowledge production.[73]

The Cartesian Obliteration of Hands-On Notions

What are the implications of this Cartesian *Unterschiebung*? The search for an indubitable foundation for the sciences has its costs: the domain reserved for epistemology becomes isolated and cannot exert any influence on the practices that brought forth this domain in the first place. With the inception of modern epistemology, scientific practice and epistemological reflection became two separate domains in Western thinking. The struggle that begins with Descartes and becomes manifest in his attempt to free a new epistemology from any contaminating ambiguities of its origins in hands-on experimenting practices culminates in a separation between modern epistemology and the modern sciences. Both domains appear to be able to develop independently of each other. We get the impression of accelerated progress, especially in the domain of the modern sciences, that can take place without any need to rethink these developments from embodied viewpoints. We can recognize this in a discrepancy between the important role that hands-on training and knowledge play in the practices of the life sciences and a conspicuous lack of sophisticated discussion of the body in these domains. In the introduction to this book, we have already seen this in the haunting body talk of the Nobel Prize–winning geneticist Barbara McClintock.

Applying the concept of *Unterschiebung* reveals that the domains of epistemology and hands-on knowledge making had in fact never been disassociated in Cartesian epistemology. However, what has become persistently obliterated in our modern thinking is the link between both domains. I use *Unterschiebung* as a tool to reconstruct the hands-on moment in Descartes's thinking and to show that embodied engagement cannot be thought away or blotted out in understanding the practices of modern sciences or in doing epistemology. Using *Unterschiebung*, we can explain a conspicuous obliviousness to hands-on perspectives in Western knowledge-making theories *and* rebut the untenable assumption that modern epistemology was born from a disembodied mind.

The second implication of the Cartesian *Unterschiebung* relates to distinctness and clarity as declared desiderata of modern epistemology. This *Unterschiebung* can be understood as a response to the modern epistemological problem of how to bring bodily gained knowledge to clarity. With the establishment of a domain where transparency reigns, we have lost sight of the body's opacity. The study of wet lab practices in the life sciences brings to light the limits of a modern project that firmly situates epistemology in the realm of cognitive clarity. Hans-Jörg Rheinberger convincingly shows in his discussion of Gaston Bachelard's and Ludwik Fleck's work that clarity is the result of a "recursive process": "Clarity is a historical product of the purifying work of the scientific mind, and not a presumed Cartesian constitution."[74] The uncritical postulation of a rationalistic hands-off stance as the starting point from which to philosophize about scientific knowledge production is what needs to be debated. In the next chapters, I take this argument a step further and show how scholars simply overlooked that Cartesian epistemology is also itself the product of a purification process

138

that Descartes initiated with his *Unterschiebung*. My analysis resonates with attempts by scholars who question the conspicuously immaterial rhetoric of modern epistemology. Applying the concept of *Unterschiebung* deepens our understanding of this bracketing and forgetting of bodily notions in relation to the process of scientific knowledge formation. As I will show, doing life sciences in a wet lab involves continuously dealing with your own body and redefining the bodily boundaries of your objects of inquiry. You cannot think bodily matters away to attain clarity.

If we bracket the dismissal of the body in Cartesian epistemology and make the substitution of a purified rationalistic construction explicit, as we have done using *epoché* and *Unterschiebung*, we can regain a hands-on perspective that provides a point of departure from within Cartesian thinking. In doing so, my reading offers a radicalized version of the so-called craftsman and scholar thesis introduced by Edgar Zilsel in 1945 and revised by Pamela Smith as the "artisan and natural philosopher" thesis questioning hands-off and theory-biased narratives of the rise of the new sciences.[75] To radicalize this thesis means to fundamentally think through, on an epistemological level, the claim that the Scientific Revolution was the result not only of theoretical reorderings but also of a new hands-on attitude toward knowledge making. I use the concept of *Unterschiebung* to understand how Descartes rhetorically erases the hands-on conditions of knowledge making. However, I take the concept in another direction than that pursued by Husserl in *The Crisis of European Sciences*, where he argues that logico-mathematical concepts—"a mathematically substructured world of idealities"—have become substituted for our everyday world that is actually given through embodied experiences and perceptions.[76] Though I concur with Husserl that rationalistic constructions and idealities must be understood as grounded in embodied practices, I endorse critical voices arguing that his idea of an underlying foundational life-world remains too much indebted to a foundationalist, ahistorical thinking.[77] Husserl's life-world concept is still redolent of a period when the historicization of the sciences was not yet applied in the same radical way to epistemology itself. In my view, the problem is not so much that we have lost sight of science's grounding in the *doxa* of life-world but that we have not yet gotten our sights on the historically contingent, hands-on practices that constitute particular scientific enterprises. Gaston Bachelard's concept of "phenomenotechnique" calls attention to the particular ways in which phenomena and theorems are produced in specific scientific knowledge-making practices.[78] Accordingly, we have yet to develop epistemologies that can account for the peculiar ways of making knowledge without reducing the hands-on character of scientific activities to "prescientific" everyday practices (Husserl's concept of a *vorwissenschaftlicher Lebenswelt*). I propose that technoscientific productivity is grounded in hands-on practices that do not simply refer back to everyday experiences but are generated in scientific settings and experimental setups, of which we have yet to explore the aesthetic, embodied dimensions. The link between life-world and scientific practices is not disputed, but that does not exempt us from the work of paying detailed attention to idiosyncrasies of scientific practices with many and various descriptions that can mirror "the pluralization process of the sciences."[79] Husserl's critique in *The Crisis* remains too general and too much indebted to a foundationalist project to radically open historical epistemology to respond to the diverse "phenomenotechnical activity" of the experimental sciences.[80]

My aim in this book is to retrieve a philosophical thinking with a methodology that combines philosophical analysis with historical and ethnographic research methods to rethink the Cartesian *Unterschiebung* in the direction of pluralistic, contextual, and historicized epistemologies. Such a multisited historioethnography goes further than a close reading of Descartes's wax meditation. It offers us an urgently needed understanding of unquestioned philosophical foundations of pioneering empirical and historical studies of scientific knowledge making.[81] It also will bring us to not yet well-explored technoscientific knowledge sites (see my afterthoughts on microbiological cleanrooms in the epilogue). And in expanding on artistic research participant observations in tissue culturing laboratories (chapter 4) and analyses of artworks (e.g., the video work by Herwig Turk, Günter Stöger, and Paulo Pereira in chapter 5), my study offers a novel approach to how to describe and understand complex hands-on/hands-off textures of today's knowledge-making practices in life science's wet labs.

Because hands-on processes are an integral part of knowledge making in the life sciences, I argue here that we need more studies that map the bodily procedures that we encounter at particular scientific sites and provide detailed—historical and ethnographic—descriptions and analyses as to how hands-on practices result in knowledge claims and give rise to epistemological reflections. The necessity of locally grounded theorizing of knowledge making is forced on us by the epistemic practices of the modern sciences. Returning to Descartes's philosophical thinking has been necessary, I argue in the following chapters, to understanding the philosophical foundations of approaches in STS that set out to radically rethink epistemology from a hands-on perspective.

From Wax *Experience* to Life Sciences
Laboratory Studies

The wax *experience* in the *Second Meditation* can be read as an investigative process that opens more than one vista when we explore it in the context of the rise of the experimental sciences and their epistemological challenges. We can discern in this wax *experience* detours and possible turns or thinking paths that Descartes as a philosopher and new scientist did not pursue, but that make out the richness of this seventeenth-century epistemological meditation. A *répétition génératrice* allows us to explore the limits and potentials of a philosophical work from within the thinking that is manifest in the text. We enter, as it were, into a movement of thought to explore what lies beyond the end points of its philosophical logic and to rediscover the fertile ground from which they emerged. We can stay very close to the text, and at the same time we can generate from the source material "unthought" dimensions or further questions. In addition, I understand a *répétition génératrice* not only as a reading method but also as a performative method that can be expanded with ethnographic approaches or reworkings that are reminiscent of reenactments but make no claim to historical congruence. We can translate the method of the generative repetition to experimental practices of knowledge making in scientific laboratories. A generative repetition also has a practical sense that is surprisingly well suited to describe the literary genre of written instructions for conducting scientific experiments. The genre of the scientific protocol knows many forms in today's instruction and research labs, ranging from didactic "foolproof" sets of instructions to train first-year students in basic procedures and concepts to more or less stable text forms that are used in research contexts. Experimental protocols in research labs can prescribe routine tasks, but they can also remain open to frequent changes as part of experimental working practices and repeated trials under changing conditions. In didactic contexts, for example, protocols are used to establish routine and familiarity and to teach textbook knowledge. Protocols are performed by repeating the described procedure, yet in research contexts protocols are also subject to continuous modifications, and experimental protocols are repeated not only to validate results but also to generate new insights. Moreover, the process of performing a scientific protocol is a complex undertaking that cannot be reduced to translating written instructions into actions but mostly involves other ways to pass on embodied knowledge via demonstrations and imitations. I will return to this point when discussing the notion of tacit knowledge in the following chapters.[82]

The method of the *répétition génératrice* also underpins the discussions in the following chapters, in which I turn from seventeenth-century meditations to debates in STS from the twentieth and twenty-first centuries. I provide a close reading of seminal studies of knowledge production in technoscientific contexts by Bruno Latour and Steve Woolgar, Karin Knorr-Cetina, and Don Ihde and explore how the hands-on character of laboratory practices in the life sciences is theorized and explicated.[83]

I relate the findings of my philosophical discussion to an anthropology of touch and embodied knowledge making. In moving from the hand to touch and bodily experience, I ask what phenomenological approaches can contribute to our current understanding of epistemic practices in the life sciences. I extend my philosophical analysis with autoethnographic accounts to develop a critique of the groundbreaking laboratory science studies, published in 1979 by Woolgar and Latour, and in 1999 by Knorr-Cetina, that have been foundational for our understanding of how (life) sciences can be studied as material and embodied laboratory cultures.[84]

The method of a generative repetition allowed me to engage in a more intimate way with the point of view of the different authors. I combine my reading of these STS classics with a practical approach that resulted in autoethnographic snapshot stories from sites similar to those studied by these authors. My hands-on perspective brings the limits and potentials of their theories of knowledge into view in a way I could not have discerned through textual analysis alone. Chapter 4 focuses on conceptual problems that arise from a disputable anti-Cartesian stance that underpins these studies. It exposes lacunae in Latour's and Knorr-Cetina's laboratory ethnographies that can be traced back to a misreading of Cartesian epistemology—one that does not radically question its supposed disembodied inception and ahistorical stance. Chapter 5 shows how *Unterschiebung* and *epoché* function as heuristics to work out the articulation of tacit knowledge that remains unsatisfactorily vague in laboratory ethnographies and in (post)phenomenological accounts in philosophy of science and technology, focusing here on the influential work of Don Ihde (1934–2024).

★　★　★　★　★　★　★　★　★　★　　　4　　　★

Revisiting Laboratory Cultures

Toward the end of the twentieth century, a growing interest in science as a material and social practice developed, replacing a predominant focus on its theories and concepts. This trend, also known as a practical turn, can be described as a more general diversion from a Western, theory-laden conception of epistemology that puts the modern knowing subject at its center. Historiography has seen a development from ahistorical analyses to increasingly historical approaches to knowledge theories in the course of the twentieth century.[1] Hans-Jörg Rheinberger provides a particularly lucid exposition of the historicization of scientific knowledge, starting from the object: "The question was no longer how knowing subjects might attain an undisguised view of their objects"—as envisioned by Descartes when transforming all bodies into *res extensae* and locating them in the seductively transparent space of geometry—"rather the question was what conditions had to be created for objects to be made into objects of empirical knowledge under variable historical conditions."[2] Historicizing scientific knowledge, Rheinberger argues here, demands close attention to processes of making and to the material means and methods "with which things are *made* into objects of knowledge."[3] With this shift in problem constellation, not only science's epistemic practices but epistemology itself becomes subject to its historicization: "Historicization of epistemology thus also means subjecting the theory of knowledge to an empirical-historical regime, *grasping* its object as itself historically variable, not based in some transcendental presupposition or a priori norm."[4] The inherent historicity of modern epistemology becomes apparent when we see under which historical conditions Descartes was able to manually forge a modern epistemological program. Pushing Rheinberger's historicization project further, I suggest, first, that the Cartesian concept of the *res extensa* is not resistant to historicization and, second, that we need to subject not only science's objects but also this act of "grasping" to an empirical-historical regime. In the previous chapter, I have shown how we can approach the *res extensa* idea from the historically conditioned hands-on perspective of early modern anatomists, thus pushing the historicization of modern epistemology to a point in Descartes's thinking that precedes the moment at which it is turned into a purely cognitive, hands-off affair. In this chapter I further develop the argument that epistemology cannot be studied independently from its historical contingent material and embodied knowledge-making practices, drawing on pioneering works in science and technology studies (STS).

Descartes's epistemology is most commonly placed within a narrative of modern innovations in physics and the mathematical mechanization of nature in the early modern period. Rheinberger noted that this disciplinary bias for so-called hard sciences paved the way for "the idea of a unitary science that would embrace everything, centred firmly in physics."[5] It also fostered an idea of modern epistemology as an ahistorical project that tries to explicate the singular and universal nature of knowledge. More recently, scholars have drawn attention to the undervalued yet constitutive role of medical and artisanal practitioners in the rise of modern experimental sciences.[6] This prompts us to rethink Descartes's theory of knowledge in the context of early modern Europe's diverse epistemic practices and to explore alternative narratives whose starting point is elsewhere. For example, as the historian Pamela Smith suggests, "the laboratories of practitioners and merchants along the canals of Amsterdam and the medical faculty of Leiden" had become "just two of the places where the new practices and new attitudes toward the nature of the experimental philosophy were disseminated and inculcated."[7] This shift in perspective resonates with the practical turn in science studies that calls attention to the diversification of the sciences and their epistemic practices and refuses to cut epistemology off its grounding in idiosyncratic scientific practices.[8] Today, it has become much more widely accepted to understand epistemology as an inherently open-ended and ongoing endeavor that "has its own permanent laboratory in the past and future history of the sciences."[9]

The general trend toward historicizing epistemology in the twentieth century is linked to the practical turn in the study of science and its history. The practical turn, understood as the study of science's social, material, and technological practices, has nowadays developed into an academic discipline in its own right. The field of science studies or STS has enriched the historical and philosophical study of science with a growing body of literature that presents detailed accounts of the *making of science*. An underlying objective of STS scholarship has been to anchor scientific knowledge within the realm of human practices and to expose the intrinsic interrelation between science and technology instead of claiming a distinct space of certain knowledge for the products of science.[10] This turn was also an anthropological turn: it was brought about by studies that have emerged since the 1970s by researchers who went out into laboratories, that is, to the sites where science is made, to study scientific knowledge making as part of human cultural and social practices and their development.

Benchwork as Embodied Practice

Bruno Latour and Steve Woolgar's *Laboratory Life*, first published in 1979, and Karin Knorr-Cetina's *Epistemic Cultures* (1999) belong to the first movement of "ethnographies of research spaces and science as practice."[11] Ethnography denotes a research method, rather than a research field or academic discipline. My analysis focuses on descriptions of life science laboratories based on data that are generated using qualitative methods such as participant observation that place the ethnographer literally at the laboratory bench.[12] *The Handbook of Science and Technology Studies*, edited by Edward J. Hackett et al., acknowledges these early laboratory ethnographies as the "foundational pillars of a new discipline."[13] These pioneering works are thought to provide the grounding for more recent work in science studies, that—as others have argued—only seldom, if ever, takes the trouble to explicate its own foundations.[14] Latour and Woolgar and Knorr-Cetina made significant contributions to the practical turn, and their early laboratory ethnographies have become classics in STS by challenging the idea that "knowledge from the lab was apolitically, asocially, transtemporally, translocally true."[15] Conspicuous by its absence from this listing of untenable ideas about the scientific laboratory is the notion of science as a disembodied, predominantly theory-driven endeavor as opposed to science as an embodied hands-on practice. Methodologically, *Laboratory Life* and *Epistemic Cultures* have been groundbreaking in tackling this notion with hands-on participatory observations of scientific laboratory practices. My analysis focuses on a key question that emerged from their pioneering work: How can we describe research spaces from a hands-on perspective as embodied spaces and life science as a manual and material practice?

An Anti-Cartesian Project?

The authors of *Laboratory Life* sought to find alternative points of departure from which to write about scientific knowledge, turning from scientists' minds (ideas, theories, concepts) to the stuff of scientists' daily material practices, as Latour summarizes in a later paper: "I was struck, in a study of a biology laboratory, by the way in which many aspects of the laboratory practice could be ordered by looking not at the scientists' brains (I was forbidden access!), at the cognitive structures (nothing special), nor at the paradigms (the same for thirty years), but at the transformation of rats and chemicals into paper."[16] This statement, in which Latour recalls his ethnographic investigations for an anthropological study of a life science research laboratory at the Salk Institute in the 1970s, is symptomatic of the ways in which modern epistemology is equated with rationalist Cartesianism: Latour refers to brains and cognitive structures to show that modern theories of knowledge have generally focused on the mind, a trend in modern epistemology that is most readily associated with Descartes. Moreover, this statement is remarkable in the way in which it envisions an escape from Cartesianism via a hands-on approach. Latour proposes a new way of theorizing science that shifts the focus from minds to matter, from thinking to handling and doing.

Latour (at least initially) understood his approach as a study that only appeared to be epistemological but turned into a sociological study of science conducted with anthropological methods. Though originally intended as an epistemological study of the endocrinologist and head of the life science research group, Roger Guillemin, who invited the anthropologist into his lab, the study, Latour claims, turned out to be something totally different.[17] It takes its departure from what humans *do* in research spaces, and Latour describes this shift in focus from heads to hands as a turn away from epistemology. Latour and Woolgar reinforce this point in the postscript to the Princeton edition of their book, published in 1986: "It is worth acknowledging Guillemin's unusual generosity in providing total access to his laboratory and his forbearance in taking in (someone he took to be) an 'epistemologist' (Dr. Jekyll) who subsequently turned into a sociologist of science (Mr. Hyde)."[18] Further on, the authors claim their study "is neither an attempt to develop an alternative epistemology nor is it an attack on philosophy."[19] Notably, Latour and Woolgar take a hands-on approach to the study of scientific knowledge production, with ethnographic means and methods as irreconcilable or diametrically opposed to what "epistemology" is. The historian of science and philosophy Hans-Jörg Rheinberger takes a different stance: he sees the practical turn as part of a reformative movement that complements historical epistemology with studies of contemporary practices conducted with anthropological means and methods. In contrast to Latour and Woolgar, and radicalizing Rheinberger's analysis, I develop a hands-on approach to science studies from *within* Cartesian epistemology with the aim to rethink current plural epistemologies as historicized and culturally shaped knowledge-making practices. I argue that modern epistemology was actually born from a hands-on perspective, though this fact has been obscured by the philosophical operation of an *Unterschiebung*, as discussed in the previous chapter.

Consequently, I argue that Latour and Woolgar's attempt to frame their hands-on approach to theorizing scientific knowledge making as an "anti-Cartesian project" is highly problematic. It shows that these practical turn scholars in fact have not taken the effort to radically rethink the foundations of modern epistemology as a historically and culturally shaped endeavor that was itself shaped by experimental hands-on practices. Latour frames his own ethnographic investigation as a counterdiscourse that approaches the sciences not from a metalevel but from a hands-on perspective.[20] Latour and Woolgar's anti-Cartesian stance is obvious in their definition of epistemology as "a particular branch of philosophy . . . which holds that the only source of knowledge are ideas of reason intrinsic to the mind."[21] What is not taken into account here is the fact that historioethnographic observations of Descartes's experimental anatomical practices show us that modern epistemology is itself constitutively and intrinsically linked to a hands-on attitude toward knowledge making, and that this linkage was only obliterated retrospectively in the philosophical operation of the Cartesian *Unterschiebung*. In this chapter I argue that the epistemological critique of pioneering practical turn scholars is in fact not radical enough, and I propose that we instead need to think *with* Descartes *beyond* the Cartesian *Unterschiebung*. What I call into question is the assumption that a hands-on approach to theorizing knowledge making in the life sciences must be conceptualized as an anti-Cartesian endeavor. Instead, I claim that we need to return to Descartes to understand epistemology in relation to the very idea of thinking about scientific knowledge making from a hands-on perspective. This entails shifting the focus of epistemology as a declared discipline of the mind to the practices that underlay, fed, and shaped Descartes's epistemological considerations. It turns out that we trace the need to think both modern science and modern epistemology from a hands-on perspective back to Descartes. As a consequence, we can situate the hands-on approach of practical turn scholars in a historical tradition that is indebted to Descartes as a hands-on practitioner *and* an epistemologist.

The research described in both *Laboratory Life* and *Epistemic Cultures* was (at least partially) conducted in modern biological research centers. Knorr-Cetina's study in particular is characterized by a recognition of the handiwork involved in knowledge making in experimental laboratories: "Sensory performance and action go together, especially when, as in molecular biology, almost all experimental work is *manual work*."[22] The authors of both studies explicitly chose to work from a hands-on perspective in order to foster an embodied understanding of life science as a human and material practice. However, a closer look at the philosophical basis of these foundational laboratory ethnographies will show why framing the hands-on approach as a counterdiscourse to Cartesian epistemology makes it difficult to describe the embodied dimension of life science's research spaces and life sciences as hands-on practice. Only when we retrieve the hands-on moment as a point of departure *within* modern epistemological thinking will we be able to articulate the specific ways in which we are confronted with our bodies in today's life science research laboratories. More provocatively put, it is necessary to rethink Descartes's epistemological project in light of his work as a new scientist and anatomical practitioner and the new philosopher as a practical turn thinker *avant la lettre* to understand the

148

epistemic importance of hands-on experiences in the practices of the life sciences today. The pioneering laboratory ethnographies remain, however, indebted to an unquestioned Cartesian thinking by understanding themselves as a counterdiscourse and reaction against Cartesian epistemology as a purely rationalist philosophical project. Keeping in mind that modern epistemology has never been solely a hands-off affair, I argue for a philosophical underpinning of the hands-on approach to laboratory ethnographies from *within* Cartesian thinking.

Hands-On Stories from Life Science Laboratories

Methodologically, my discussion draws not only on textual analysis but also on observations I made over more than a year working in research and teaching laboratories. My multisited historioethnography draws on "snapshot stories."[23] I have gathered these, predominantly through participant observation, during fieldwork at the Gorlaeus Laboratories of Leiden University, where the life science teaching and research laboratories are housed.[24] The snapshots are written as "I was there" testimonies, but I also refer to them as stories here in a specific sense. I employ storytelling as a method to portray lived-through moments in laboratory settings. Knorr-Cetina and others who have studied the stories that travel around in laboratory culture argue that, in keeping relevant experiences alive, they form a salient feature of epistemic cultures in (molecular biology) laboratories. In *Epistemic Cultures*, Knorr-Cetina defines stories as "scenarios of former experiences that participants have had directly or have heard about."[25] Stories are "another means of enhancing experience in the laboratories studied, because they preserve the scenic, phenomenal aspects of events."[26] She points out that "stories emulate experience," that is, "they keep some part of the experiential context in the picture" instead of reducing experience "to abstract rules or instructions" (as in protocols).[27] Knorr-Cetina argues that stories convey phenomenal meaning contexts perceptible by the senses and through immediate experience of laboratory culture.[28] I present my observations in the form of ethnographic snapshot stories to articulate and investigate in writing how I became cognizant of my bodily presence when I participated in coursework intended to train future benchwork scientists. My stories explore how I experienced the handling of materials and instruments involved in experimental work in a molecular genetics laboratory and how these hands-on experiences gave rise to epistemological reflections.

149

I had already spent several weeks at the molecular biology laboratory at Leiden University when a box of those little cone-shaped plastic containers that I had been using on a daily basis caught my eye. While I was preparing some samples for a gel electrophoresis, the slogan on the box of centrifuge tubes hit me. Unnoticed for weeks, on the shelf right above my bench, it read "EPPENDORF—In Touch with Life." I looked down at my hands—one holding a micropipette, the other clasping a small plastic vessel, and got to wondering. . . . What precisely does it mean to be in touch with life in today's biological laboratories? What kind of hands-on work is done at the bench? How exactly is touch involved in life sciences research? How does touch relate to life? How does touch relate to knowing in the practice of science?

Readers who are familiar with the practical turn in the historical, social, and philosophical study of science may recognize certain figures of speech in this personal narrative of my "discovery" of Eppendorf's corporate slogan "In Touch with Life" on a box of tubes above my bench (figure 4.1). My anecdote echoes the wonder of the main protagonist of Latour and Woolgar's *Laboratory Life*—the anthropological observer who has never been in a laboratory before and who employs his sense of wonder as a tool to articulate the mundane practices of the sciences and to explore their epistemological relevance.[29] But before I focus on a detailed analysis of this "stranger device," I want to highlight the intrinsic historicity in Latour and Woolgar's pioneering anthropological approach and their rhetoric of curiosity.

150 The shift from a focus on the history of scientific ideas and theories to processes and practices also entailed a heightened interest in rhetorical strategies and the metaphors that are used to tell stories about new findings. Lorraine Daston effectively describes how these narratives change in form over the course of history.[30] Inspired by her historical analysis, my story deliberately borrows terms and phrasings from other classical narratives in the history of scientific discoveries that have been employed as "strategies of originality." In her essays on the cognitive passions in scientific attention, Daston points at the interplay of wonder, astonishment, and curiosity that is called on when speaking about a phenomenon worthy of scientific inquiry.[31] For Isaac Newton and the generation before him, she states, astonishment excited curiosity by rousing the philosopher from his

4.1 ∗ A box of Eppendorf tubes above "my bench" at the molecular genetics research laboratory, Leiden Institute of Chemistry, Netherlands. Photo by the author.

idle dreams and guiding his attention to a scrupulous examination of the phenomena in question.[32] The way in which astonishment and curiosity are perceived today seems to be still indebted to a shift in meaning that took place in the early seventeenth century. Daston describes the career of these passions during the following three centuries, meticulously depicting how curiosity (*Neugier*) gained status, eventually becoming the hallmark of the diligent naturalist. Quite the reverse happened to astonishment (*Staunen*). In the course of the following centuries, astonishment lost its meaningfulness as a philosophical passion as it became increasingly associated with the illiterate crowd: "Once celebrated as a philosophical passion par excellence, around 1750 astonishment became emblematic for genuine ignorance."[33] Wonder, on the other hand, maintained its positive sense as long as it was experienced in moderation; Daston elaborates by citing Descartes, who was perhaps the clearest on the delicate balance to be struck between just enough and too much wonder.[34] Daston writes that for

151

Descartes, "astonishment differs in degree from wonder—'astonishment is an excess of wonder'—but their cognitive effects are diametrically opposed: Whereas wonder stimulates attentive inquiry, astonishment inhibits it, and is therefore, Descartes asserted, always bad."[35] The wax passage in the *Second Meditation* illustrates this beautifully: Descartes brings into play a rhetoric of curious interest that leads to a scrupulous examination of a mundane object, a piece of wax, that through this process is turned into an epistemic object. The transformations that wax undergoes in the experimenter's hands make the protagonist wonder about what we can know for certain. If we focus our attention on the process of questioning, rather than on the outcome of Descartes's meditations on the transformational states of a piece of wax, we can see that kneading and wondering play a crucial role in the forging of a new epistemological program, as I have argued in chapter 2.

As Daston points out, however, it was only later that naturalists employed a more explicit rhetoric of the extraordinary to call attention to what appears, at first sight, too mundane to be worth investigating. In her contribution to the history of rationality, she describes how the linkage of these passions in the seventeenth century entailed a bias among natural philosophers toward rare, exotic, or mysterious phenomena. The particular linkage between these passions prompted naturalists who, subsequently, wanted to examine more mundane, everyday, or common occurrences to present them as extraordinary in order to arouse curiosity and the accompanying attentiveness.[36] Daston cites the telling advice of Robert Hooke (1635–1703) on how to discern the ordinary in nature:

> In the making of all kinds of Observations or Experiments there ought to be a huge deal of Circumspection, to take notice of every least perceivable Circumstance. . . . And an Observer should endeavour to look upon such Experiments and Observations that are more common, and to which he has been more accustom'd, as if they were the greatest Rarity, and to imagine himself a Person of some other Country or Calling, that he had never heard of, or seen, the like before: And to this end, to consider over those Phenomena and Effects, which being accustom'd to, he would be very apt to run over and slight, to see whether a more serious considering of them will not discover a significancy in those things which because usual were neglected.[37]

Daston's analysis shows how the first experimental naturalists made use of literary strategies of alienation. They introduced narrative conventions that are later powerfully employed by laboratory ethnographers for the study of experimental sciences in modern laboratories.

In Touch with Laboratory Life

The very same strategy of alienation that Hooke had used was strikingly brought into play some three hundred years later in Latour and Woolgar's *Laboratory Life*. Indeed, it seems as if Bruno Latour acted on this very advice when he, together with Steve Woolgar, constructed the fictional character of an observer to relate Latour's observations on daily occurrences in a laboratory. Their ethnographic approach to science is reminiscent of the narrative conventions used by the first experimental naturalists, one of whom was Descartes. Yet, in their narrative, Latour and Woolgar shifted from the scientific wonder at natural phenomena to a wonder at the way in which these phenomena are explored in scientific laboratories, thus broaching for the first time the examination of the ordinary in scientific knowledge making as itself an object of scientific inquiry. To put it differently, Latour and Woolgar "discover," as it were, that the modern site of scientific findings, the laboratory (and not "nature"), is in itself a prolific source of wonder that deserves scientific attention. Redirecting the focus onto mundane aspects of experimental sciences has caused debates and raised many methodological questions among historians, philosophers, and sociologists of science. And other issues and events, not discussed here, caused quite a stir, eventually culminating in the 1990s in the so-called science wars.[38]

Latour reflected on his use of a strategy of alienation and claimed that imagining a fictional investigator is crucial for creating a situation that represents "the maximum possible distance between the inside and outside of natural sciences and then also represents the best possibility for really explaining science."[39] He notes that literary means become pivotal in achieving a distanced viewpoint when using participatory observations as a research method: "If we cannot stage such a situation experimentally [placing an observer, ignorant of the Western conception of science into a scientific context], we can approximate it through literature."[40] Latour defies the criticism that only natural scientists, being experts in their specific fields of knowledge, are capable of reflecting appropriately on the process of scientific knowledge production. He argues that an expert account may forestall genuine reflection "if it simply restates the account it is supposed to explain."[41] In contrast, employing the fictional figure of a competent layman has the methodological advantage of observing scientists as if they were

153

strangers while doing science. Latour claims the imagined observer "could account for elementary concepts like 'observing,' 'explaining,' 'studying' or 'recording' in entirely new terms."[42] The impossibility of a "Pure Visitor's" viewpoint has earlier and famously been discussed by Ernest Gellner (1925–95), a leading figure in the social sciences in Europe.[43]

Moreover, it is worthwhile to note that Latour's discursive method is not new at all; his literary means have their own history in science, as Daston demonstrated. I made use of literary devices in my snapshot story that have a long history in writing about scientific discoveries. Indeed, Latour falls back on the very rhetorical devices that were used by the first naturalists. The seventeenth-century founders of our modern conception of science constructed their stories of discovery with similar means to attract scientific attention through wonder. Situating Latour and Woolgar's laboratory study within a broader history of scientific attention makes the inherently historical dimension of an anthropology of science apparent.[44] These narratives form a frame of reference for my personal account, which is both indebted to the work of these authors and intent on investigating the intrinsic entanglement between an anthropological mode of inquiry and the historicization of epistemology.

Constructing a Fictional Observer

One further point needs to be made on Latour and Woolgar's construction of a fictional observer. Obviously, Latour's choice for a character that only exists within the text was not just some methodological trick for looking afresh at what scientists do. Latour creates an alter ego on paper, an observer who has only a textual existence, because his study is all about texts and paperwork. He chose his method deliberately because it aptly allows him to describe how "nature" is transformed into text in the laboratory as a literary activity.

Latour and Woolgar wrote a compelling book not least because its form ingeniously reflects its thesis. Their study of practices of "literary inscription" performs this practice itself in an exemplary fashion, inscribing Latour's observations into the text by transforming his person into a fictional character.[45] Indeed, this is the very same "deflating strategy" they used to show how things were turned into paper in the laboratory.[46] Their imagined anthropologist observes in amusement how much effort it takes to achieve this translation, turning the observational spotlight on activities that are usually blacked out in narratives of scientific discovery. This care-

154

fully constructed point of view brings to light mundane aspects of scientific knowledge production that hitherto remained outside the scope of epistemologists. Yet, this fictional observer does more than just expose. Despite his deflated, two-dimensional existence on paper, this figure casts long shadows that obliterate other minutiae. Latour abstracts from the very beginning of his story from his physical presence in the lab by turning his embodied eyes (and hands—Latour assisted with benchwork tasks as a lab technician) into a "hands-off" third-person observer/narrator. The literary figure, in fact, implies a turn to a disembodied account of science. What is sacrificed in order to "present the laboratory as a system of literary inscriptions" is the hands-on perspective.[47] This is the way in which his body became implicated in the described "transformation of rats and chemicals" and other *mani*pulations.

Latour and Woolgar point out that the material dimension of the laboratory "very rarely receives mention" and describe how laboratory space is inhabited by a range of inscription devices.[48] The hands-on (wet) work at the bench is mentioned as a means toward an end of literary production (publication of papers).[49] Yet the laboratory as embodied space and benchwork as bodily activity are not described from their constructed point of view of the literary observer. A more detailed discussion of Latour's ethnographic method shows how he constructs, so to speak, a "disembodied" ethnography from a hands-on perspective when he elides any "body talk" from his material account of scientific practices.[50]

My analysis exposes how the embodied dimension of laboratory work is systematically written out of *Laboratory Life*. The obliteration of bodily issues in fact makes manifest how these early laboratory ethnographers struggle with the Cartesian legacy. They fail to see how the Cartesian *Unterschiebung* works: first, how it obliterates the fact that hands-on engagements are inherent in modern experimental endeavors, and, second, how the Cartesian *Unterschiebung* establishes a "disembodied" thinking about scientific practices. Searching out the hands-on moment *within* Cartesian thinking demonstrates that we need to think scientific practice from a hands-on perspective, and thus from an embodied perspective. A critical reading enables us to see how *Laboratory Life* remains indebted to disembodied conceptions of knowledge making as it fails to radically rethink Descartes's double movement (see figure 2.2) from a constructed viewpoint that aims to efface the observer's own embodied standpoint. In contrast, I argue *from within* Cartesian thinking for the need to include the articulation of bodily issues in accounts of life science wet-lab cultures and for the need to think *about* these epistemic practices as embodied practices.

It is rather striking to see how the ethnographer-in-the-role-of-technician becomes marginalized by the literary figure of the observer in Latour and Woolgar's account. This is emblematic of the way in which hands-on experiences and manual training are written into the margins of the text in general. Latour conducts participant observations in the role of technician, but he appears to use his first hands-on experiences at the bench not so much as primary source material for his descriptions of laboratory activity but rather as a strategic position from which he can observe what other scientists do and how their social and literary activities lead to the construction of facts. "The majority of the material which informs our discussion," the authors explain, "was gathered *in situ* monitoring scientists' activity in one setting."[51] From this statement we gain the impression that participatory observation is first and for all used to monitor the action of others and is not intended to include self-reflexive analysis about how it feels to become trained in the manual laboratory work that Latour conducted in his role as technician. His own hands-on experiences at the bench, apart from the activity of writing that forms the common denominator of the research activity of the observer and the observed, are not described as part of the process of making scientific knowledge. It seems as if it is precisely the *transparency* of the technician's role that was most appealing to Latour. The role of technician provided him with an insider's perspective on laboratory life (in situ observations) that was retrospectively turned into the outsider's gaze of a disembodied observer.

RESEARCH INTERNSHIP (SECOND WEEK): TRAIN
FROM LEIDEN TO AMSTERDAM

I am sitting in the train back to Amsterdam. My head is still spinning from all the impressions I have gathered during my first two weeks in the laboratory of the molecular genetics research group at Leiden Institute of Chemistry. I am exhausted and euphoric at the same time. Today, I had for the first time the feeling that I could grasp what I was doing and as a result could start to anticipate my next move. It was such a relief not to stumble around anymore; the whole first week I felt like a foreign body, an obstacle barring others from moving around swiftly between their benches and other equipment installed throughout the lab space. I recognized many of the tools, like pipettes, centrifuges, heating blocks, and vortex mixers, from the practical

courses I had followed, but what I learned the first week was that enter-
ing a new lab meant first and foremost to become familiar with its specific
topographic situation and to tune myself into this new landscape. It was
not only that I did not know in the beginning where to find what, or how
to use certain instruments; I also had the feeling that other staff members
apprehensively observed my movements, afraid that I would bump into
their setups and mix up or touch their samples. Even after all the course-
work I had passed, I felt awkwardly out of place in this space. Certainly, I
had anticipated this in an intellectual sense and prepared myself for a "Snow-
ian culture shock."[52] *I expected to be confronted with questions about what*
qualifies as a relevant scientific question and what are taken to be appropri-
ate scientific methods when crossing the border between the humanities and
the natural sciences, but I had not anticipated that I would feel misplaced in
such a literal, bodily sense.

What struck me when reading *Laboratory Life* was that the place that
Latour actually occupied as a technician within the space of the laboratory,
his own spatial presence, his own bodily being in situ, his own actions at
the bench, are not subject to his reflections. In *Laboratory Life*, being in situ
functions primarily as a point of view from which to monitor the actions
of others.[53] The *observer's* point of view, the authors explain, functions as a
"principle of organization" to gain a better understanding of how order (in
form of "systematic and tidied research reports") is created from apparent
chaos.[54] The technician's spatial position is turned into a geometric point,
becoming a coordinate—an organizing principle; the embodied perspective
is conceptualized as a position with no extension in space; the technician's
standpoint becomes a conceptual *viewpoint*. The geometrization of the situated being of the ethnographer-as-technician remains indebted to a Cartesian geometrization of bodies.[55] The in situ method employed in *Laboratory
Life* can be criticized from a phenomenological understanding of space. Andrew Pickering's summary of writing on space by phenomenologists such
as Edmund Husserl, Martin Heidegger, and Maurice Merleau-Ponty makes
this explicit: "Our sense of space derives from our necessarily embodied
actions in the world, and, as it were, always refers back to our bodies as
specific, marked places in the world. Our sense of neutral, unmarked space,
the argument goes, is a parasitic construction laid as an organizing principle
upon that prior sense."[56] The geometric viewpoint is an organizing principle that blocks out the possibility of articulating the embodied dimension
of laboratory spaces. We could also say that the in situ method used here is

itself not conceptualized as a situated approach to science studies.[57] From the point of view of the mythical observer, the authors introduce a different level of analysis that obscures and obliterates the embodied perspective of the manually engaged technician/epistemologist.

Laboratory Life focuses on detailed descriptions of writing activities and pays only cursory attention to other kinds of hands-on engagement involved in doing life science research (such as handling instruments, laboratory animals, and chemicals). The book provides a detailed map of the laboratory but no further elaborations on the bodywork conducted within these spaces (the administrative offices and the bench space) or the specific ways in which researchers move about in the different spatial settings. How does our experience of special controlled laboratory spaces shape our thinking, and how do these spaces shape and define bodies?[58] We can situate questions about the laboratory as a specific site of bodywork and embodied space within a broader debate that entailed a paradigm shift in the thinking about the relation between mind and body. Within the cognitive sciences, an embodied cognition paradigm, which has been expanded into a conception of cognition as embodied, embedded, extended, and enactive, has gained wide acceptance. It fosters a thinking that departs not from the mind-body dichotomy but from the presupposition that intelligence is inherently dependent on a moving body inseparable from its environment.[59] In my view, this calls for more than the general body theory or body philosophy as Merleau-Ponty develops it in *Phenomenology of Perception*, first published in French in 1945.[60] It also calls for detailed accounts of the specific bodywork involved in particular scientific practices. This shift from brain to hands, in which thinking becomes grounded in doing, resonates with Latour and Woolgar's declared objective of gaining a better understanding of scientific productivity by approaching the sciences from a hands-on perspective instead of inquiring into science's mental products (theories, concepts, ideas). However, what is important here is that Andreas Blum and colleagues emphasize embodiment as a temporal process that unfolds in time.[61] Yet the mobilization of a geometric viewpoint necessarily entails an ahistorical stance that does not take into account the historical dimension of the laboratory space as a training space in which being in situ always involves being trained and developing bodily skills (I will elaborate on this point in chapter 5). Latour has provided us with detailed accounts of the historicity of scientific facts.[62] But how scientists *become* scientists in laboratory space is not part of the story of *Laboratory Life*.

158

One passage in *Laboratory Life* gives the hands-on practicing technician a voice with which he reflects on his benchwork experience. Despite its apparent triviality, this passage provides us with a vivid example of mindful hands in modern biology labs:

> One of the most difficult tasks was the dilution and addition of doses to the beakers. He had to remember in which beaker he had to put the doses, and made a note, for example, that he had to put dose 4 in beaker 12. But he found that he had forgotten to make a note of the time interval. With pipette half lifted, he found himself wondering whether he had already put dose 4 in beaker 12. He blushed, trying to remember whether he had made a note of when he had made a note! He panicked and pushed the piston of the pasteur pipette into beaker 12. But maybe he had now put twice the dose into the beaker. If so, the reading would be wrong. He crossed out the figure. The observer's lack of training meant that he continued in this fashion.[63]

The authors conclude that the incompetence of the participant observer/ technician provides valuable insights into the "wealth of *invisible* skills [that] underpin material inscription."[64] Two points in this account drew my attention. First, note how Latour and Woolgar readily frame their description of a manual activity as part of scientific knowledge production in terms of invisibility: they describe adding correct amounts of doses with a pipette into beakers as an "invisible" skill. Hands-on work can only be articulated when it becomes framed as that which is not visible, that which goes unnoticed, is hidden, unseen, or out of sight. Second, it is significant that the authors describe a feeling of awkwardness and that they hint at a heuristic potential of such awkward experiences.

The pioneering study *Laboratory Life* is an exemplary and innovative attempt to demonstrate the epistemological value of a hands-on perspective on science that critically questions assumptions inherent in traditional accounts in history and philosophy of science. However, we can discern in Latour and Woolgar's account an epistemological bias against manual work that it seems to share with those very accounts it aims to criticize. About ten years after the publication of *Laboratory Life*, Steven Shapin published a seminal article on seventeenth-century new experimental science and the role of technicians.[65] It is illuminating to reread Shapin's analysis in the

159

context of the earlier discussion here regarding how a technician's role is methodologically employed for an anthropology of science. In "The Invisible Technician" Shapin states: "The transparency of technicians' roles in making scientific knowledge reflects both historical and modern attitudes toward the value of skilled work."[66] In *Laboratory Life*, Latour and Woolgar readily make use of the invisible technician's role in making scientific knowledge without questioning the prejudice toward manual work that is implicit in the way in which the technician remains unnoticed. They describe a division of labor between scientists ("reading, writing, typing") and technicians ("cutting, sewing, mixing, shaking, screwing, marking") according to which the handling of equipment and manual work with materials other than written documents are almost exclusively ascribed to the latter while both those working at the bench ("known as technicians") and those working mostly at desks in another section of the laboratory ("individuals referred to as doctors") are portrayed as "compulsive and almost maniac writers."[67] This study describes how a literary activity is also part of the technician's work, but it provides hardly any descriptions of hands-on benchwork in the tasks that scientists perform. A distinction between "mere" manual work and literary activity is maintained, and though Latour and Woolgar find it imperative to describe the latter in its materiality, the account of investigative activity given here appears to favor an image of laboratory life science as a discipline of the pen rather than a discipline of mindful hands and implicated bodies. My point, then, is to draw attention to the fact that we can discern an epistemological bias in the way in which manual work and the bodily dimension are elided from the analytical level, while they play a constitutive role on the practical level. *Laboratory Life* is, in a certain sense, a paradoxical study. We can describe it as a pioneering attempt to write an epistemological study from a hands-on perspective: for almost two years Latour was invited to work part-time as a technician in the laboratory of an endocrinology research group, a position that put him "in situ" within the laboratory space and "hands-on" at the bench. Besides this part-time employment from 1975 to 1977, Latour gained access as an observer who was provided with his own office space, the possibility of attending most discussions, and access to archives and other written output from the laboratory.[68] The declared aim of his anthropological study was to "retrieve some of the craft character of scientific activity through in situ observations of scientific practice."[69] Yet it is this hands-on character of the scientific field under investigation that remains undescribed. Endocrinol-

160

ogy is a branch of internal medicine and a subfield of physiology. It seeks to understand the inner workings of living bodies by studying the endocrine system, consisting of glands that are responsible for the secretion of hormones that regulate various bodily functions, including metabolism, growth, and development but also mood. How do the researchers gain access to these inner bodily processes? Latour and Woolgar provide no detailed descriptions of the hands-on skills like pipetting or handling and preparing laboratory animals and body parts for treatments or dissections that characterize benchwork in endocrinology research.

The focus is on disclosing how science amounts to a predominantly literary activity, with a particular interest in the materiality of writing processes and inscription devices, but Latour and Woolgar's book provides no further accounts of the bodily activity and manual skills involved in what they described as transforming rats into papers. What is problematic, however, is the persistent rhetoric of transparency and invisibility that *Laboratory Life* shares with traditional historical and philosophical accounts when reflecting on matters of bodily presence and manual engagement in science.

Latour and Woolgar provide a *methodological* motivation for their use of an in situ concept in a "disembodied" sense. On the one hand, their book underscores the importance of "being on the spot" in studying scientific practices: they emphasize the importance of an in situ approach in gaining insight into the "craft character of science."[70] In the postscript to the Princeton edition of *Laboratory Life*, Latour and Woolgar argue that "in situ observation provides more direct access to events in the laboratory than, for example, interview responses. In both cases, the general idea is that more is to be gained from being on the spot than from attempting interpretation from a secondary perspective. The in situ monitoring of contemporaneous scientific activity portrays the scientist located firmly at the laboratory bench and treats with some scepticism the kind of representations provided by the scientists."[71] However, they later take issue with the very suggestion that doing in situ observation provides a more direct account of how science is done. In other words, they question whether in situ observations can be simply linked to ideas of participatory immediacy. They argue that this method allows researchers to immerse themselves in the daily activities of the lab, but the question of how to achieve an analytical distance in reporting these observations remains an issue. As a solution, they chose to place "the burden of observational experience on the shoulders of a mythical 'observer.'"[72]

Latour and Woolgar discuss this methodological problem in terms of the dilemma of achieving the right balance between closeness and distance. Analytical distance and reflexivity are achieved by turning a hands-on perspective into hands-off observations. Thus, the in situ notion is a heuristic to gain a perspective on scientific activity that provides insight into the micro-processes that underlie what Latour and Woolgar call a retrospective characterization of scientific activity in terms of "logical reasoning."[73] But, at the same time, the authors warn against a naive realism implied in the idea that "*being* there" amounts to knowing what "really" happens. Latour and Woolgar claim that they have circumvented this dilemma by abstracting from embodied presence with the literary tool of the "stranger device." They react here to a critique of ethnography that questions the idea implicit in participant observation that embodied presence entails the possibility of an immediate and authentic account, as if physically being at the spot implies that all strangeness vanishes and full congruence between "what is" and "what is experienced" can be achieved. Latour and Woolgar imply a link between the physical, embodied presence and the methodological problem of implying "direct," "undisturbed," "immediate" observation. The problem that an in situ, or hands-on, approach can lead to a false impression of "immediacy," "authenticity," or "directness" implicit in the idea of "being present at the spot as it happens" creates, for Latour and Woolgar, the need for strategies of estrangement that lead to quite "disembodied" accounts of the scientific activity of both the observer and the observed.

However, the real problem may be that the authors so readily convert a hands-on in situ position into notions of transparency. Should we not rather think of being in situ as a position that starts out with being a "strange body"—not merely a stranger device but in a literal sense a foreign body?[74] The embodied sense of being in situ is precisely marked out by its opacity, by the difficulty of tuning into the environment, and by its resistance to becoming invisible. A stranger's body failing to blend in chimes with Michael Lynch's account of how he could only *become* an observer who did not stand out but was merged into the laboratory setting *through* bodily training.[75] I propose that Latour and Woolgar's methodological discussion, in which bodily presence functions as a "strategy of authentication" that implies direct access, is flawed because it presupposes that embodied experience is tied exclusively to concepts of congruence, intimacy, and closeness.[76] I also sense here the assumption that embodied presence hinders, rather than facilitates, attempts to gain analytical distance. For me, the much more interesting question is: How can we articulate the resistance of bodies to becoming invisible, on both the practical and the epistemological level of analysis? And how can we employ our body's *opacity* as a methodological tool?

From a phenomenological perspective, we can locate a sense of alienation at the core of bodily experience. In the act of touching and being touched, I experience my hand both as my own and as a thing that I can examine as something external to me. I can probe my right hand with my left hand as if it were an object lying on the table before me and sense that *my* hand is touched by my other hand. As discussed in chapter 3, this phenomenon of alienated experience and an experience of "ownness" locate notions of closeness and distance, intimacy and alienation right at the core of bodily experience. Jenny Slatman's phenomenological analysis in *Our Strange Body* describes how Cartesian epistemology culminates in a mind-body distinction by capitalizing on this possibility of making an alienated body experience absolute on a conceptual level.[77] With this hypostasizing operation Descartes achieves a philosophical position in which all bodies are turned, on an analytical level, into potentially intelligible entities within a *cognitio mathematica* ideal of the sciences. This, in turn, makes the bodies as *objects* accessible on a practical level for the experimental grasp of the "new science." What is sacrificed in this philosophical operation is the possibility of theorizing knowledge making as an embodied activity and of describing embodied experiences. We can discern in *Laboratory Life* a similar trajectory of thought: the authors begin from a hands-on perspective, but then write any reference to embodied dimensions of scientific activity (as experienced by the hands-on practicing observer) out of their epistemological account of knowledge making. I do not criticize the authors of *Laboratory Life* for turning a blind eye to issues of embodiment and the specific bodywork involved in doing research in the life sciences—selectivity is, after all, a scientific virtue and necessity—but my objective is to show why they *cannot* articulate this problematic due to their underlying philosophical assumptions.[78] I have argued that it is precisely because these early laboratory ethnographers did not see that it is historically and philosophically untenable to think their hands-on approach *in opposition* to Descartes's thinking that they remain indebted to an unquestioned conception of Cartesian epistemology. This becomes manifest in the impossibility of accounting for the epistemological relevance of bodywork from within the theoretical and methodological framework of *Laboratory Life*. The merits of *Laboratory Life* lie in the rigorous way in which the authors argue their point with a well-chosen selection of material to expose the centrality of literary activity

163

in doing science. But they force this point at the cost of untold body stories buried in paperwork.

Embodiment and the Disembodiment of Life

This essay, then, is a meditation; an attempt to explore—tentatively and a bit quixotically—the history of the biological gaze, focusing in particular on the different ways in which that gaze has become increasingly and seemingly inevitably enmeshed in actual touching, in taking the object into hand, in trespassing on and transforming the very thing we look at.

Evelyn Fox Keller, "The Biological Gaze"

Today cultures of living, reproducing cells in petri dishes form an indispensable and ubiquitous research instrument in the life and biomedical sciences. Hannah Landecker states in "New Times for Biology: Nerve Cultures and the Advent of Cellular Life in Vitro" and *Culturing Life: How Cells Became Technologies* that the introduction of a technique for culturing cells from complex multicellular organisms outside of the body must be viewed as part of a larger shift in twentieth-century laboratory practice from in vivo to in vitro experimentation. This move from in vivo to in vitro experimentation marks not only a shift in lab practices but also an epistemological shift because the advent of cellular life in vitro called into question life's dependence on the body. Landecker describes Ross Harrison's experimentation with nerve cell cultures that took place from 1907 to 1910.[79] Harrison tried to observe the development of a nerve cell live, as it grew. To make this internal life process visible, he had to isolate the cell from the body. His experiments differed significantly from experimentation with isolated body parts that were kept alive for a short time. As Landecker asserts, "It was a deeply held assumption that pieces isolated from the body were in fact dying; and though they might be kept isolated for a period of time, they were fully expected to perish relatively rapidly."[80]

Harrison's experimental setup broke with the underlying assumption of these experiments. He did not want to study temporary survival but instead designed his experiments to observe ongoing life outside of the body. These experiments violated the primacy of an "intact" context out of the desire to see internal changes as they take place, even if this meant isolating them from the body. The way in which Harrison combined already existing methods to make a growing nerve cell visible called into question what had been considered the right means of producing suitable and reliable experimental knowledge. His intention was also at odds with the prevailing histological methods and techniques that were used to gather information on internal development and change in organisms. In histology, change was represented using a series of different specimens that were prepared for microscopic research by staining, fixing, and slicing. Internal change could thus only be inferred indirectly through observations made on dead specimens.

Yet, the handling of dead material in research on life processes was not seen as problematic, but rather was considered to be of practical advantage. What was most crucial, however, was that these techniques were believed to leave the context of the body more or less intact. In contrast, Harrison threatened to transgress a vital boundary with his attempts to isolate a living cell. According to Landecker, his experiments were not so much disputed for the way in which they challenged existing views on the relation between the body and its parts; his practice was highly controversial because it questioned the significance of the inner context, or internal milieu, for the growth and reproduction of cells in complex organisms.

With his concept of a *milieu intérieur*, the nineteenth-century physiologist Claude Bernard declared the internal fluids and tissues of the body to be indispensable for the survival of a cell. Histological methods supported this understanding. Harrison's experimentation, however, contradicted the prevailing view both practically and conceptually. He displaced internal life processes such as cell growth from the inside to the outside. Thus, he not only took the body apart but also turned it inside out. By doing so, he questioned the then dominant conviction that research on life phenomena has to take place in the context of the body. Landecker shows in her analysis how Harrison's nerve culture experiments transformed the opaque interior 165 into a transparent exterior. His research was cutting edge in a literal sense: he transgressed the boundaries of a living body that were considered to be vital. Harrison's attempt to grow living cells in a glass tube brought about a new method to study life in the laboratory, and his first experimentation on nerve cells gave birth to a new concept in life science research: cellular life in vitro. Landecker's history of cell culturing can be situated within the larger history of a biomedical tradition that sets out to make bodies transparent and a twentieth-century approach to the body that calls for its fragmentation and its concomitant demand for a constant renegotiation of bodily boundaries.

With the advent of vitro technologies, research on the phenomena of life appears to have discarded the body as a necessary context within which to study living processes. But is that really so? Evelyn Fox Keller ruminates on manifold bodily boundaries in organic bodies: "Biology recognises many bodies, corresponding to many skins: in higher organisms, there is the multicellular body contained within an outer integument; in all organisms, cellular bodies are contained by cell membranes; and in eukaryotic organisms nuclear bodies are contained by nuclear membranes."[81] Keller argues for the cell membrane as a relevant bodily limit in research practices. She emphasizes the importance of the skin, or membrane, in the theory and practice of the life sciences and states, "Given the dire effect that physical erasure of this boundary would have on the survival of the organism or cell, it scarcely seems necessary to elaborate on the inappropriateness of its conceptual erasure."[82] In fact, the concept of the internal context did not lose value, but in vitro technologies made it necessary to revise the accepted boundaries between the inside and the outside of the living "unit."

This discussion illustrates that there exist no stable bodily boundaries in the life sciences. Depending on the research question, boundaries have to be continually drawn anew. In addition, biologists stress that the cell membrane cannot be conceptualized as a preexistent, fixed borderline. The membrane is permeable and continuously in motion, so the boundary work of the body must be seen as dynamic and active. It is important to note that the boundaries of biology's research bodies do not necessarily coincide with the bodily limits of living organisms. With the possibility of culturing cells in vitro, internal processes became exteriorized. Hence, the bodily boundaries in research practices have moved inward, from organism to cell, and from in vivo to in vitro. But, most important, they keep shifting in and out. It is crucial to keep in mind that in vivo experimentation was not replaced, or displaced, by tissue culture.

Landecker's statement is highly significant in this respect: "Indeed it was not until 40 years later that tissue culture began to be used in the widespread manner that led to its contemporary ubiquity in the laboratories of biomedical research. However, the conceptual and practical shift signaled by its establishment in the first decade of the twentieth century indicates the appearance of a new way of thinking about, seeing and experimenting upon the cells of complex organisms."[83] "The body," she concludes "was not replaced by the cell, nor reduced to it; rather, this technique of fragmentation substituted an artificial apparatus for the body and generated new views of the autonomy and activity of cellular life. As a result, understanding of both the cell and the body was at this time fundamentally altered, as were ideas about their relationship to each other."[84]

Landecker signals that the body has not lost its significance in laboratory research on living processes. Rather, the substitution of the body for an artificial apparatus in the lab has turned the body into a technoscientific problem. As internal life processes are taken out of the body and placed into a glass tube, the body and its boundaries are made explicit as they must be dealt with on a conceptual as well as a practical level. What constitutes the body and its boundaries is a question of ongoing concern in laboratory practice: the missing bodily context must be compensated for by an artificially controlled environment that protects the cell cultures from external influences and provides them with the right temperature, nutrition, fluids, and adhesion base. Cell life in vitro is not a life form in itself but depends on a highly advanced compensation technology.

Sterility forms an important constituent of these compensation technologies that make cell proliferation in vitro possible. A sterile environment ensures that cells will not be contaminated by other organisms, such as bacteria that grow much faster and are lethal to the cells. Cell cultures do not have an immune system and are therefore very prone to any such contamination. What makes the concept of sterility so intriguing is that it calls attention not only to the shifting bodily boundaries of the research object but also to the researcher's own body and the vital part that it plays in compensating for the missing bodily context. I noticed that researchers in the lab were constantly concerned with the possible *interference* of their own body—with the potential for their body to be a contaminating vector. The "disembodiment" of living processes has led to an embodied practice that requires an acute awareness of one's own body in the course of producing knowledge in the lab. In most labs where research is conducted on cells isolated from humans or animals, researchers use a sterile hood. The need for sterility makes it clear that the postulated borderline between the investigator and the investigated is not preexistent but the result of a laborious process. The shift from in vivo to in vitro in the laboratory practices of the life sciences has exposed the body and its limit in a very particular sense as a central issue in life science research. This should also alert us to a misleading "gene talk" dominated by disembodied metaphors such as "life as a code," a script, a master program, or a blueprint that others have already deconstructed.[85]

167

Umbilical Cord and Blood Cell Isolation

SYMBIOTICA—AN ARTISTIC LABORATORY
DEDICATED TO RESEARCH AND HANDS-ON
ENGAGEMENT WITH THE LIFE SCIENCES,
SCHOOL OF HUMAN BIOLOGY AND ANATOMY,
UNIVERSITY OF WESTERN AUSTRALIA, PERTH

Monday afternoon. I had been on "standby" for weeks, when I received Guy's call. He was excited and told me that the baby had been born and that he expected the delivery of the umbilical blood and cord any minute. I hurried to the tissue engineering laboratories. Guy had eagerly been waiting for this moment. He had told me some time ago about his plans to harvest stem cells from the freshly cut umbilical cord of the child of his friend and colleague, the bio artist K. He had asked me whether I wanted to assist him with cell isolation and to document the experiment with my camera. He was especially excited about this very promising protocol for stem cell isolation from umbilical blood. His hopes were high that this protocol had more chances of success than the less sophisticated procedure that we would use for isolating cells from the cord tissues. It was a strange and exciting idea that we were going to work with a freshly cut umbilical cord, the body part of the child of his friend that not even an hour earlier had connected mother to child. A body part from the depth of her uterus and of vital importance, a lifeline, pumping air and blood, and providing the embryo with all her nutrient needs, connecting the blood circulation from mother to child. A body part that immediately becomes superfluous once the baby is born. Yet, the umbilical cord appeared to me as the material manifestation of their intimate relationship, which enters into a new phase of symbiosis the moment the mother starts breastfeeding her newborn child, and when both parents start providing the baby with all its needs, milk, warmth, comfort, care.

A big white box has already been delivered when I arrive at the building. Guy lets me in and we enter the laboratory. First we have to put on one of the special lab coats that are only to be used in this laboratory space and are not allowed to be worn outside of the tissue culture laboratory. Guy opens the lid of the box. I see a bloody, pale, and fleshy string inside a Plexiglas box. It reminds me of a snake or some inner organ at the butcher's. Next to it is a flask filled with a thick red fluid. Before we unpack the box, I take a picture of its contents. We are going to perform two different protocols, hoping that at least one of them will work out and will result in growing cells in vitro. One protocol

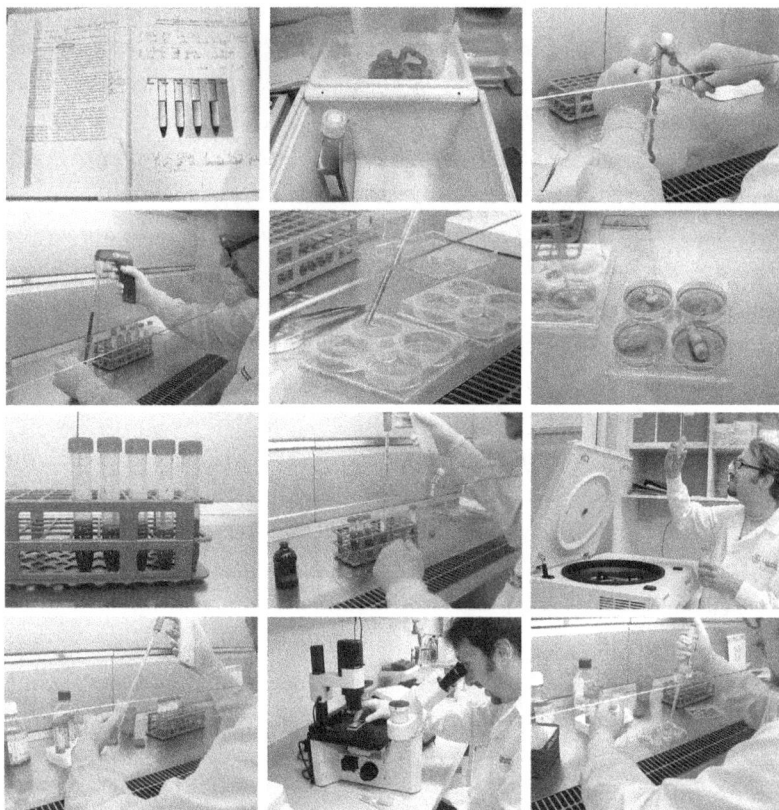

4.2 ★ Guy Ben-Ary, artist in residence at SymbioticA, performing an umbilical cord and blood isolation experiment in December 2006 at the tissue culture laboratories, School of Human Biology and Anatomy, University of Western Australia, Perth. Photos by the author and courtesy of the artist.

169

describes all the steps of how to isolate stem cells from umbilical cord blood. The other one is a more crude procedure for harvesting cells from primary tissue and is less likely to give positive results.

We start with the preparations for the protocol for stem cell isolation from umbilical cord blood. We wash our hands, and Guy puts on gloves. He puts all the materials on the bench into the laminar flow cabinet: a tray, a plastic bag with tubes, bottles with solutions, an automatic pipette gun, and several sealed sterile pipettes. He explains to me how important it is not to forget anything and to have everything well prepared beforehand since he should stay put at the sterile hood once he begins with the protocol. He sits down in front of the opening of the sterile hood, extends his arms into the interior, and puts several tubes into the tray. He opens the sterile wrap of the pipette without touching its tip and inserts it into the pipette gun. Next he unscrews the lid of one of the tubes with one hand. With the other hand he transfers the umbilical cord blood from the big flask into the tube. He repeats this with the rest of the tubes and casts the pipette away. He puts a fresh pipette into the pipette gun, inserts it into one of the other bottles, and adds the same amount of fluid to every tube. He is careful to leave the tubes and bottles open only as long as necessary and to touch only the inner sides of the bottle and flasks with the tip of his pipette. When he is finished, he takes the tray with the tubes from the sterile hood. Finally, he puts the tubes in the centrifuge on full power for thirty minutes [figure 4.2].

In the meanwhile, we start with the primary tissue cell isolation protocol. The other week Ionat Zurr and Boo Chapple had taught me how to culture cells from a cell line, so I had already worked in a laminar flow cabinet once before. They had explained to me that I had to be careful not to contaminate the cells. I had learned that it was of vital importance to work in a sterile manner with living cells. With no immune system, those "bare" cells are very vulnerable because the air, my hands, and my body are full of microbes that could harm them. If I touched, for example, the inner side of the lid of the flask in which we grew the cells, microbes from my fingertips would enter the flask, and they would grow so much faster using up all nutrients that the cells would have no chance of survival. Culturing cells from cell lines was a strange, rather anaesthetic experience because I could not see, feel, or smell the cells. That time we had taken a vial of frozen cells from a human cell line out of the freezer. I could not see with my naked eye that the transparent frozen fluid contained living cells. So when I sat at the bench and carefully pipetted the cells from the little microcentrifuge tube into a flask, I only knew that I was inserting potentially living cells into a flask in which they would be fed with a nutritious solution, bovine serum. A few days later, Boo showed me how to observe the cells under the microscope. She showed me how to manually adjust the microscope to focus on the cells and pointed out what I had to look at and how I could distinguish between moving, that is, free-floating dead cells and immobile cells that had attached themselves to the wall of the flask and had started to grow into a thin layer of tissue. It took me some time to handle the focusing knob gently enough to get the right focus and to recognize the cells.

I put on gloves and take all the materials and solutions that I need from the bench and place them into the laminar flow cabinet. Sterilized instruments, that is, tweezers, pinchers, scalpels, and surgical scissors to cut up the *umbilical cord and to mince the tissue as fine as possible to loosen the stem cells and detach them from other tissues. And petri dishes, a tray, sterilized pipettes, and a pipette gun. I sit down at the bench and extend my arms into the opening. It feels a bit uneasy to work in a laminar flow cabinet. I repeatedly bang my head against the glass front when I try to get a closer look at what I am doing. Guy had cut the umbilical cord into smaller chunks. It took him some time to cut through the thick, tough cord. He puts the chunks into a tube filled with some sort of cleanser. After shaking them, he puts the washed pieces on a petri dish. I pick up a piece of the cord. The texture feels thick and stiff. I feel a bit clumsy, afraid to let the slippery piece slip from my grip. I take up some scissors with my other hand and try to cut the cord. It is difficult to cut through the tough flesh. It feels weird to pull and tug at this body part. I wonder, whose body part is it anyway? Is it from the mother or the child? Where does the mother's body end, and where does the baby's body start? And what about the cells that we isolate from it? Finally, I manage to cut off a small piece and start to mince it with a scalpel. This does not work very well. I take up the scissors instead; they do not seem to work better, and it is difficult to hold onto the slimy substance without letting it slip from the dish. Finally, I fill the wells of a four-well plate with the bovine serum. I clumsily unwrap the sterile pipette and insert it into the pipette gun, afraid to contaminate its tip by brushing it against my hand or the bench. While I hold the pipette with one hand, I try to unscrew the flask of the bovine serum with the other. It looked so easy when Guy did it. Finally, I manage to get it open with one hand. I put the lid upside down on the bench, proud that I remembered not to let its inner rim touch the bench. I carefully insert the pipette into the bottle and suck up the right amount of fluid. I fill the wells without letting the tip touch the plastic, just in case. Next, I transfer the cut-up pieces of the cord from the petri dish into the wells and close the lid. That's it, Guy says. I pass* *him the plate through the opening, and he puts it into an incubator. During the next few days we will come back to change the nutrient solution and to check whether some cells have attached themselves to the bottom of the well and then whether they keep growing. This would be it then: life outside the body, cellular life, in vitro life—in any case, handle-me-carefully-do-not-touch-me life.*[86]

With the practical turn, unified accounts of science have rightly been challenged, and scholars have called instead for accounts that recognize the plurality of the sciences and their diverse epistemic practices.[87] My philosophical inquiry into the possibilities of grounding Cartesian epistemology in hands-on practices feeds into this line of research that sets out to historicize epistemology by writing knowledge-*making* histories rather than a history of ideas.

How can we not talk about bodies and bodywork at scientific sites that are all but dedicated to bodily matters: to the manipulation and fragmentation of living bodies, to the constant attempt to redraw bodily boundaries, to question the "malleability of living substance," and to explore "the plasticity of living matter"?[88] When "the study of vital phenomena became an experimental science," as Keller observed, epistemological accounts must engage with hands-on experiences.[89] The exteriorization of life in vitro involves continuous boundary work by many hands.[90] Moreover, practices in the life sciences today need to be understood from within the modern biomedical project to make bodies transparent. How do technologies that increasingly facilitate the transparency of bodies under investigation go together with an increasing attention to the opacity and obtrusive presence of the investigators' bodies? In the next section I further explore how today's biomedical practices involved in making bodies "transparent" actually prompt us to rethink them from the perspective of the body's opacity.

When researchers today investigate our inner bodily workings, what do they actually touch in their laboratories? What does hands-on research mean in a contemporary biology lab? These knowledge-making practices have raised many questions: How do scientists' own bodies become implicated in their investigation of living processes? In what sense is modern physiology today still a bodily invested endeavor? How are scientists in *touch* with life in today's life science laboratories? What kind of bodies are implicated and engendered in biomedical and life science laboratories?[91]

In the seventeenth century, the emerging modern sciences of life involved hands-on examinations of, and experimentations with, living and dead bodies and body parts. The craft character of modern life science has its history in the laboratories of medical practitioners and the institutionalization of these new practices in medical faculties at universities in seventeenth-century Europe. This craft character is expressed in a hands-on attitude in which the bodies of the investigators themselves became implicated in the experimental exploration of bodies and living processes.[92]

In her essay "The Biological Gaze," Evelyn Fox Keller points to the "indispensable role the hand has come to play" in experimental biology.[93] She distinguishes a modern "interventionist" biological approach from a period of preexperimental inquiries during which natural historians investigated living phenomena with the naked eye. "Surely," she writes, "if there ever was a non-intrusive, non-transgressive, biology, that was it," and she further elaborates on the natural historians' gaze: "The naked eye, with the relatively gentle ways of touching that went with the scientific study of biological form, left the mysteries of vital phenomena unexplored."[94] She then opposes the natural historians' gentle gaze to the experimenter's probing touch: "Touching implies intervention, manipulation and control."[95]

This remark encapsulates what modern experimental life science is all about. It is especially interesting given the intrinsic relation between touch and life. Touch is essential to living beings, as Jenny Slatman has pointed out when she writes: "Living beings are capable of surviving even when they lack a sense of smell, sight, taste or hearing; but they cannot stay alive without the sense of touch. Therefore, if there were something like 'the sense of life,' it would be the sense of touch."[96] The importance of manual training in biological sciences has, however, long remained understudied. Keller credits Ian Hacking, a philosopher of science, as "one of the few" who deliberated on the importance of hands-on training in the case of modern biological microscopy.[97] Thinking about life, or living beings, thus always implicates thinking about touch; thinking about the life sciences therefore calls for a thinking that allows us to start with touching, hands-on investigations, and *mani*pulations.

In Touch with Epistemic Cultures

EXPERIMENT: CLONING AND COMPLEMENTATION IN YEAST; SECOND-YEAR PRACTICAL COURSE, "GENE TECHNOLOGY, LIFE SCIENCE AND TECHNOLOGY STUDIES," GORLAEUS LABORATORIES, LEIDEN UNIVERSITY

Just reading the protocol is a funny experience. The aim of the experiment is to complement a gene in a yeast strain, which means that we would insert a gene into the genome of unicellular fungi in which the corresponding gene had been deleted. To do so involves the use of a range of basic gene technologies and the handling of two of the most studied model organisms in molecular biology: the eukaryotic model yeast Saccharomyces cerevisiae *and the prokaryotic model bacterium* Escherichia coli.

The experiment would teach us how to clone a yeast gene that is involved in DNA *repair; how to insert the genes into another species, that is, into bacteria; how to select the successfully cloned genes; and how to express them in this foreign body of the bacterium* E. coli. *Next we would learn how to amplify the genes in* E. coli, *then isolate the plasmids containing the gene. In the*

173

next step, these bacterial plasmids would be transformed into yeast strains. And finally we would check whether the transformation was successful with a UV test, which shows if the genetically complemented yeast strain has become UV damage–resistant.

The benchwork included working with model organisms and with genes inside and outside of a living organism, a concept that was still hard for me to grasp. The idea that genes are the "basic unit of life," but on the other hand that isolated genes in the lab are just a molecule, a chemical substance that can be isolated, purified, replicated, cut, rejoined (ligation), and again inserted into a genome of a living organism where it would resume its task and function as nothing had happened was mind-boggling. But what I was most curious about was not so much the ethical implications of genetic manipulation or the ramifications of this manipulation for our understanding of, and intervention into, natural processes, but, in a very literal sense, what gene manipulation means. What I wanted to know was: What kind of handiwork is involved in doing gene technology? How does it feel to handle a gene inside, or outside, of an organism?

What was most striking to me was that all the handling of bacterial cells, or yeast cells, fragments of DNA, or plasmid DNA most of the time meant juggling around minuscule plastic tubes (microcentrifuge tubes): putting them into trays, on ice, into heating blocks, centrifuges, and fridges. Most of the time the microcentrifuge tubes contained minuscule amounts of fluids that needed to be added or carefully removed, transferred into other tubes, and inserted into gels with microliter pipettes. I had difficulty keeping track of when I was handling living cells, "bare molecules," chemicals, or just sterile water. What I remember most vividly was that most manipulations involved a careful "regime of touch" to keep samples free of contamination. For example, I had to be careful not to pollute my samples with my DNA, or with the bacteria from my hands. I also had to protect my body against DNA damage when working with carcinogenic chemicals, such as the colorant for gels: ethidium bromide. And I also needed to wear gloves when I tested the UV sensitivity of my yeast cells because the UV would cause DNA damage not only in my samples but also in the DNA of my hands.

The ethnographic study *Epistemic Cultures* by Karin Knorr-Cetina was published in 1999, some twenty years after *Laboratory Life*.[98] Knorr-Cetina draws on in situ observations to describe the epistemic culture of a molecular biology lab as an intrinsically hands-on and bodily endeavor. The objective of the book *Epistemic Cultures* is to exemplify the need for many different epistemological theories that can account for disciplinary particularities; the study thus gives insights into diverse epistemic cultures, which have been described meticulously in their specific forms "through direct observation."[99] *Epistemic Cultures* therefore emphasizes not only a historicization of epistemology but also an epistemological pluralism. Importantly, Knorr-Cetina's call for a pluralism shifts the perspective from disciplines that have traditionally been favored by epistemologists—physics, mathematics, and astronomy—to the hands-on practices of experimental life sciences.

Knorr-Cetina's book provides a rich image of the "bench work style of molecular biology."[100] Its descriptions of bench scientists respond to Steven Shapin's call for an understanding of the laboratory "on the model of a workshop" and scientists "on the model of technicians."[101] Knorr-Cetina also uses the term *workshop* to define a contemporary molecular biology lab: "This bench laboratory is always activated; it is an actual space in which research tasks are performed continuously and simultaneously. The laboratory has become a *workshop* and a *nursery* with specific goals and activities."[102] We can link the preference for workshop terminology to historical arguments that foreground bodywork.[103] These arguments also resonate with Knorr-Cetina's discussion of how "bodily presence and activities were essential" to the laboratories under study.[104]

Knorr-Cetina divides her account into descriptions of two kinds: one focusing on the transformation of material objects and the manipulation of bodies and the other focusing on the scientist's body and the different ways in which it is implicated in the experimental work. The objective of my analysis of *Epistemic Cultures* is twofold. First, I show that Knorr-Cetina convincingly demonstrates the necessity of studying the practices of the life sciences from a hands-on perspective in order to expose their hands-on character. Second, I critically assess Knorr-Cetina's account to disclose how *Epistemic Cultures* remains trapped in Cartesian body concepts when the author attempts to articulate the embodied practices on a conceptual level.

Knorr-Cetina begins by providing an impression of the various ways in which material objects—bodies—are subjected to all kinds of interventions, handlings, and treatments:

> Objects on the level of experimental work are subject to almost any imaginable intrusion and usurpation. They are smashed into fragments, evaporated into gas, dissolved in acids, reduced to extractions, run over columns, mixed with countless other substances, purified, washed, spun round, placed in a centrifuge, inhibited,

precipitated, exposed to high voltage, heated, frozen, and reconstituted. Cells are grown on a lawn of bacteria, raised in media, incubated, inoculated, counted, transfected, pipetted, submerged in liquid nitrogen, and frozen away. Animals are raised and fed in cages, infused with solutions, injected with diverse materials, cut open to extract parts and tissues, weighed, cleaned and controlled, superovulated, vasectomized, and mated.[105]

I recognized many of the listed experimental activities from protocols from my life science and technology practical courses, from my internship in the molecular genetics laboratory, and from my participation in a laboratory animal workshop. What most of these operations that involve a range of manual skills have in common is that they all need to be conducted under more or less strict regimes of sterility. Knorr-Cetina affirms that issues of contamination and sterility are central to molecular biological benchwork: "Extreme precautions with regard to hygiene and sterility are part and parcel of all laboratory work."[106] The importance of issues of sterility and contamination is repeatedly foregrounded in *Epistemic Cultures*, yet the author does not comment on these issues in her discussion of the experimenter's body.[107] In the next chapter I will elaborate on the peculiar hands-on/hands-off experiences involved in performing protocols that demand sterile working techniques and how this central role of sterility in life sciences epistemic cultures cannot be grasped when we keep thinking about bodies in terms of transparency.

After describing molecular biology as a scientific practice that revolves around materiality and manipulation, Knorr-Cetina switches from the experimental level to an analytical level, where she differentiates between three kinds of bodies that make up the scientist's body or the experimenter's body: the "acting body," the "sensory body," and the "experiencing body." She stresses the "object-oriented interactive quality of molecular biological practice," but it is notable that material objects and scientists' bodies are described separately rather than in interaction.[108] The tendency to describe the one preferably in a passive voice and the other in an active voice also points at an indebtedness of Knorr-Cetina's study to an oppositional subject-object thinking commonly associated with Cartesian dualism. A closer examination of Knorr-Cetina's body concepts shows how she struggles to articulate the embodied dimension of life science laboratory practices from a hands-on perspective: when leaping from experimental observation to the conceptual level, bodies are recurrently rendered in Cartesian dualistic notions. This leads to strange juxtapositions and to inconsistent or even self-contradictory accounts.

This is illustrated by the programmatic assertion with which Knorr-Cetina opens her analysis of the scientist's body. "By the scientist's body," she writes, "I mean the body without a mind."[109] "If the mind were included," she elaborates, "hardly anyone would deny the presence of the body. The *body*, as I use the term, refers to bodily functions and perhaps the hard wiring of intelligence, but not conscious thinking."[110] Note how an attempt to "re-embody" the sciences leads on a conceptual level to first "dis-mind" the scientist's body. Speaking about the body here only seems possible when utilizing an analytical framework that remains trapped in an unmitigated Cartesian dualism.

This is also manifest in Knorr-Cetina's analysis of what has led, in her view, to a "disembodiment" of science. In the passage about bodily functions, she identifies two factors: first, "the inclusion, into research, of technical instruments that outperformed, and replaced, sensory bodily functions," and, second, "the derogatory attitude that important scientists have developed toward the sensory body." She refers here to seventeenth-century experimentalists and consolidates, rather than questions, a persistent image in the history and philosophy of science that, as I have argued in the preceding chapters, cannot be ascribed to a sudden *obsoleteness* of the body. On the contrary, historians of science have called attention to the *increasing importance* of bodily matters in the experimental sciences. We can signal a need to replace a thinking in "disembodied" terms with other rhetorical repertoires, such as "mindful hands" and better articulations of the relationship between hands-on and hands-off experiences.

Body Talk

A closer look at the body concepts that Knorr-Cetina introduced in the 1990s confirms this first impression but at the same time shows how her observations of life science laboratory practices resonate with my historical and aesthetic analysis in the previous chapters. Knorr-Cetina starts out by discussing the role of the "sensory body." She emphasizes the importance of the sensory body to experimental work in molecular biology, and—at the same time—she claims its utter irrelevance "as a primary research tool," arguing that molecular biologists have turned away from investigating objects in their natural environment with their senses (perceiving/observing, listening, hearing, smelling, and tasting). Her detailed observations of experimental research in molecular biology laboratories reverberate strikingly with my rendition of emerging new practices in seventeenth-century natural philosophy when she writes, "The intervening technology of benchwork in molecular biology implies a shift toward the manual or instrumental manipulation of objects and away from . . . naturalistic observations of events."[111] Others have convincingly shown that Galileo's experiments do not seek to describe phenomena in the course of nature, and that the use of experimental setups in Galileo's investigation of falling bodies comes close to the idea of a "virtual laboratory" as the locus of the new scientists' thought experiments.[112] In chapter 1 I have traced this historical development as a deviation from Aristotelian sense-based empiricism toward an experimental empiricism.

Knorr-Cetina posits that the senses play a role in a host of secondary ways—for example, in "visually inspecting" materials and traces of materials and "in checking and determining the status of experimental processes."[113] Her study offers examples of the required "visual know-how," but to my surprise she fails to comment on the conspicuous "invisibility" of molecular biology's subject matter. And despite the explicit attention to the highly manual character of the contemporary life sciences and the descriptions of "manual know-how" involved in performing routine tasks that call for highly skilled lab hands, Knorr-Cetina does not broach the issue of the peculiar intangibility of molecular biology's epistemic culture.

A problematic use of Cartesian dualistic notions becomes apparent when Knorr-Cetina describes how the "sensory body" interacts closely at these sites with the "acting body."[114] The term *acting body* is quite ambiguous. On the one hand, it seems to refer to the body of the bench scientist who performs protocols and conducts experiments (e.g., manually mixing, measuring, pipetting, shaking, and plating substances). On the other hand, Knorr-Cetina defines the acting body as one "that works without conscious reflection or codified instruction."[115] This narrows the definition down to bodywork that is acted out without explicit attention to one's movement and that is not verbalized in protocols or manual instructions but rather is presumed to be known and routinized by the experienced experimenter (e.g., the correct handling of a pipette). It is rather striking that the acting body is defined as being "perhaps the first and most original of all automats" and further described as "an information-processing machinery."[116] On the other hand, Knorr-Cetina stresses that bench scientists "tend to treat themselves more like intelligent materials than silent thinking machines."[117] Note here the unreflective use of a metaphor that immediately brings to mind Descartes's mechanistic explanations of living bodies and echoes the "thinking machine" computer language that has been heavily criticized by phenomenologists for its problematic Cartesian rhetoric.[118] Moreover, discussing the body in terms of black-boxed information-processing tools provides no suitable model to describe bodily interactions in the lab because it directs attention to inward cognitive processes, which is at odds with the objective of *Epistemic Cultures* to focus on the "object-centered interactive quality of molecular biological practice."[119]

We can also observe an explicit use of mind-body dichotomies in a passage discussing the observation that many scientists believe in the importance of doing (parts of the) experimental work themselves. This is interpreted as "giving priority to the body" over the mind because "the body is trusted to pick up and process what the mind cannot anticipate."[120] With the concept of the *acting body*, Knorr-Cetina wants to describe specifically that which is more to a body than simply carrying out an action. When she defines the body of the scientist as a black-boxed "information-processing tool," she calls attention to knowledge that is generated by placing oneself in the situation and doing, seeing, and feeling things for oneself.[121] Because this knowledge does not pertain to the cognitive realm, it eludes verbalization and codification. Descartes's ideas about the importance of performing experiments himself and his conception of experimental work as a solitary activity resonate strikingly with Knorr-Cetina's theory about the body functioning as an epistemological warrant for true findings: "To have performed the relevant tasks of an inquiry oneself—or at least to have seen them done—is the capital on which trust in the results is based. Results not seen directly or not produced through embodied action cannot be properly evaluated and are prone to misinterpretation."[122] Knorr-Cetina singles out "performing" and "witnessing" as essential features of the acting body and thus stresses the important role of the body in processes of knowledge making in the molecular biology laboratory. At last, it seems as if the author is forced to backtrack from the opening definition of *Epistemic Cultures*, in which she rigorously excluded all notions pertaining to the realm of the mind from the body.

Knorr-Cetina signals a conspicuous "absence of discourse concerning embodied behavior" and relates this to the popular phenomenological idea that "the acting body works best when it is a silent part of the empirical machinery of research." She concludes her analysis by portraying the scientist's body as a "silent instrument."[123] We will recognize this motif in

179

other studies that put the body at center stage yet grant a leading part only to muted, tacit, silenced, invisible, and self-effacing bodies.

Before we further scrutinize this persistent motif of unobtrusive bodies and invisible hands, however, let us turn to the last of the three bodies that Knorr-Cetina distinguishes in the practice of bench science—"the experienced body."[124] The experienced body, that is, the body that is produced by the use of the sensory and acting body in experimental situations, comprises "the bodily archive of manual and instrumental knowledge that is not written down and only clumsily expressed."[125] This knowledge is "inscribed in the body of the scientist and tends to be lost when the scientist leaves the laboratory."[126] This last body is also described as a black box, since the experiences it incorporates are not (or only insufficiently) externalized in written or oral form. In this context, Knorr-Cetina critically discusses the term *tacit knowledge* that has been coined to refer to embodied knowledge in expert settings. The term is problematic because it applies only to the expert's unarticulated knowledge, indicating that it "is derived from the model of a thinking knower."[127] The "silent body" that she presents as an alternative model is based on the expertise of manually and sensorially skilled bench workers. However, it does not become clear in what sense the criticized tacit body actually differs from the concept of the "silent body," since Knorr-Cetina also defines the embodied knowledge of silent bodies as that which exceeds expression in words (written or spoken).

This overview of Knorr-Cetina's rather confusing "body talk" makes one thing apparent above all: that body talk which remains entrenched within Cartesian dualisms is prone to inconsistencies and conceptual problems that obviously cannot be solved from within this analytical framework. The hands-on approach that discloses the hands-on character of the life sciences appears to be trapped in modeling descriptions of bodies and bodily activity in Cartesian dualistic terms and displays a tendency to silence the bodies that it makes visible. We can trace the reasons for this problematic body talk back to the Cartesian *Unterschiebung* discussed in more detail in chapter 3. Instead of proposing a counterdiscourse to Cartesian epistemology that leads to conceptual inconsistencies, I have argued that we can retrieve *within* Descartes's thinking and doing the inherently hands-on character of the creation of scientific knowledge. A recurrent point of interest in this context is the discrepancy between the ahistorical thrust of an unquestioned Cartesianism and the historical dimension of hands-on practice as a learned practice.

RESEARCH INTERNSHIP (FIRST WEEK): MOLECULAR
GENETICS RESEARCH LABORATORY, LEIDEN INSTITUTE
OF CHEMISTRY, LEIDEN UNIVERSITY

Today is the third day of my first week in the molecular genetics laboratory. I am exhausted. It is not that the work itself is so tiring. I am surprised to see how much I enjoy the manual work, the precise preparations, the repetitive movements, the meticulous attention to clean working procedures and working together with others. It is, rather, the continuous anxiety about asking stupid questions, making a wrong move, handling an instrument in the wrong way, or, in fact, making any mistakes at all. I feel strangely guilty about not being "well prepared" enough to be able to function smoothly without keeping others from the work they are doing. It is only a few weeks later that it dawns on me that no one actually expected me to become an invisible wheel of the laboratory research machinery the moment I entered the lab. On the contrary, I slowly realize that this internship is all about learning in practice rather than being examined on what I already know and if I am capable of applying my knowledge appropriately. The strange nervousness remains during the whole course of my internship. I am not able to put my finger on the source of my uneasiness. Why do I find it so difficult to adjust my thinking to a confident hands-on attitude? I start to wonder: Where did I get this inhibiting idea that knowledge is something preexistent, something I should already be in command of at this site?

Shapin broaches this issue when linking his historical considerations on the Scientific Revolution, the role of technicians, and the meaning of hands-on notions in seventeenth-century discourse to contemporary practices.[128] He discusses the tendency in some contemporary laboratories to strictly distinguish between knowledgeable minds and skilled hands:

> In such places scientists think, while technicians carry out the manipulations dictated by the results of scientists' thoughts. Manipulative skill and hands-on experience are evidently not highly valued in such settings. Science here is seen largely as a reflective and rational activity. By contrast, there are many laboratories in which the distinction between skilled and knowledgeable people, and correspondingly between technicians and scientists, is blurred. It is believed in laboratories of this sort that genuine understanding is properly based upon the direct experience that can come only from working with

181

the materials and processes under study. Here there may be talk of the value of "good hands" or of the importance of a "feeling for" organisms, data, or apparatus.[129]

Shapin's article polemicizes the lack of attention paid to technicians' work in history, philosophy, and sociology of science. He argues that "the price of technicians' continued invisibility is an impoverished understanding of the nature of scientific practice" and relates this to the fact that "the predominant biases in the Western academic world have traditionally portrayed science as a formal and wholly rational enterprise carried out by reflective individual thinkers."[130] To counter these biases he proposes that the workshop and technician provide more fruitful models for understanding how science is made in laboratories.[131]

A critique of an epistemological bias against hands-on work implicit in the opposition between technician and scientist and the increasing interest in the hands-on training of a scientist feed into the contemporary turn toward historicizing epistemology. Studies that emphasize the pedagogical dimension of the laboratory as an instruction site describe what it takes to *become* a scientist. They complement the early laboratory ethnographies in which the construction of scientific facts was the focal point. *Laboratory Life* exposes the intrinsic historicity of scientific facts by telling the stories of their construction. *Epistemic Cultures* extends this historical approach by showing that not only scientific facts but also scientists have a history. No one is born a scientist; thus the historical dimension encompasses not only the making of facts but also the making of the "fact makers." The intrinsic historicity of science is exposed and expanded on in *Epistemic Cultures* with descriptions of processes by which embodied "epistemic subjects" are formed.[132]

182   The central role that hands-on training plays in the life sciences is a key theme in the analysis of molecular biological laboratory practices. Knorr-Cetina shows that the laboratory is a training site—a crucial setting that "life scientists in the making" must pass through in the course of an academic career. In fact, she points out that it is only in the final step of becoming a laboratory leader that scientists arrive at the point where they are able to delegate the benchwork (= bodywork) entirely to others, such as analysts and technicians. Becoming a leader is also all-dependent on "'owning one's own lab'" and—at least in principle—on maintaining "close ties with the work in the laboratory."[133]

Moreover, one should keep in mind that the laboratory work is to a great extent performed by other scientists in training, such as master's students,

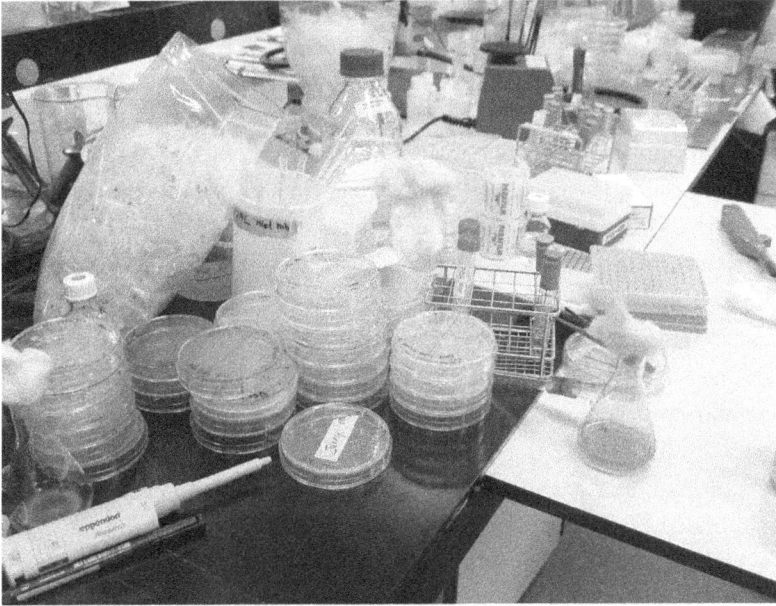

4.3 ★ A workbench, with Eppendorf pipette (*lower left*) and a bag of used tips and tubes, at the molecular genetics research laboratory, Leiden Institute of Chemistry, Netherlands. Photo by the author.

doctoral students, and postdocs, most of whom have to undergo extensive hands-on training at the bench to reach that stage of their career (figure 4.3). This training comprises, in general, about three years of laboratory practicums at an undergraduate level and then at least two to four more years of hands-on research work at a postgraduate level and finally more years at the postdoc level in order to gain a "repertoire of expertise."[134] Postdoctoral training, in particular, does not imply an increasing distance from the bench; on the contrary, it was initially designed to cultivate hands-on knowledge by laying the emphasis on "non-text-based practices and skills."[135] Group discussions on relevant articles, however, are also part of the core curriculum of scientists in training and form an essential part of scientific daily life, as do reading and writing papers, as Latour and Woolgar rightly point out in *Laboratory Life*. From my experience, I would agree that early career researchers in training are tested in the laboratory for mental and manual learning agility with an "emphasis on hands-on technique, rather than rote book-learning."[136]

Knorr-Cetina describes, for example, how students have to undergo a six-week observation period in which they work under the supervision of a researcher before they are offered the possibility of earning their master of science degree with a research internship in that laboratory. In short, to *become* a molecular biologist one must "pass through career stages, from the incipient scientist at the student level to the full-fledged scientific person."[137] This "training up" in laboratories is an essential part of becoming a professional researcher, Knorr-Cetina argues.[138] The pedagogical dimension also remains crucial in a later stage because senior researchers often supervise incipient scientists.[139] The laboratory is simultaneously a site at which scientific knowledge is produced and a training site for emerging scientists. Thus, in the laboratory, "knowledge is simultaneously taught and created."[140]

Life scientists are never simply "just there" in the laboratory producing knowledge about living phenomena; they are continuously at a stage of "becoming a scientist" or—at least in some way—engaged in instructing or mentoring future scientists who need to pass through intensive hands-on training. The two processes, pedagogy and research, are intrinsically intertwined, yet only a few scholars have stressed that "teaching and research are mutually reliant" and "that modern scientific knowledge is tied to teaching and training."[141] One critique leveled at ethnographic laboratory studies is that they have not, thus far, paid attention to this pedagogical dimension. Cyrus Mody and David Kaiser point out "that knowledge-*making* is the primary focus of these studies; their protagonists' roles of teachers and/or students are subordinated to (or invisible beside) their roles as researchers."[142] Note the recurrent rhetoric of invisibility here not only in relation to the embodied dimension of scientific activity but also in relation to its pedagogical dimension. This is no coincidence. It can also be linked to the philosophical operation of the Cartesian *Unterschiebung* and the ensuing need to sacrifice pedagogical aspects of knowledge making in favor of an ahistorical epistemology.

184

Awkward Bodies and Hands-on Training

INTRODUCTION AND BIOCHEMISTRY PRACTICUM I: INSTRUCTION LABORATORY FOR UNDERGRADUATE STUDENTS, LIFE SCIENCE AND TECHNOLOGY STUDIES AND BIOCHEMISTRY, GORLAEUS LABORATORIES, LEIDEN UNIVERSITY

I pushed open the door leading to the instruction laboratory. I was confronted with a row of laboratory benches that structured the space into four rows with four working tables each. On each bench, basic instruments were laid out:

three adjustable automatic microliter pipettes, boxes with blue and yellow pipette tips, a cooling box, some tube holders for tubes and microcentrifuge tubes, and a waterproof marker. Lab coats were piled up on one table. Here I was. This was it. Finally, I had made it to the heart of the life sciences—the space where life scientists were made.

I still remember that I felt quite tense during this whole first practicum. I was overwhelmed by all the new information and—in contrast with my fellow students who came fresh from high school, where most of them had just majored in either chemistry or biology—I was lacking a framework within which I could place all the terms and concepts. Though it came as a surprise to realize that I had much less trouble on account of my lack of background knowledge during the lectures in cell biology than during the practicums. It was only during the latter that I felt so self-conscious and awkward. I realized that lectures and extensive homework from the introductory handbook The Cell *felt rather familiar from my studies in humanities. Studying* The Cell *reminded me of learning a new language that allowed you, step by step, to enter a whole new world. During the practicums, though, I felt conspicuously ill at ease. I was not used to being trained in manual tasks and had to put my rusty math into practice by mixing and measuring buffers and other, often colorless and scentless, solutions.*

From the 1990s onward, the embodied work of scientists (i.e., the "fact makers") has drawn more attention in the field of STS. Attentive readers will recognize in my snapshot story rhetorical similarities to the opening paragraph of Park Doing's seminal article "'Lab Hands' and the 'Scarlet O': Epistemic Politics and (Scientific) Labor," in which Doing intertwines narratives about his own experience, working as an operator in a physics laboratory while completing his master's and doctoral degrees in STS.[143] Doing employs literary strategies that combine the previously discussed rhetoric of wonder with an Indiana Jones–type feeling of entering a sacred tomb and discovering a hidden treasure. He investigated how hands-on notions function in a physics laboratory as identity markers and how they are employed by different lab users (scientists and operators) to present static and dynamic models of expertise. He further shows how the meaning of "hands-on" can shift during different stages of lab development. More STS scholars have paid acute attention to scientific laboratories as specific sites of embodied perception. "We can discern," writes Cyrus Mody in an article on sound and hearing, "the whole physical presence of laboratory workers, not just their eyes—how they comport themselves, how they inhabit specially constructed lab spaces, how they interact with instruments and artifacts, how they shape and move their bodies to be perceived and disciplined by the gaze of others, and how their bodily experiences (their illnesses and exertions) are insinuated into their craft."[144] Mody's observations can be situated within a broader trend in STS that set out "to explicate how embodied practice generates scientific knowledge" and to describe how bodies are implicated in making scientific knowledge.[145] More explicit attention has been paid to hands-on training and the pedagogical dimension of scientific practice, for example, in a review article on scientific training in which hands-on techniques and skill acquisition are theorized as proper aspects of scientific activities, in the third edition of *The Handbook of Science and Technology* (2008).[146]

However, scientific sites should not be viewed as spaces in which bodies recede from view and, therefore, that we need ethnographies to "retrieve" these entities that have slipped so far beyond reach that a concerted effort is required to recover them. Instead, I propose to shift the attention from science's tacit bodies and "invisible hands."[147] The goal is to explore ways of writing about weighty, unwieldy, awkward, and cumbersome bodies— bodies that stand in the way, bodies that force themselves on us within life science's specially constructed, contamination-controlled environments. Bodies are not transparent. In particular, studies that take into account pedagogical dimensions pay attention to the ways that bodies resist becoming invisible and to the effort it takes to make them fade into the background.

The STS scholars Cyrus Mody and David Kaiser point out that the pedagogical dimension has been "an important but underemphasized ingredient in many STS narratives."[148] Mody and Kaiser state that only a "few STS scholars have analyzed the experimental workplace as a pedagogical site that simultaneously fosters knowledge creation and the training/disciplining of knowing subjects," which is remarkable, given that it is "only after intense pedagogical inculcation . . . [that] new recruits develop the 'disciplined seeing' or 'hands' of accomplished practitioners."[149] As a consequence, they call for elevating "pedagogy to a central analytic category."[150] They also point to a shift that has taken place in scientific higher education from text-based training to laboratory-based instruction that emphasizes hands-on techniques.[151]

Notably, Mody and Kaiser open their contribution on the integral role of hands-on training in and for the creation of scientific knowledge with a discussion of the heuristic potential of awkwardness. They refer here to another STS classic, Harry Collins's thought experiment "The Awkward

Student," in which a student pretends not to understand the teacher's instructions for a mathematical operation by performing the instructions in a "correct" but nonsensical way.[152] Collins's example demonstrates the "soft interpretative character" intrinsic to all and even the "hard" sciences, such as mathematics; by applying this example to protocols rather than formulas, or other numerical processes, Mody and Kaiser show how this thought experiment provides us with an example of the "interpretative flexibility inherent in experimental practice." They write: "No description of an experimenting setup can ever be complete enough that it will be safe from 'awkward' misreadings by replicators."[153] Both Collins and Mody and Kaiser discuss awkwardness as a heuristic method for illuminating "miscommunications" and "misreadings" that offer "a picture of science as a human, temporally and culturally situated endeavour."[154]

Mody and Kaiser mobilize "The Awkward Student" experiment to point to the significance of the pedagogical dimension, but they do not further develop the heuristic potential of "awkwardness" in relation to embodied skills when they refer to "misreadings" rather than to "misperformances." Yet, is it not precisely a certain awkwardness, in the sense of an initial ungraceful, ungainly comportment, that characterizes the bodily behavior of apprentices or recruits in new laboratory settings? It is not accidental that Latour's only remark about his own hands-on experience revolves around awkwardness and clumsiness, though he appears to use this ethnoautobiographic story to emphasize first and foremost his outsider's gaze.[155] Interestingly, after reading Mody and Kaiser's argument, one can see the ethnographer's awkwardness that is expressed in physical signs of his embarrassment ("he blushes") in a different light: his initial ineptness at performing the task correctly and efficiently reveals a core characteristic of bench science's epistemic cultures, rather than illustrating the ethnographer's distance from the practice under observation.

More important, in the methodological section, Mody and Kaiser draw parallels between ethnographers of science and technology and "students and trainees that are usually the newest entrants to laboratory settings."[156] As they point out, "Often, they have the same awkward questions as the ethnographer and many of the same difficulties in adjusting to local practices, even the same insider/outsider's critical perspective."[157] Mody and Kaiser see in the similarities between pedagogy and ethnography a methodological advantage: ethnographers should make use of the existing methods used to acquaint new entrants with local laboratory working practices. It is significant that Mody and Kaiser single out this similarity, as

187

becoming trained at the bench is an integral part of the creation of knowledge in the life sciences.

Awkwardness is a recurrent motif in stories that science ethnographers tell about the bench training they had to undergo in order to *become* an observer of laboratory practices.[158] Their discussions show how the in situ concept is linked to the notion of hands-on: methodologically, participant observations imply an in situ and embodied approach, being at the spot and *doing* hands-on observations. The notion of hands-on thus adds a dynamic dimension to the static notion in situ, and because we also already associate it with practical learning, the pedagogical dimension is inherent in the term *hands-on*. The method of participant observation implies at least some aspect of firsthand training: observing laboratory activity from a hands-on perspective necessarily involves, to a certain extent, "becoming skilled" and "learning a craft." Therefore, participant observers can capitalize on a core feature of epistemic practices in the life sciences. It is especially this intricate intertwinement of doing science and reflecting on scientific practices with hands-on methods that we can trace back to Descartes's philosophical practice and that we can explore from *within* Descartes's thinking.

However, the historical perspective of how thinking is shaped in hands-on learning processes came to be eclipsed by the Cartesian *Unterschiebung* of ahistorical epistemology, or first philosophy, founded on a *cognitio mathematica* ideal of the sciences. Turning bodies into *res extensae* meant sacrificing the materiality, opacity, and historicity of bodies to the ultimate ideal of geometric transparency on an analytical level. In this same process the heuristic potential and epistemological value of notions, such as "awkwardness," that cannot be raised to clarity and distinctness became obscured too. This discussion points in the direction of alternative epistemological projects that refrain from engaging in an act of thought that leaps to disembodied and purified conceptions of scientific knowledge making, and that instead explore how a hands-on approach enables us to retrieve more obscure notions of cumbersome and recalcitrant bodies to gain a deeper understanding of epistemic practices in life science wet laboratories.

188

In Touch with Life

On Molecular Vision and Microliter Technologies

What do I need to work? A work space in a laboratory with an approval for gene technological experiments, three pipettes to pipette volumes ranging from 0 to 1000 μl, the rest is luxury. That is at least what it looks like. . . .

    The molecular biologist, also called "molli," handles most of the time tiny amounts of mainly transparent, colorless fluids.

Cornel Mühlhardt, *Der Experimentator Molekularbiologie/Genomics*

On the first day of my research internship in a molecular genetics laboratory, the senior researcher welcomed me into her office, a small and narrow room with two desks along each wall. The desks on the right held a computer and were cluttered with papers; the others were clean and empty. A coffee machine was placed on a table in the back. A door connected the office to an adjoining research laboratory. After some short introductory remarks about the research of her group, I was ushered into the laboratory. Almost immediately, she handed me over to the analyst, who showed me my bench space. This was it. Here I was, ready to go "hands-on." Finally, I had made it to the heart of a molecular biology laboratory—a bench with my own set of pipettes, tips, and tube holders. One of the first things I had learned at Leiden University's laboratories was that the life sciences' firm grasp on molecular life is made possible by some small but ubiquitous tools: micropipettes and microcentrifuge tubes.

Researchers in the history and philosophy of science observed the rise of a "molecular vision of life" in the twentieth century and "defined the locus of life phenomena principally at the sub-microscopic region between 10^{-6} and 10^{-7} cm."[1] Many instruments were vital to the molecular approach or the new biology's more general focus on the subcellular level. The historian and philosopher of biology Hans-Jörg Rheinberger identifies, for example, several biophysical, biochemical, and genetic technologies that played a central role in the early development of molecular biology: X-ray crystallography, radioactive tracing, and analytical and preparative ultracentrifugation. The ultracentrifuge, for instance, was instrumental in creating new representational spaces in the life sciences.[2] What does this venture into subcellular realms mean in terms of hands-on handling of materials in molecular biology laboratories? How does a molecular vision and the new biology's "emphasis on the ultimate minuteness of biological entities" affect our sense of life while exploring, manipulating, and representing the living?[3] How *do* we get in contact with life at twenty-first-century sites of experimental knowledge production? The molecular vision pushed research not only into biology's invisible dimensions at the submicroscopic level but certainly also into regions that lie beyond our grasp. What kinds of "techniques of touch" go along with the life sciences' visualization, magnification, and observation technologies? How can we understand the intangible dimensions of the life sciences from a hands-on perspective?

Evelyn Fox Keller has already pointed out that a biological gaze or—for that matter—a molecular gaze must always be understood as intrinsically reliant on the haptic sense; others have also shown that life science's visualization technologies must be understood as the result of extensive manipulations rather than documentations of hands-off observations.[4] How is molecular life handled in the lab, what kind of technologies of touch do we encounter in these practices, and how do they shape our sense of life? What are the daily tools used in life science labs that extend researchers' grasp to intangibly small organisms and vital phenomena on the cellular level and below?

Tangible Impressions from the Bench:
Calibrating Pipettes

INTRODUCTION PRACTICUM (FIRST YEAR):
LIFE SCIENCE AND TECHNOLOGY STUDIES
AND BIOCHEMISTRY, GORLAEUS LABORATORIES,
LEIDEN UNIVERSITY

The first tool we were introduced to on our first day was the automatic pipette. The assistant gave a demonstration on how to operate this liquid-handling device before we were allowed to lay our hands on it. She stressed that a microliter pipette is a precision instrument and that it is impossible to perform a good experiment without a good pipette: a pipette, we learned, should always be handled meticulously and with respect.[5] I picked up an Eppendorf pipette from our table; the ergonomically shaped plastic casing felt light, and the tool lay comfortably in my palm. With its elongated form, a pointed nose on one side, and a push-in button on the other, it looked almost like a ballpoint pen. We were instructed to push the button on the plunger gently with our thumb until we felt a light resistance and only then to push all the way through. One of the most important rules was to never use a "bare" pipette—always fix a tip to its nose before use. I reached to grasp one of those light blue tips to make my instrument ready for action, but I was instructed to wait and watch. This is how you place a tip onto your pipette: open the right tip box (blue for 1,000 μl pipettes, yellow for 200 μl and all smaller sizes), hold the pipette vertically and with the nose pointing downward, then push the nose firmly into one of the tips in the placeholder to fix the right-sized tip to the nose. Finally, close the tip box to keep the remaining tips clean and sterile.[6]

191

Proper handling, I learned, required some delicacy. This was not only about gently and carefully handling a precision tool. Besides exerting the right amount of pressure, the art of pipetting implied knowing precisely which parts of the device to touch and, most important, which not to touch. Never touch the tips with your hands, I was told, even when they are still clean. Also be careful that the tip does not graze against the bench, your sleeve, or any other surfaces. In the course of this introductory class, I began to get a sense of the invisible microbial world that would decisively shape my behavior within laboratory spaces.

For our first task, however, it was not microbial cleanliness that mattered but precision and another important component of lab work: repetition. Our first task at the bench was to check the instrument's accuracy. In the course of the experiment, students who complained that their tools were not working correctly were repeatedly sent back to their bench to try again. After a while, I realized that calibrating was not just about testing the working of the instruments; perhaps even more important, it was about testing our own hands-on performances. Our first experiment was an exercise in learning how to handle a pipette by correlating the readings of the pipette volumes with the readings of another standard [see figure I.1]. My bench partner and I shared a set of three pipettes that could measure incredibly small volumes up to two millionths of a liter. The smallest pipette was used for 2 to 20 μl, the medium-sized for 20 to 200 μl, and the largest one for 200 to 1,000 μl. We learned to always use the smallest pipette possible to obtain the best results. With every pipette, we measured several volumes of purified water (Milli-Q water) by weighing small droplets of different sizes on a scale.[7]

My partner put an empty weighing tray on the platform of the scale and reset the display to zero. I adjusted the pipette to the right volume by scrolling with my thumb on the wheel at the upper end of the pipette until the volume display showed 2 μl. To load my pipette, I pushed the plunger until the first point of resistance, then I inserted the tip of the pipette vertically into a small vessel of purified water until the tip was immersed just enough to cover the end. I slowly released the button to take up the desired amount of water, while keeping the tip immersed in the liquid. To check whether I had loaded my pipette, I squinted at the transparent tip, where I could only just discern a faint waterline. Finally, I released the droplet of water into the empty weighing tray by pointing the tip against its inner side and pushing the button slowly all the way down. (Later, in the research laboratory, I learned to load and release my pipette in a more controlled manner by placing my elbow on the bench and slightly pressing the tip diagonally against the inner wall of the tube to steady my pipetting hand.) We noted the indicated weight and proceeded with the next volumes to be measured with the smallest pipette: 5 μl, 7.5 μl, 10 μl, 15 μl, 20 μl. We repeated the exercise with the medium-sized pipette and performed eight measurements and weighings ranging from 20 to 200 μl. Finally, we tested the third pipette with volumes between 100 and 1,000 μl.

The scale at which we operated with these liquid measurement tools, ranging from raindrops to barely visible droplets, lay beyond my imaginative powers. And though I had major difficulties in grasping a concept such as two millionths of a liter, my hands were doing their job just fine. I enjoyed the fine-motor movement and the meticulous attention that this task required. I was not used to working with such concentration while doing something as apparently simple as weighing different-sized droplets of water. Though the amounts increased in size, the manual operation stayed the same. After a while my movements became almost rhythmic, and I took pleasure in the meditative aspect of this repetitive task.

It was no coincidence that we began our first day by calibrating pipettes. Pipetting is one of the most frequent tasks in life science laboratories, and the right handling of the microliter pipette is critical in performing experiments successfully and in managing the repetitive task accurately and without strain. Pipettes are used for all kinds of experimental procedures that involve the handling of living cells and their internal components. They are used to transfer cells onto plates; to mix, freeze, and feed cells; and to prepare cells for centrifugation and to kill them. They are used to prepare molecules for rapid replication, to load gels with DNA molecules, and many more mundane laboratory activities that are involved in in vivo and in vitro experimentation, all of which come down to handling small volumes of mostly transparent solutions. Pipettes are not mere liquid measuring instruments; they are key players in the manipulation of living and nonliving substances in today's life science laboratories.

In *An Epistemology of the Concrete: Twentieth-Century Histories of Life*, Rheinberger points to the central role that instruments have played since the seventeenth century in our understanding of the experimental sciences. He discerns a shift in historiography: first instruments were primarily described and conceptualized as "ideally transparent media" that were praised for extending or enhancing our senses and were employed to "isolate, purify, and quantify" experimental perception.[8] While instruments did play a crucial role in telling the story of the rise of modern science, as "transparent media" they became invisible within the grand narrative of the scientific revolution as "a revolution of the mind" to which experimental innovation was merely ancillary.[9] Rheinberger calls attention to the material opacity of instruments and objects rather than their ability to become transparent, or invisible, in historical accounts of scientific innovations.

Not only instruments but also experiments have long played a subordinate role in the history and philosophy of science. In his essay on the rise of experimentation and instrument use, the philosopher of science Michael

Heidelberger diagnoses an "invisibility of experiment" in philosophical discussion up to the 1980s, since before this time experiments were discussed only in terms of their deficiencies. He states that the philosophy of science remained blind to experiments because ideology had long ascribed to experiment the role of merely verifying theory.[10] Rheinberger argues for a change in attitude toward experimentation and instruments, describing experimental systems not as instruments of verification but as material settings for the exploration of yet unknown phenomena.[11] The philosopher of science Tim Lenoir sees a link between developments in the history and philosophy of science and the increasing ascendancy of the biological sciences: this rise at the end of the twentieth century, he observes, has been accompanied by a "growing preference for practice-oriented as opposed to theory-dominated accounts of knowledge production."[12] Traditional theory-laden accounts tended to frame science's material realities—its instruments and experimental settings—in terms of transparency and invisibility, if they were mentioned at all. The practical turn in history and philosophy of science and the rise of STS as an academic field has led since the 1980s to a rapidly growing body of historical and ethnographic studies focusing on scientific instruments and technologies and their role within experimental innovation and the creation of scientific knowledge.[13] As Rheinberger points out, historical literature has called into question in a fundamental way "the much vaunted transparency" of scientific instruments. Many case studies foreground the importance of historical and local contexts, given that "as a rule instruments neither work nor produce insights by themselves."[14] Today, historians and philosophers of science argue for a theory of knowledge formation that does not transcend the historical contingencies of the science's objects, instruments, and concepts.[15] Not theories, but material histories, are seen as viable starting points for any further theoretical research into the creation of scientific knowledge.

Scientific Toolmaking in Historical Perspective: Crafting the Life Sciences

With the rise of microliter technology adapted for accuracy and exactness on unprecedented small scales, new research fields emerged that could penetrate ever further into the molecular world. Because they enable the convenient and precise handling of very small liquid volumes, microliter technologies played a significant role in the rapid progress of molecular biology. The micropipette and its counterpart the microcentrifuge tube became key players in the molecularization of life that profoundly transformed laboratory culture.[16] Omnipresent in virtually every modern biological laboratory, the handheld automatic pipette forms a mundane yet indispensable operating tool among the surrounding high-tech apparatus. In 1957, the prototype of the most used handheld instrument in today's life science laboratories, the automatic adjustable micropipette, was defined as a "device for the fast and exact pipetting of small liquid volumes."[17] Not even half a century later, the tool that evolved from this first piston strike pipette, now known as the Marburg pipette, was described as an icon of modern biological technology and biomedical research, "a mainstay in laboratory work" and "a lifelong friend" to bench workers.[18]

Contemporary witnesses to the invention of what later became known as the Eppendorf pipette emphasize the importance of this precision instrument (figures 4.3, 5.1, and 5.2). In an interview in 1999, one of two former directors of the German medical equipment company Eppendorf that brought the first Eppendorf pipette onto the market observed that the new research fields of molecular biology and gene technology would not have even been possible without the invention of the pipette; moreover, both directors stressed the importance of the miniaturization of technology—the rise of microliter technology—for the development of what is today designated as the life sciences.[19] Despite its decisive influence, the piston strike pipette has not received much attention by historians of science. Birgit Pfeiffer's dissertation on the making of the so-called Marburg pipette and its inventor, the German physiological chemist Heinrich Schnitger, is a rare exception. Pfeiffer traced the history of an early prototype of the automatic piston strike pipette, drawing on archival material and interviews with contemporary witnesses. She concluded that the scientific focus on life's ever-smaller constituents would have been unthinkable without the invention of microliter technologies, of which the automatic pipette is an emblematic example.[20] Building on Pfeiffer's account, we can describe the invention of the Marburg pipette as a story of tedious tasks and creative craftmanship.

The former directors of Eppendorf, the company that later became one of the leading enterprises developing, producing, and distributing devices for use in life science research, described in the interview how in 1958 they developed the first Eppendorf piston strike pipette in collaboration with the inventor of the Marburg pipette for use in clinical chemistry.[21] In the 1960s, the Eppendorf pipette was integrated into a microliter kit for enzyme assays that allowed physicians to conduct fast and reliable blood analyses with very small samples at their office, an innovation that was especially

195

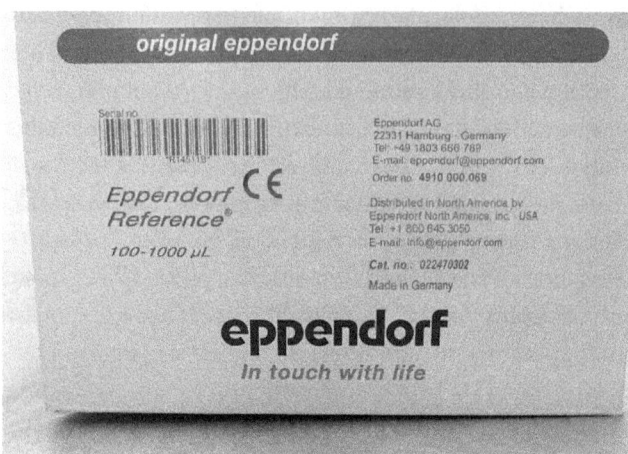

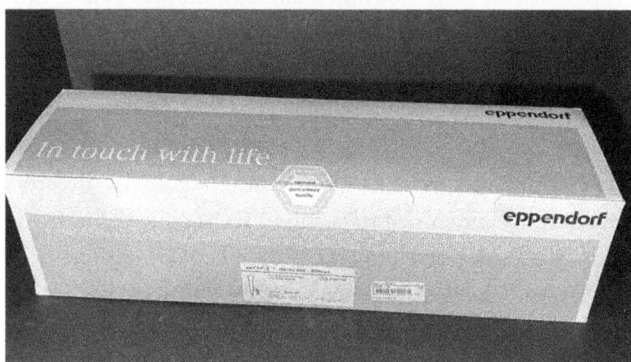

5.1 and 5.2 * Eppendorf AG corporate slogan "In Touch with Life" on a box of pipettes and a box of pipette tips. © Eppendorf AG. Image sources: https://labstuff.eu/images/product_images /info_images/IMG_0358_0.JPG; https://i.ebayimg.com/images /g/2qYAAOSwNTRho1Af/s-l1600.png.

welcomed by pediatricians. From the 1970s onward, large central laboratories emerged that took over these blood analyses, and the total volume of sales has shifted from clinical chemistry (15 percent) to molecular biology and cell technology (90 percent).[22]

Today, the Eppendorf pipette, the Eppendorf microcentrifuge tube, and the Eppendorf centrifuge are still three of the company's best-known products. Since the 1970s, the piston strike pipette has been copied and further developed by other companies the world over in response to the special demands of various research fields. The automatic microliter pipette and the disposable microcentrifuge tube, which allow for the precise and uncontaminated handling of very small liquid volumes, can, without exaggeration, be considered the most widely used instruments in biotechnological and biomedical research. Microliter technologies have become indispensable for many practices centered on "the representation, definition, measurement, analysis, production and circulation of molecules."[23]

The handling of instruments has increasingly been discussed in philosophical, historical, and sociological reflections as an integral aspect of the creation of scientific knowledge in the experimental (life) sciences.[24] Consequently, epistemological reflections became both more grounded in material practices and also historicized. This epistemological interest in instrument use has been accompanied by a more explicit interest in hands-on work and the training that takes place in scientific settings.[25] This development falls in line with (post)phenomenological attempts to integrate an embodied dimension into more political and sociological approaches to the study of scientific knowledge production.[26] Intrigued by the fact that hands-on benchwork in a molecular biology laboratory revolves, to a great extent, around pipetting, I will take a closer look at the use and history of the automatic microliter pipette, a basic instrument that can be found today in virtually every life science laboratory.

The Piston Strike Pipette: A Story of Tedious Tasks and Gifted Tinkering

In the context of the material and practical turn in science studies, the chronicle of the invention and history of the Marburg pipette is illuminating in several respects. Written in the form of a traditional discovery story that traces the original circumstances of a brilliant idea, Pfeiffer's narrative remains largely descriptive, yet it forms a fruitful source for further reflections on the implications of studying the life sciences from a hands-on perspective. What stands out in this historical tale of a revolutionary invention is that the story revolves mainly around practical nuisances, tiring handiwork, and creative tinkering. Pfeiffer's account shows that reinventing pipetting with an automatic piston strike device was a manual, rather than an intellectual, breakthrough with major implications for the expansive research area of the life sciences. She traces the invention of one of the first prototypes for the revolutionary automatic piston strike pipettes back to Heinrich Schnitger, who worked at the Marburg Institute of Physiological Chemistry in post–World War II Germany.[27] Schnitger, who developed this device while working closely with the German medical equipment company Eppendorf, had studied medicine and had written his dissertation on the automatic measurement of blood coagulation, a procedure for which he had developed a new device.[28] After his graduation in 1956, he joined a research group working with experimental systems in the area of enzymology, first on a research stipend and as a scientific assistant beginning in 1957. From 1961 onward, he continued to work as a researcher (*wissenschaftlicher Mitarbeiter*) at the same institute until his early death in 1964. The head of Schnitger's research group, Theodor Bücher, had come to Marburg in 1953, where he was actively involved in rebuilding the research institute after World War II until accepting a professorship in Munich in 1963. In the 1950s and early 1960s, his research group conducted pioneering work in the field of enzymology with the use of column chromatography. Experimental work in enzymology was very complex and costly, especially the isolation and purification of enzymes and substrates, which required many hands for many hours. One way to make the procedure more efficient and less time-consuming was to radically scale the samples down in order to work with microscopic and even nanoscale volumes.

Bücher had quickly recognized his assistant's talent for solving technical problems and granted the young Schnitger a good deal of freedom in pursuing his own technical interests. For instance, Bücher allowed him to set up his own workshop with advanced equipment, including a high-class workbench and other precision tools for instrument makers. In 1957, Schnitger worked with a doctoral student in what was then the new research area of nucleotide chromatography, attempting to further develop and refine column chromatography, a method that at that time was very labor-intensive and time-consuming. Their research involved a separation process for determining nucleotide concentrations, a wearisome task that involved many hours of pipetting.[29] Pipetting was a tedious and cumbersome process. Schnitger and the other research assistants had to use glass pipettes to transfer or measure out small quantities of liquid. The desired amount of liquid could be sucked up either orally or with an ancillary device of an elastic bulb attached to the slender tubes of glass. Some of these pipettes delivered the desired volume with free drainage, whereas others required that the last drop be either blown out or washed out with an appropriate solvent. Working with pipettes was thus not only tiresome but at times also dangerous, since toxic or infectious fluids could come into contact with a worker's lips and mouth if a pipette was not handled properly.

One of the most labor-intensive aspects of this work with glass pipettes was their cleaning.[30] After use, the glass pipettes would be collected in a

198

watery solution until they were cleaned either with a detergent or, in case of severe contamination, with an appropriate acid. If this was done carefully, the pipettes could be washed in a laboratory dishwasher. However, when it was necessary to sterilize the pipettes, they needed to be preheated to prevent damage to the surface that could negatively affect the calibration of the instruments. Laboratory researchers also had to make their pipettes themselves. Self-blown glass pipettes are still used in the laboratory for specific tasks, though most pipetting work is now done with automatic micropipettes. Two research assistants from Bücher's group recall that instrument making monopolized a large part of their time in the laboratory: "Glassblowing, also the making of pipette tips, was one of our main tasks. We bought glass pipes. They could be purchased from Firma Kroge. We then moved them over the Bunsen burner and we drew [the glass] very thin, as we wanted them. Yes, we did everything: instruments, except for the *Photometer*, we could purchase that, but all other equipment (for example, distiller, apparatus) we had to build ourselves."[31] The research assistants used Peterson pipettes that lacked precision and were very liable to break: "You had to take heed not to get the stuff into your mouth. You had to keep sucking. And you had to feel out when you encounter resistance [from the fluid]. That was all but ideal."[32] These anecdotes provide a vivid sense of how researchers were bodily implicated in laboratory work not only in the early times of experimental sciences in seventeenth-century workshops and anatomical theaters but also in modern biochemical laboratories (on this, see also chapter 4).

What is of specific concern to the following discussion is that bodily engagement and instrument handling in life science laboratories today inevitably revolve around questions of sterility. The proliferation of knowledge in the field of bacteriology at the end of the nineteenth century and throughout the first half of the twentieth century was accompanied by an increasing awareness of microbial contamination and the need for sterile working procedures and equipment, sophisticated decontamination, and advanced barrier technologies not only in medical settings, where the discovery that death could be caused by contamination revolutionized the practice of surgery and its operation sites, but also in (micro) biological and biochemical research laboratories.[33] The laboratory became, more and more, a place in which the nature of bodily engagement at the bench could not go unnoticed. Laboratory workers' bodies became a constant risk factor, subject to contamination control and risk management procedures. I will return to this point regarding the regime of "touch control" and the modern laboratory as a site of bodily regulation later in this chapter.

One of Bücher's former research assistants, Hans-Jürgen Hohorst, who later became a professor of physiological chemistry, commented on the varied backgrounds of the researchers in Bücher's group. Some had a background in chemistry, some in medicine, and some in both disciplines. In an interview, Hohorst remembers that Schnitger had originally been trained as a craftsman (or journeyman). He points to the craftsman-academic tradition (*Handwerker-Akademiker*) at the Kaiser Wilhelm Institute for Cell Physiology where Bücher had been trained by the Nobel Prize winner Otto Warburg. Warburg had developed new methods in photometry in close collaboration with trained craftsmen. Hohorst points out that Bücher followed this tradition when he employed Schnitger, "who was a phenomenal bricoleur [*Bastler*], though not academically trained."[34] In the interview, Hohorst emphasizes the technical knowledge and talent of his former colleague Schnitger, but he does not remember that Schnitger had received any academic training. Hohorst was at that time studying chemistry but also held a medical degree, like Schnitger, who had received his doctorate in 1956 before joining Bücher's research group.[35] Hohorst's mistaken memory of Schnitger as an ingenious tinkerer without an academic education is significant here; it shows how levels of hands-on engagement are employed as identity markers in labor relations.[36] In his study on the changing notions of "lab hands" in a physics research laboratory, Park Doing convincingly describes how stories about talented handymen in laboratory contexts can intentionally or unintentionally be employed to cordon off "lab hands" from "workshop hands," thus indicating hierarchical differences between knowledge agents at scientific research sites.[37]

Obviously, Schnitger's hands-on knowledge and interest in problems and solutions arising from handiwork at the bench did not qualify him as a well-trained scientist in the eyes of this scientific staff member, who viewed him as just a talented tinkerer. This characterization resonates with Doing's observations on how "being hands-on" can be associated with a knowledge model that views ability as "embodied, inherent, and located in the hands (and not in the head)" and conjures up associations with primitivism, intuition, and ad hoc improvisation, a model that is played out against knowledge models that view ability as something acquired by learning and study (as in case of an academically trained scientist).[38]

Willi Bender, a former apprentice at the physiological chemistry workshop, remembers the close collaboration between workshop and laboratory.

When asked whether he was also involved in the problems the scientists faced in the laboratory, he answers:

> Yes. Sometimes you had to come to the laboratory, too, that is still the case today. With Professor Bücher it was still something special. He was very fond of the workshop, he came frequently and asked questions. He was quite skilled in handiwork, too. We had to come up to see what the instruments were used for. That was always quite interesting. But, surely, he was also very impatient [he laughs]. Everything had to be done quickly, he was always in the workshop, checking on us, how far we were. But I think he did not only do that in the workshop, but with everyone.[39]

Wilhelm Bergmann, the electrical engineer, and one of the two later chief executive officers at Eppendorf, remembers the outstanding quality of Bücher's experimental workshop:

> In Marburg he had an experimental workshop [*Versuchswerkstatt*] which was remarkably good for the time. Professor Bücher wanted it that way. He took charge of this institution on the condition that he would get a brand-new experimental workshop that had to be top quality. And Schnitger, when he started there as a young man, he did use the workshop, but he also immediately built his own next door in the cellar. There was a master, a first-class trained fine mechanics master, who essentially ought to assist Klingenberg and Schnitger. Schnitger really cut the ground from under his feet. He said: "What you are doing is all well intended, but actually you don't have a clue." And then he insisted that they let him work in a cellar space—one with bricked-up windows, you have no doubt heard about that— he wanted to be independent of daylight conditions. He furnished this walled-up space, that remained cool and operational during day and night or winter and summer times, with apparatus—he had, by the way a small "Hommel" machine: that is a special refining machine—the devil knows where he knew that one from, they used to have them in submarines; he had bought or ordered this one, and with it he could build stuff himself.[40]

When reading the interviews, what struck me most was how these laboratory stories resonate with early modern studies that foreground the role of craft in the development of modern experimental science. Tracing the history of the most-used handheld tool in the contemporary life sciences leads us back and forth between the laboratory and the workshop, a meandering storyline that illustrates the intricate intertwinement of craft and science. The interviews also illustrate the resistance of histories of technoscientific discovery to being shaped into the linear form in which a theory or idea leads directly to success or scientific progress. Rather, they foreground the importance of historical attention to sites, materials, and mundanities. Birgit Pfeiffer's source material calls to mind early modern studies that emphasize the importance of handiwork and craft knowledge for the rise of modern experimental sciences.[41]

Based on contemporary witness accounts, Pfeiffer concludes that Schnitger apparently created the prototype of the Marburg pipette without any support from the institute's experimental workshop.[42] One of his former assistants, Roland Scholz, who in 1957 worked as a PhD student with Schnitger, vividly recalls Schnitger's annoyance at the dull and time-wasting task of pipetting that was involved in identifying and measuring the nucleotide contents of fractions which were eluted from chromatography columns. He remembers how Schnitger left the laboratory after yet another week of tedious pipetting with glass pipettes for three days without any notice, only to come back and resume his work as if nothing had happened. However, Schnitger now used a little device he had put together himself, which he could operate with one hand: it consisted of a glass tuberculin syringe to the tip of which he had attached an elongated plastic tube with a piston. With the aid of a spring, the piston returned immediately and automatically to its original position after having been pushed down, a mechanism that allowed him to conduct his work much faster.[43] Pfeiffer notes in her dissertation that Bücher immediately recognized the importance of this innovation that would significantly simplify chromatography technology on nano and molecular scale. Schnitger's *Urpipette* became a prototype: in the following weeks, together with the head of the experimental workshop and his apprentices, Schnitger built fifty pipettes for Bücher's scientific staff members.

The later patented Marburg pipette consisted of a pumping mechanism: a plunger that fits closely within a tube in which it moves up and down against a spring. When the plunger is pushed down, the spring moves a piston. The piston stroke length is defined by a calibrated stop, which means that exactly the same amount of air is always pressed out of the pipette's tip. A replaceable tip attaches to the end of the cylinder, the plunger is depressed, and the plastic tip is immersed in the sample solution. The liquid enters the tip when the plunger is released. To drain the solution, the plunger is pressed down to a second stop, which ensures that any residues are ejected from the tip. Due to the piston stroke mechanism and the exchangeable tip, the solution never touches the plunger. A high accuracy of measurement was ensured by the double stop: one that defined the piston stroke length while keeping the air volume small and thus constant, and another that ensured the ejection of any solution residues.[44] This innovation was revolutionary not just because it made possible the pipetting of small liquid volumes in an exact and reproducible manner; it also guaranteed fast pipetting of different solutions successively—and one-handed! The instrument was also easy to clean and reassemble (unfortunately, the original prototype has not been conserved).[45]

When Schnitger developed the removable tip for the Marburg pipette, the story goes, he locked himself up in his workshop and worked around the clock for a week, keeping himself awake with drugs, to develop the first tips made from Teflon. This material had recently been invented for use in space research and possessed the required characteristics, such as water resistance and chemical inertia. With its extremely high melting point, however, Teflon was difficult to process; moreover, it was expensive, which made it an unsuitable material for mass production. In the early 1960s, the Teflon tips were replaced by tips made from polypropylene, which had the same required characteristics but was also cheap, easy to process, and, importantly, transparent, allowing the user to see what was being pipetted.

In this period, Wilhelm Bergmann came to Marburg every two weeks to develop the pipette further for mass production with Schnitger. In 1962, Eppendorf eventually brought the first disposable polypropylene tips onto the market—an innovation that, as Pfeiffer writes, was fully in line with the emerging trend toward a "throwaway society" (*Wegwerf-Gesellschaft*). The disposable tips also laid the foundations for applications in areas of cell culturing and gene technology, such as tissue engineering and the revolutionary polymerase chain reaction (PCR) for which sterile tips are essential.[46]

The invention of the automatic microliter pipette is an indispensable part of the history of "technologies of living substances."[47] The Eppendorf pipette and the ever more specialized liquid handling devices that were developed in its wake were vital for the emergence of new research areas and activities that bring us "in touch with life"—the Eppendorf company's corporate slogan. This slogan hints at the extent to which researching life is a matter of hands-on work and how researchers are bodily implicated in life science research. Yet, the gripping catchphrase obscures rather than highlights, perhaps even intentionally, how life science research has become ever more intricately intertwined with technologies of contamination control.

The story of the Marburg pipette makes the close relationship between the workshop and the laboratory palpable. Pfeiffer's account of the period in which the pipetting of small amounts of liquids was reinvented shows that understanding hands-on benchwork often requires that we trace activities back and forth between (technical) workshops and research laboratories, blurring the conceptual boundaries between the work of a technician's mindful hands and a handyman scientist. When we follow the development from the glass pipette to the Eppendorf micropipette, it is important to emphasize that this is not a linear history; one tool did not replace the other. Glass pipettes are still widely used in biological laboratories and innovative research areas such as gene technology, for instance, in protocols for the purification of genomic DNA. When I participated in the practical course for first-year life science and technology students, we were still instructed in the art of pipette making and learned how to heat the front end of a glass pipette with a Bunsen burner and pull to form a needlelike tip using forceps, a technique that requires some skill to make very fine tips without breaking the glass or clogging the opening. It needs to be practiced to be executed fast and efficiently—a manual skill that I had not quite mastered during my coursework.

Lab practices in STS studies have critically examined the taken-for-granted character of instruments that led to accounts in which instruments function primarily as transparent media in the hands of scientists with revolutionary ideas. The historical account of the close collaboration between Bücher's research group and the Eppendorf company also illustrates, for instance, how innovation must be understood in the context of what has been described as "the emergence of the industrial research laboratory and the scientification of industrial production," a trend that calls into question the "'ivory tower' role of the university" and that explicates further developments in the latter half of the twentieth century with a triple helix model.[48] Innovation is not simply born from a scientist's ingenious mind but is firmly grounded within university-industry-government relations.

The true protagonists of Pfeiffer's historical portrait—the Marburg pipette and the automatic microliter pipette that emerged in its wake—have apparently remained historiographically invisible thus far. However, from the historical material emerges an image of pipettes' persistent materiality, with pipetting depicted as part of a manual practice that revolves around notions of resistance and recalcitrance rather than effortlessness.

We could say that in Schnitger's hand the micropipette becomes an "epistemic thing."[49] It allows us to trace a molecular vision of life by following the development of a material object, "rather than pursuing the development of concepts, disciplines, institutions, or individual researchers," as Rheinberger has argued with Gaston Bachelard's concept of phenomenotechnique in mind.[50] The micropipette as an epistemic thing indeed requires us to abandon commonly used classifications in the history and philosophy of science, for it appears almost impossible to confine this study to any one of the fields, disciplines, or research areas of enzymology, chromatography, physiological chemistry, clinical biochemistry, biotechnology, molecular biology, or the life sciences more broadly. Rheinberger's description of epistemic objects provides us with a strikingly apt vocabulary and conceptual framework with which to understand the birth of the micropipette and its later success:

> The force and the reason of epistemic objects lies thus in the conjectures of what they might become, all while what they are going to be cannot be anticipated. . . . These entities for the very same reason do not belong to the realm of objectivity in the sense of representing something independent from our manipulations. But they do not belong to the realm of deliberate construction either. The mode of existence peculiar to such entities derives precisely from their resistance, resilience, and recalcitrance rather than from their malleability in the framework of our constructive and purposive ends.[51]

In his analysis, Rheinberger focuses on life science's odd entities rather than on its instruments, though it has become almost impossible to differentiate between these two categories in the age of biotechnology. The Marburg pipette resulted from a concrete local problem in an enzymologist's research laboratory. Of course, Schnitger had a purposively designed device in mind when he set out to find a solution for the extreme nuisance he experienced in the course of his daily work, but in light of the later applications of this device across many disciplines and its role in the emergence of new fields, we cannot speak in this case of "deliberate construction."

The micropipette became the right tool for the job.[52] This occurred in a range of research fields that Schnitger could not have imagined when he sunk his teeth into finding a handy solution to the tedious task of pipetting with glass pipettes. The Marburg pipette is a striking example of how instruments come into being and how popular tools do not merely simplify the work in a specific field, but how they can open up new research fields and

205

instigate new conceptions of the field itself.[53] Rheinberger further describes the dynamic character of epistemic things, which migrate from the center of attention to its margins when they become a stable part of an experimental system.[54] It is on this stabilized stage that former epistemic things can turn most easily into "transparent media," a process we can witness in the case of the Eppendorf pipette and which is illustrated, for instance, by the fact that the micropipette has not received much historiographical attention. Adele E. Clarke and Joan H. Fujimura remind us that Simon Schaffer had already described this stabilization, using the term *transparency* to indicate that when instruments recede from the focus of attention, they become viewed as "reliable transmitters of nature's messages."[55] Understood in this light, instruments are, as it were, transparent vessels through which it is possible to see right into nature's secrets.

The process by which instruments become incorporated into an experimental situation has also been described as a memory loss. Technologies that no longer attract the experimenter's attention and whose histories have become forgotten have also been described as black boxes and defined as "that which no longer needs to be considered, those things whose contents have become a matter of indifference."[56] The concept of a black box appears rather static, but scholars have stressed that blackboxing must be viewed as a dynamic process that must be studied in its use-context.[57] Latour further describes blackboxing of technologies as a paradoxical effect. He mobilizes a rhetoric of opacity for an effect that has mainly been circumscribed in terms of transparency, namely, "the way scientific and technical work is made invisible by its own success."[58] Latour describes the capacity of scientific objects and technologies to fade out of focus as a process of alienation: "When a machine runs efficiently, when a matter of fact is settled, one need focus only on its inputs and outputs and not on its internal complexity. Thus, paradoxically, the more science and technology succeed, the more opaque and obscure they become."[59] We can observe here how notions of transparency and opacity, in a seemingly self-contradictory way, become interchangeable in describing the successful incorporation of tools into experimental arrangements and daily practices. Opacity is related here to notions of indifference; what is black-boxed is that which works but no longer demands attention. However, my exploration of opacity points in a different direction, namely: something that is opaque is that which cannot simply be overlooked or ignored but continues to stand in the way.

Thinking about instrumentation within a conceptual framework that revolves around notions of transparency makes it impossible to tell the story of the Marburg pipette, or that of its later development into today's "icon" of biotechnological and biomedical research. Rather than histories of apparent effortlessness, we have to make do here with mundane histories of tedious resistance that do not let us ignore the recalcitrant materiality of the experimental sciences, which has, especially in the case of molecular biology and genetic technologies, become obscured behind immaterial predicates associated with the genetic code and literary activities, such as copying, pasting, transcribing, and translating. To gain a deeper understanding of the laboratory life of the life sciences, we need not only to reflect on the concepts of the life sciences but also to explore from a hands-on perspective the material idiosyncrasy of experimental contexts.[60]

Rheinberger views experimental settings as sites of specific intersections. He provides, in *An Epistemology of the Concrete*, a historical analysis of apparatuses to show "that instruments in the biological sciences generate opaque, very diversely configured *intersections* between themselves and the objects they are used to investigate."[61] With the notion of "intersection," Rheinberger sets out to draw attention to the "surfaces on which apparatus and object make contact."[62] His main question concerns the making and negotiation of what we could call touch zones: "How are these planes and points of contact between the animate and the inanimate as it were, between organisms and technical apparatus formed?"[63] Rheinberger understands this project as being part of the enterprise to historicize epistemology. He is interested not in the unification but rather in the diversification of epistemology: his account focuses on specific problems of the life sciences where intersections mark the plane of contact at which life and technology meet. He writes: "Ever since the life sciences took it upon themselves to turn organisms inside out it has been incumbent on them to define the limits of life. And ever since the beginnings of this empirical adventure they have been haunted by the question of the legitimacy of drawing these 'minor' lines of demarcation"; therefore, we need to pay attention to these "minutiae," he argues, for their impact may be much greater for the formation of other frontiers between knowledge domains than one can anticipate.[64]

An epistemology of the concrete needs to concern itself with "these manifold, often unspectacular intersections and interfaces in their own right."[65] It is "the specific materiality of each particular object of the biological sciences carved out in this manner" and "the many small, provisional borders between what still may be considered biological nature and what may have be conceived of as an artifact" that give rise to epistemological questions.[66] Redrawing the boundaries between life and nonlife amounts to a task that must be performed daily on the practical level where objects and instruments meet and must be considered an integral part of the life sciences. Rheinberger's "intersections" designate the boundary surface between apparatus and object of inquiry. The sites where life and technology intersect are critical: "since in general the living entity is wet and soft and the technological one dry and hard special precautions have to be taken to ensure their compatibility."[67] Intersections need to be mastered and maintained. This at times cumbersome work plays out on a daily basis at the bench, but its importance must not be underestimated: "The scientific and cultural—indeed the aesthetic and ethical—significance of experimental systems turns on what is achieved at this intersection."[68] Thinking with Rheinberger, we could say that hands-on and "wet" work at the bench requires skill, expertise, and care to maintain boundary surfaces that bring the life scientist in touch with life. The intersections are manipulation sites: to grasp "molecular" and "cellular" life means to skillfully handle the appropriate precision tools that form the contact points with inanimate and animate material objects. These intersection sites between the inanimate and animate, between living and nonliving technologies, are generally subordinated to sterile regimes. The life science laboratory thus becomes a complex site of touch and no-touch zones. Hence, to be in touch with life at the bench becomes a delicate hands-on/hands-off affair.

A Choreography of Benchwork

The video work *setting 04_0006* (figures 5.3–5.8) opens with a shot of an empty scene: the viewer sees a corner that is dominated by a grid pattern formed by white tiles and black joints that cover both walls and the bench. The static setting is disturbed when hands enter the scene from the right. The hands are clad in white, semitransparent gloves. As they extend into the space, the white sleeves of a lab coat become visible. We see them move back and forth as the hands reach out to grasp imaginary objects. The rest of the body is cut off by the picture frame. The hands stay in constant motion, floating in front of the rigid structure of tiled walls. They move purposefully at a steady speed. Their rhythmic movements describe a pattern, as if dancing. The ephemeral nature of their movements is accentuated by filmic means. The movements of several pairs of hands are just slightly delayed as if one set of hands echoes the movements of another. The pairs are superimposed on each other and cross-fading, imparting a translucent, semitransparent quality. It is impossible to tell whether it is the same pair of hands that is depicted, as the extremities are rendered anonymous by uniform lab gear.

The moving hands describe distinctive patterns. The right hand, for instance, appears to hold an invisible, slender object in its palm: four fingers secure a sticklike instrument with a plunger on top that is pushed down by the right thumb at infrequent intervals. The right and left hand continuously approach each other, the right hand hovering above another invisible small object clasped by the left index and thumb. I can discern repetitious gestures, familiar to me from daily benchwork in a biological laboratory: the delicate motions of pipetting with the right hand, while opening and closing small tubes with the left. The bench appears to be filled with invisible objects, as the continuous grasping movements of the left hand suggest. The hands move constantly, apparently performing the same protocol time and again. The rows of tiles throw into relief this ghostly ballet of delicate finger movements. The stark surroundings that form a background of precisely quantifiable units contrast strikingly with the vivid movements of the fluttering hands. The setting is suggestive of precision, measurement, and plain purity that exudes an atmosphere of sterile cleanliness. Yet, the dancing pair of gloved hands gives way to an excess of meaning that escapes the firm grip of the grid, floating on, elusive and ephemeral.

In the exhibitions *peripheral vision I* (Museu das Comunicações, Lisbon, 2007) and *Say it isn't so* (Weserburg Museum für moderne Kunst, Bremen, 2010), the video installation *setting 04_0006* was shown as a wall projection next to a photo series (figure 5.9).[69] The video work by the artist Herwig Turk and videographer Günter Stöger in collaboration with the Portuguese cell biologist Paulo Pereira is complementary to the series *agents*, consisting of photo portraits of laboratory instruments that Turk had encountered at the Institute for Biomedical Research in Light and Image at the University of Coimbra, Portugal, where Pereira works as a researcher at the Center for Ophthalmology.[70] The photos portray laboratory equipment against the same sterile black-and-white background used in *setting 04_0006*. Together the photo series and the video work seem to provide an impression of science's material dimensions, that is, the scientist's tools and hands that generally remain invisible in scientific accounts. The makers have chosen their subjects for a selective visibility of the laboratory environment: Turk's series *agents* (figure 5.9), for instance, features laboratory instruments that are portrayed as aesthetic objects, rather than as commodities in daily use in the laboratory, by blending out all operating or manipulating hands. By contrast, Turk and Stöger's video work *setting 04_0006* and Turk's later version *hands on (version 3)* from 2014 blend out all instruments, equipment, and materials, and instead spotlight the scientists' hands in action.[71]

209

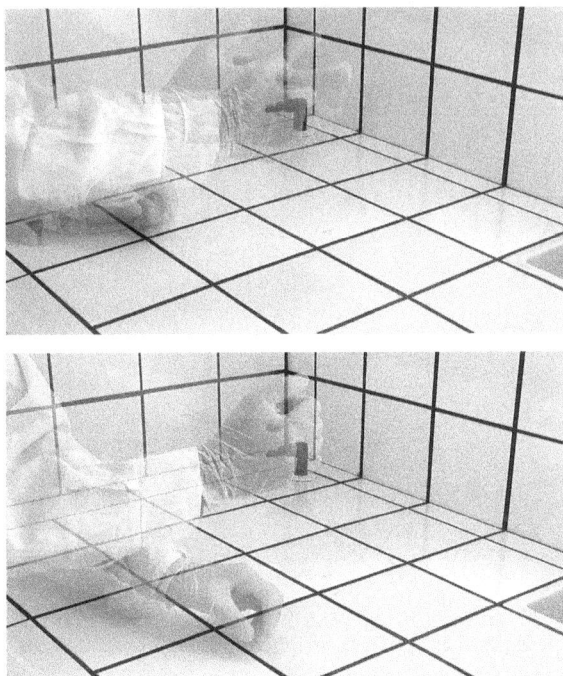

5.3–5.8 ∗ Stills from *setting 04_0006*, © 2006. Video format, PAL 16:9; 6 min., 21 sec. Concept realization: Herwig Turk (Lisbon/Vienna), Günter Stöger (Vienna/Berlin), Dr. Paulo Pereira (scientific supervisor; Coimbra), Beatriz Cantinho (performance supervisor; Lisbon). Video stills courtesy of Herwig Turk.

The art historian Ingeborg Reichle has interpreted the video work in terms of visualizing science's invisible hands and staging "implicit manual knowledge."[72] Her reading of the video work echoes Klaus Hentschel's critique of a hagiographic image of science that rendered most experimental handiwork invisible, as historians of science tended to be mainly interested in the intellectual work of a few geniuses while neglecting all other hands involved in the creation of scientific knowledge.[73] Making these invisible hands visible, Hentschel asserts, is long overdue in the context of modern historiography's declared interest in studying science as forms of cultural and social practice and not in terms of discovery stories of universal laws and objective entities. In her visual analysis, Reichle introduces the term *gestural knowledge* to articulate the hands-on engagements imperative to the experimental sciences, a concept that resonates with the work of Michael

211

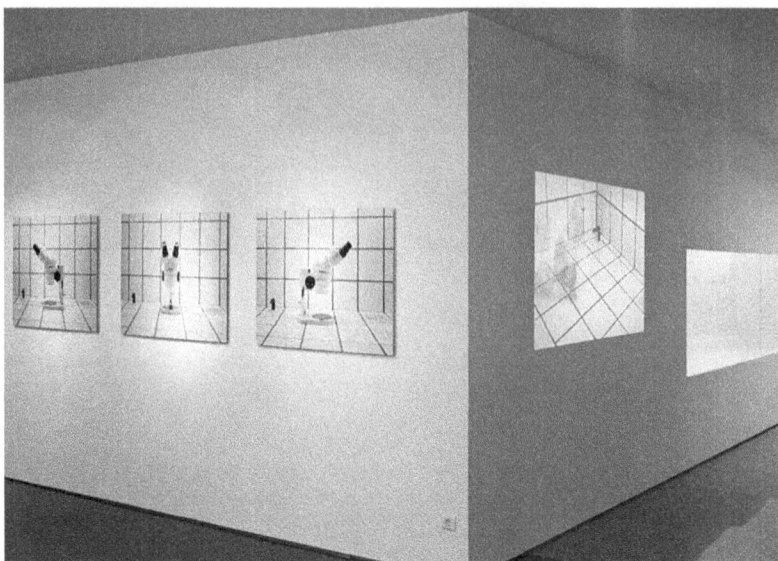

5.9 * Installation view, *agents* (photo series, *at left*) and *setting 04_0006* (video projection, *at right*), at *peripheral vision I*, exhibition at Museu das Comunicações, Lisbon, 2007. Photos © Herwig Turk.

Polanyi, who coined the term *tacit knowledge* and was one of the first to call attention to the importance of skill and manual knowledge for the experimenter.[74] In this reading of the video work, the gloved hands that we see moving about are readily inscribed within a discourse that revolves around science's unseen and tacit dimensions. Reichle's interpretation moves from a descriptive to an analytical level when she gives a reading of the artwork that focuses precisely on the invisibility of what the video work in fact makes visible: the acting hands of the scientist apparently first need to be silenced and rendered imperceptible in order to articulate their epistemic significance. The question arises as to whether we can interpret this artistic visualization of lab hands differently and whether we can raise epistemological issues without readily leaping to a rhetoric of invisibility.

Reichle's interpretation is based mainly on a visual analysis of *setting 04_0006*. In contrast, I offer an interpretation that emphasizes the process of creating the video work, which I will describe not as a visualization or an artistic way of rendering the invisible visible but as a hands-on/hands-off experiment. Reichle provides a detailed description of how the video work was made and the protocol that is performed by the scientist. My descriptions of the working procedure are based mainly on her accounts and a personal conservation with the artist, Herwig Turk.[75]

We could say, for example, that *setting 04_0006* not so much stages science's "invisible hands" but rather investigates the very idea of scientific instruments as "transparent media" with an artistic experiment in which a scientist is asked to handle invisible tools. The video portrays a scientist who performs a protocol empty-handed by executing a series of manual operations with imaginary tools and materials. Reichle's interpretation of the artwork lays the emphasis on a critical engagement with the sciences and with conceptualizations of scientific practices. Central to her interpretation is the aesthetic contrast between the stark background of a grid of white tiles that is readily associated with ideals linked to scientific claims, such as precision, repeatability, measurability, objectivity, and purity, as opposed to the unique, personal, ephemeral, and blurred movements of the hands. The repetitious hand movements demonstrate the precision with which routine processes are executed, yet the constant repetition makes us, at the same moment, aware of "small differences" and "deviations" that "raise some doubt whether day-to-day laboratory practice can indeed always meet the requirements of scientific claims."[76] Though her observations are illuminating, I think this work not only appeals to a critical engagement with the sciences but also engages the viewer in more ways. For instance, the artists bring into play absurdity as an artistic explorative strategy. The artwork responds in a ludic way to the lack in the traditional history and philosophy of natural sciences that centered on "ideas" and mental products rather than the material culture of the experimental laboratory, as Reichle also pointed out.[77] The video *setting 04_0006* makes it visually palpable that scientific instruments were long considered to be mere auxiliary means and thus played no, or indeed only a marginal, role in research of natural sciences' material culture.[78] The performance takes the form of a "real" experiment with which thinking about instruments as invisible tools can be literally explored as an empty-handed practice.

The video was recorded in a laboratory at the medical faculty of the University of Coimbra in Portugal. Turk and Pereira asked a scientist, Rosa Christina Fernandes, to perform a protocol in which cellular proteins are prepared for a technique that is widely used in biochemistry, molecular biology, and other life science and biomedical laboratories. This method, known as SDS-PAGE, serves to separate proteins with electrophoresis according to their electrophoretic mobility. The scientist enacts a protocol for the isolation of proteins from cells for further separation

213

and characterization of the proteins using SDS-PAGE. Before running the SDS-PAGE gel, the samples are centrifuged and divided for further analysis. In practice, this entails much pipetting of small amounts of fluid into microcentrifuge tubes. Turk and Pereira proceeded in a similar fashion for the other work, *hands on (version 3)*, for which they asked four researchers to perform an analytical technique called Western blot that makes use of a similar biochemical method. The video installation *hands on (version 3)*, which displays the naked hands of four different researchers, consists of two video projections that show alternately the different actors side by side. For both works, *hands on (version 3)* and *setting 04_0006*, the artists asked researchers to enact the manual movements of a protocol, but without any of the tools or materials involved in this routine task. In doing so, the artists turn the protocols into a script for a performance. Pereira describes the impact of the vanished tools as follows: "As seen from the inside-of-the-laboratory perspective, the scientist has lost her tools. The objects are no longer present, but trained memory is still able to reconstitute a series of movements. Because of its highly functional nature—this is not a symbolic language—the movements lacks [*sic*] objects, or rather the objects act as extensions of the scientist's hands."[79]

Let us take a closer look at the particular hands-on experience that is generated in this experiment that stages a routine bench performance as a "hands-off instruments" event. What becomes apparent are the constraints of notions such as tacit and gestural knowledge that remain too general to explore and articulate the particularities of today's hands-on experiences in life science laboratories. For *setting 04_0006*, the bench scientist, Rosa Fernandes, is asked to execute a task that she usually performs almost automatically. Habitually, she does not consciously move her hand toward the tubes or think about how to position her fingers when grasping a pipette. Rather, she concentrates on the amount of fluid she is measuring and on accurately filling the row of tubes. Grasping and holding the different instruments as well as handling the different materials when preparing an SDS-PAGE gel is a matter of routine, a task her hands knowingly perform without her having to reflect on her bodily movements. The video not only puts the spotlight on the bodily dimension of scientific work, showing that knowledge is hand-made in a medical laboratory, but in this artistic experiment, Fernandes herself has to focus her attention on her body. To create bodily awareness, Pereira and Turk collaborated with a choreographer who called the scientist's attention to the movements of her hands and arms when enacting the protocol. All tools and materials normally used for this protocol become at once signifiers of bodily motions: for example, the choreographer made Fernandes aware of slowing down her motions and speeding them up depending on the weight of the objects she pretends to handle. The size and form of the instruments define the distance between right and left hand and the position of her fingers. Fernandes now consciously enacts what previously had been a matter of unreflective action.[80]

In this hands-on/hands-off experiment, Fernandes is confronted with her body, which demands her attention: her hands, first moving about unnoticed, now materialize, so to speak, before her eyes when she tries to perform the protocol properly. Performing the protocol with "invisible instruments" affects her bodily experience: she experiences her own body here not as a "transparent source of perception" but as an "object of awareness" that must be monitored and whose parts must be consciously and conscientiously moved about.[81] The artistic research experiment directs our attention to the persistent presence of the handling hands that no longer recede into the unmentioned, or unnoticed, margins, and thus challenges us to enrich and refine our conceptual vocabulary concerning embodied practices.

This is, of course, especially true for healing hands in biomedical practices. A striking series of photographs by Eric Avery, titled *Hands Healing* (1977), portrays surgeons' hands in touch with living bodies during daily practice in the operating theater. *Hands Healing* was exhibited in a corridor of the Hershey Hospital, Hershey, Pennsylvania, where the surgeons portrayed in the photographs work. In his book *Emergent Forms of Life and the Anthropological Voice*, Michael Fischer describes the effect of Avery's photographs on a surgeon who, when seeing this familiar scenery through the artist's lens, was deeply disturbed and unable to perform surgery for some time.[82] The visual presence of her own hands in the photos affected her relationship to her work. Fischer's analysis shows us how photography "can disrupt the detachment of the medical gaze and can intervene in and reframe our self-reflexivity."[83]

In *Bodies in Formation*, Rachel Prentice discusses how the concept of "good hands" in training surgeons not only refers to precise manual motoric skills but also encompasses "the much broader combination of craft skill and situational awareness that makes up surgical ability."[84] Today's surgical settings that rely on advanced digital technologies have been further explored as sites of specific embodied hands-on skills. One example is the video work *Da Vinci* (2012) by the Italian artist Yuri Ancarani, which features a surgical robot and was part of the Venice Biennale in 2013.[85] Another is the collaborative research project Making Clinical Sense: A Comparative

Study of How Doctors Learn in Digital Times (2016–22), led by the medical anthropologist and STS scholar Anna Harris.[86]

(Living) Materials, Instruments, and Skillful Hands

Aseptic technique should be to you not merely a way to open your cultures sterilely, but a way of thinking and acting always in the lab.

Kathy Barker, *At the Bench: A Laboratory Navigator*

The video work *setting 04_0006* brings into focus the intersection between instruments and researchers' hands. By removing instruments and samples from the scene, the video work makes palpable that these scientific instruments do not work on their own but are always part of a complex hands-on situation that encompasses research objects, instruments, and the embodied researcher. It is precisely the complexity of the experimental situation that has become the focal point in studies that began to challenge a "theoretical determinism" and instead defined "the hands-on concrete practices of science" as the "major new problem areas."[87] In *The Right Tools for the Job*, the editors Clarke and Fujimura argue that nothing can be left out of the situation if we want to get a better understanding of scientific practice. In contrast to Rheinberger, who zooms in on contact points where the animate and inanimate intersect, Clarke and Fujimura make a case for addressing the complexity of the situation as a whole: "Only by identifying and paying serious scholarly attention to all of the elements in the research situation can we eventually hope to understand the processes of doing science."[88] They discuss how scholars have drawn attention to coconstructing processes in which all elements are "mutually articulated through interactions," pointing out that all elements are "situationally constructed," though it is possible to differentiate between the elements on an analytical level.[89] Clarke and Fujimura use the notion of a "situation" in a deliberately broad sense to refer to workplaces, scientists, theories, instruments, and skills but also work organizations and, for example, audiences.[90] To gain a better understanding of the different elements of the situation, it is necessary to move from the empirical situation to the analytical realm.

The heuristic value of the artistic experiments that resulted in the video works *setting 04_0006* and *hands on (version 3)* lies in the fact that the artworks do not retreat to an analytical level in order to explore constituent elements of an experiment. The collaborative work of artists, scientists, and a choreographer explores what Clarke and Fujimura see as the main concern of an epistemology that does not submit to a "hegemony of theory" and instead asks *"empirical* questions about complex interweaving phenomena that can be quite dicey to specify, much less to study."[91]

What the video works in fact do is to investigate the complexity of an experimental laboratory situation by deliberately blending out specific elements, thus bringing other elements from the periphery to the center of attention. Yet this making visible/invisible does not take place on an analytical level; it is put into practice in an *empirical* situation. The artworks turn an inquiry into experimental work into an aesthetic exploration; they visually and palpably explore a routine laboratory hands-on situation and create an aesthetic play with hands-on and hands-off experiences. The artists capture with this hands-on/hands-off experiment idiosyncrasies of benchwork that resonate with my own benchwork experiences, but which have not yet attracted much scholarly attention. What struck me most when I was working in life science teaching and research laboratories was the peculiar hands-on/hands-off character of the benchwork. The gloved hands in *setting 04_0006* are emblematic of the paradoxical situation in a life science laboratory. The laboratory is a site of highly manual research activities, as Knorr-Cetina has shown.[92] It also, however, is a working space that is characterized by more or less strictly sterile hands-off regimes that must ensure that materials and instruments are processed without being touched by (bare) hands. Hands-on work and bodily relations are constituents of experimental situations, yet we have to think about them in many medical and scientific contexts against the background of sterile regimes and contamination control technologies that put these popular notions beyond commonsense understandings.

The first time I encountered sterile working regimes in a laboratory was during my field trip in November 2006 to SymbioticA—An Artistic Laboratory Dedicated to Research and Hands-On Engagement with the Life Sciences, at the University of Western Australia, where I was introduced to tissue engineering techniques by the artists Ionat Zurr and Boo Chapple. Ionat Zurr showed me how to passage living cells into another flask and refresh the nutrient medium, a task that mainly comes down to pipetting.[93] I remember how I first was introduced to sterile working principles and the difficulties I had internalizing them when I worked for the first time with living cells in a laminar flow cabinet. It felt as if my hands and my body persistently moved from the periphery to the center of my attention when I tried to pick up instruments, open and close tubes and flasks, and pipette cells and solutions without contaminating my samples. The invisible touch of my hands became a vital concern that required continuous monitoring of my movements, alerting me in a rather counterintuitive manner to all the contact points between my hands, the instruments, and other objects and materials. Working in a sterile hood is tiring, even for experienced bench workers. It is telling how experienced laboratory bench workers and medical practitioners (e.g., surgeons and dentists) talk about sterile working techniques becoming second nature to them. This manner of speaking not only indicates that trained practitioners develop a routine of working in a sterile manner but also shows that this routine, though internalized, remains other than their "first" nature.

The recent trend toward rethinking scientific practices as embodied or bodily endeavors must go beyond efforts to simply expose embodiment as that which preconditions and structures all knowing in the first place. Instead of grounding scientific doings in everyday situations, I argue that we must look closely at the particular ways embodied practices manifest themselves at particular sites in our knowledge society, such as molecular biological laboratories. What is needed, in my view, is not yet another attempt to lay bare the basic structures of embodied relations in scientific and everyday situations but to develop an observational and conceptual vocabulary with which we can describe and analyze the idiosyncrasies of life science's specific knowledge sites. I built hereby on the work of scholars who explore the particular and peculiar bodily engagements that characterize specific medical or scientific practices.[94] My analysis takes its starting point from body-instrument relations. This has been a main objective in the pioneering work of Don Ihde, a philosopher of technology who applied body philosophies and phenomenological methods to describe technoscientific contexts.[95] I will return to Ihde's work later in this chapter. It provides a shift of perspective: from Rheinberger's epistemic objects and intersections, and Clarke and Fujimura's tools, to hands-on experimenters.

Rheinberger differentiates between epistemic things and experimental systems in which these objects are embedded. With experimental systems he denotes "a broader field of material scientific culture and practice, including the realm of instrumentation and inscription devices as well as the model organism to which these objects are generally connected, and the fluctuating concept to which they are bound."[96] He emphasizes the historical and organic nature of experimental systems as the "smallest integral working units of research" when he writes that such arrangements are not simply there from the beginning, but that it is a laborious process to set up and maintain such "systems of manipulation."[97] What is implicit in his definition of experimental systems is that they "contain" not only the scientific objects but also the experimenters. These systems are designed, as Rheinberger explains, "to give unknown answers to questions that the experimenters themselves are not yet clearly able to ask."[98] Systems are not set up *by* experimenters as material answers to theoretical questions; rather, experimenters are part of experimental systems that must be seen as "vehicles for materializing questions."[99] Experimental systems are "not simply experimental devices that generate answers" as if they could be designed and used merely to test "properly delineated conception."[100] Any experimental situation is a complex entity. Experimental systems do not test for or against a theory thought up by a scientist. Instead, scientists must be understood as the part of the experimental system that actuates processes of simplification: "It is only in the process of making one's way through a complex experimental landscape that scientifically meaningful simple things get delineated; in a non-Cartesian epistemology, they are not given from the beginning. They are the inescapably historical product of a purification procedure."[101] Rheinberger emphasizes here the complex historical nature of science's "simple things" and calls for an epistemology that can account for the processes of purification that generate "simple truths."

However, describing this cleansing process with which clarity can be obtained as "the purifying work of the scientific mind," and not as embodied labor, is reductionist.[102] Moreover, when Rheinberger calls for understanding clarity as a historical product in anti-Cartesian terms, he conveniently overlooks that Cartesian epistemology is also itself the product of a purification process that Descartes initiated with his *Unterschiebung*. The Cartesian *Unterschiebung* actuated a conception of modern epistemology that centered on the result of an epistemological simplification and obscured the hands-on complexity of its inception (see chapter 3).

Tubes and Touch

GENE TECHNOLOGY PRACTICAL COURSE (SECOND YEAR),
LIFE SCIENCE AND TECHNOLOGY STUDIES
AND BIOCHEMISTRY, GORLAEUS LABORATORIES,
LEIDEN UNIVERSITY

During my gene technology course, most of the handiwork I performed involved pipetting cells and solutions from tubes into slots in gels, onto plates, and into other tubes. We had to be careful at all times not to contaminate our

219

samples, either with our own DNA or with proteins or microorganisms from our hands. Most times we worked without gloves. We only used gloves when pouring gels to which we had to add carcinogenic chemicals. We had to be careful not to touch our samples when plating out our cells on agar plates, adding solutions to tubes, or inserting toothpicks with cells on them into growth media. We used microcentrifuge tubes to mix cells with other solutions, to centrifuge cells, to warm cells in heating blocks, to cool them down on ice, or to freeze samples for later use. We worked with living cells and with their inner components: isolating, duplicating, cutting DNA molecules, loading them into gels, and reinserting them into the genomes of living microorganisms (mostly bacteria or yeasts).

For all these processes we juggled around tiny plastic vessels, filled with mostly transparent solutions. In the beginning the whole tube threatened to flip out of my hand, and I also tended to get cramps in my tightly clenched fingers when opening it with my thumb. If one and the same tube had to be opened and closed repeatedly, I had to be very careful not to spill any of the fluids by moving too jerkily; I needed to train the fine sensorimotor behaviors of my hands to attune my hand-eye coordination to handling this new tool in co-ordination with the microliter pipette. The tubes had to be sterile to prevent bacteria from the air or from our hands growing in the samples. We had to work carefully and meticulously adhere to sterile working principles when replicating DNA with a PCR machine, for instance, for which we had to use special minuscule tubes and tips that were kept under stricter sterile conditions than the tubes and tips on our bench. The analyst pointed out that the minuscule PCR tubes must be handled very conscientiously: "Never ever touch its rim or the interior of the lid." It took some time until it became a routine operation to handle these tiny objects correctly. This task involved, at least in the beginning, an increased awareness of what had become exposed to my touch. It could happen easily that my thumb grazed the rim unconsciously, the contact being so slight that I didn't sense it. Though we did not need to be afraid that we could replicate DNA from our own hands by chance, since fragments of our DNA would not in general attach to the primers we used, we still had to keep in mind that our hands contain nucleases that could cleave the DNA strings of our sample.[103]

The name Eppendorf tube, or more fondly *eppy* (German and English) or *epje* (Dutch), is today often used generically in laboratories to indicate microcentrifuge tubes that are designed to hold very small amounts of liquids and which fit into a centrifuge. These small containers with conical bottoms are found today in virtually all areas of biomedical, pharmaceutical, and life science research. They can be closed with a lid that is attached to the tube by a small, stiff plastic hinge. The lid snaps tightly closed if its round brim is pushed firmly into the opening of the tube, and the small lip on one side allows it to be flipped open again with one thumb, while the tiny, nearly weightless tube is secured between the other fingers of the same hand. Handling the centrifuge cups with their snap-tight caps is one of the most routine but nonetheless rather delicate tasks I had to learn in the laboratory.

Working with PCR and other gene technologies in a molecular biology laboratory comes down to manually mastering at least the basic research tools that make the investigations of cellular and molecular life processes possible. While learning to handle new instruments, I learned how to adapt my hands to this task. We can describe this process in which my hands become the right tools for the job as a process of coconstruction.[104] But it also can be viewed as a "world-enclosing" activity, as Don Ihde suggests in his book *Instrumental Realism*.[105]

As life became molecularized in the modern laboratory, and as cellular life in vitro became one of the most powerful and ubiquitous tools in biomedical, pharmaceutical, and biological research, getting in touch with life became a matter of mastering the use of microliter tools and the most basic technical skill required in a molecular biology lab, that is, good sterile technique. Descriptions of sterile technique in operating manuals provide the practitioner with instructions on how to behave and move about in laboratory space; how to handle, maintain, and clean instruments; and how to work with media and living samples. In tissue engineering and molecular genetics laboratories, I learned that sterile technique begins with myself becoming aware of the potential sources of contamination and using the proper technique to reduce the likelihood of contamination during experimental procedures. Focusing on sterility implies a shift in perspective that directs the attention from objects and instruments toward the laboratory practitioners themselves. This shift is apparent in a webinar produced by the Eppendorf company that, under the heading "It's in Your Hands," identifies the lab practitioner's own hands as a major source of contamination.[106]

After weeks of hands-on training, my *epjes* did not feel edgy anymore in the palm of my hand. Little by little, my hands lost their clumsiness as I flipped the small vessels open and closed. I could give my undivided attention now to other tasks—choosing the right chemicals, putting my tubes into the right order, keeping track of my operating schedule. In the lab I was in touch with life through these small vessels. The touch of my fingertips was not just mediated by these tiny tubes; the tubes and my hands merged, as it were, constituting a body ready to explore life on a cellular and molecular level. A phenomenological description of how I got in touch with life in this laboratory goes beyond a description of direct tactile experiences. When I grasp my microcentrifuge tube containing a mix of restriction enzymes and plasmid DNA, I am in touch with molecular life *through* the tube. The tube enhances my sense of touch, just as the microscope enhances my sense of sight, opening up the invisible and untouchable worlds of molecular life, the living cell and its constituents. The tubes are not mere containers of cells, DNA, and other media, and they do not simply "protect" living and nonliving samples from the touch of my hands; rather, it is through the physical activity of pipetting that I enter the world of "molecularized" life. The pipette and tubes function as means through which I can perceive and act on otherwise imperceptible environments and contexts. They form a "special class of artefacts" that are normally not perceived themselves as objects but "are capable of engaging in 'symbiotic' relationships with the human body," which become "embodied" in perceptual acts that are technologically mediated.[107]

Hands-On Benchwork as Incorporation Processes

Philosophers of technology have tried to understand instrument use in scientific contexts as "instrument-mediated perception."[108] The embodied relationship between scientists and instruments for scientific investigations takes center stage in this approach. Don Ihde addressed similar questions as we already encountered in seventeenth-century experimentalism: "How can we examine phenomena that we cannot perceive with our senses?" and "How do we embody tools and techniques that give us access to worlds that lie beyond ordinary sensory perception?" More recently, works by scholars of embodied cognition, including Shaun Gallagher's groundbreaking study *How the Body Shapes the Mind* (2005), have examined complex forms of body awareness in instrument-use contexts.

The work of the French phenomenologist Maurice Merleau-Ponty, who reevaluated the role of the body and its epistemological significance in his most famous work, *Phenomenology of Perception*, first published in French in 1945, has been of major importance to pioneers of this approach.[109] The following discussion of Merleau-Ponty's body philosophy focuses on central concepts in descriptions of instrument-embodiment relations and hands-on learning processes. Merleau-Ponty borrowed the technical term *body schema*, used by psychologists of his time, to describe a dynamic, potentially open understanding of body space. Hence, the confines of one's body

222

space do not have to coincide with the biological body, as Jenny Slatman explains in *Our Strange Body*: "Just as the 'extended phenotype' comprises more than the actual organism, the boundaries of the body schema do not coincide with the boundaries of the organism."[110] Merleau-Ponty's *schema corporel* defines bodily boundaries in a manner that is conditional on the situation in which one finds oneself. When I became trained at the bench in routinely handling micropipettes and microcentrifuge tubes, the contours of my body schema expanded. Over the course of time, my body schema encompassed, as it were, part of the bench and the instruments I handled. A well-versed hands-on practitioner does not handle tools; she incorporates them: "A movement is learned when the body has understood it, that is, when it has incorporated it into its 'world'."[111] To handle pipettes and tubes routinely means "to be transplanted into them, or conversely, to incorporate them into the bulk of our own body."[112] In the language of Merleau-Ponty, at the end of my internship, after countless hours of pipetting, I had incorporated the pipette and microcentrifuge tube. For me to use a tool or to master a technology meant to integrate it into my bodily space: the measurement instruments became part of my body schema. This process of incorporation can also be described as an expansion of one's body: "Habit expresses our power of dilating our being-in-the-world, or changing our existence by appropriating fresh instruments."[113]

Merleau-Ponty pays extra attention to operations that become routine because this process cannot be described with cognitive models of bodily perception. It is precisely the moment when I am no longer aware of my body that I master an operation routinely. When a new task becomes a habit, I no longer tend to notice the operations of my limbs or the tool at hand. Merleau-Ponty employs the term *body schema* to indicate that one's being toward the world (*être au monde*), which forms the precondition of any knowing of the world, is grounded in prereflexive engagements. Merleau-Ponty uses the expression *être au monde* to describe how one relates to the world through bodily movement, or what he describes as the situated spatiality of a body. In Slatman's words: "The boundaries of our body, determined by the body schema, comprise both the body and its situation."[114] The body schema expresses that one's body cannot be understood as an object moving about in space. When I move my hand toward my lined-up Eppendorf tubes, I do not calculate the distance that my hand must overcome to reach the objects on the bench. I do not experience the movement of my hands as a displacement of body parts from one location in space to another. Phenomenologically speaking, I do not conceive of my body as

223

an object positioned among other objects in the laboratory. The situation that I find myself in is not simply a configuration of objects positioned in space; rather, my body encompasses the space that constitutes my world. In other words, I could say that I "embody" a situation that I find myself in.

Merleau-Ponty uses the concept of body schema to distinguish between habitual bodily movements and consciously enacted bodily movements. I perceive my reaching hands not as objects that are moved about in space by cognitive acts. The successful process of learning to operate a tool is described with the term *incorporation*, indicating a loss of awareness. Working with living substances in this lab did not involve any direct sensual experience of living matter; I touched only nonliving objects, such as microcentrifuges and pipettes. This does not mean, however, that I did not engage bodily with life on a molecular scale. In this description, an aesthetics of immediacy makes way for an aesthetics of technological mediation as an embodied activity.

Merleau-Ponty's model of incorporation has been expanded on in interdisciplinary studies in philosophy, psychology, and technology by Don Ihde and the psychologist and philosopher Shaun Gallagher. Their elaborations on the incorporation model show that it provides valuable insights into the embodied dimensions of hands-on work at the bench. However, this model remains too general to account for the specific embodied experience I had in the molecular laboratory that takes into account issues of sterility. Moreover, I noticed how these adaptations of the incorporation model remain constrained to framing embodiment in terms of a rhetoric of transparency. This makes it difficult to describe how I experienced myself in the laboratory not as a transparent source of perception but as a continuous source of concern, to use Richard Shusterman's phrasing, or even as an obstacle.[115]

224   The last section of this chapter will take a critical look at the conceptual frameworks introduced by Ihde and Gallagher. Both authors have been influential in shifting the perspective from mind-biased epistemologies to embodiment relations, but a close reading of their work reveals a problematic tendency to frame bodywork in terms of transparency. The body epistemologies they offer struggle to articulate our bodies' persistent presence. Somehow, here, too, bodies tend to fade away. This discussion leads to some final reflections in the epilogue on technoscientific spaces designed to deal with scientists' bodily presence and their resistance to being erased from scenes of knowledge making.

Gallagher and Ihde introduce the concepts of "experiential transparency" and "instrumental transparency" to further describe the process of learning to handle tools as an incorporation, or extension, of the body schema.[116] Gallagher, a psychologist, focuses on the body that incorporates, that is, on the question of whether and how the perception of one's body changes during the process of learning. Ihde, in contrast, lays the emphasis on the way in which one's perception of objects changes when one successfully incorporates a new technology. These two approaches are already anticipated in Merleau-Ponty's terminology that defines the dynamic structure of the body schema at once as integration (what we could call Gallagher's "inward perspective") and expansion (Ihde's "outward perspective"). A critical examination shows how Ihde's work grants partial opacity to instrument relations, but it remains focused on how I experience my own body in body-instrument relations mostly as a "transparent source of my perception or action and not," as Shusterman critically observes, "as an object of awareness."[117] Shusterman's work on somaesthetics offers a critique on this persistent bodily absence in philosophical accounts.

Tool Incorporation as Instrumental
Transparency

In his book *Instrumental Realism* (1991), Don Ihde draws on Merleau-Ponty's "body epistemology" to describe "how we utilize technologies and how such use transforms what it is we experience *through* such technologies."[118] Ihde wants to provide a description of body-instrument relations that does not highlight the subject body but rather the incorporated objects. His philosophy is explicitly aimed at extending phenomenology into the realm of technologically mediated experience with examples from scientific instrumentation. He describes scientific perception as a highly specialized mode of perception that is mainly, though not exclusively, mediated through technologies.[119] Ihde's *phenomenology of technics* lays out a program that differentiates between three body-instrument-world relations: embodied, hermeneutic, and alterity relations. They occupy a continuum, though the extremes are quite distinct, ranging from "transparency" to "opacity."

The most basic way of relating to technologies that mediate our sensory perceptual world are embodiment relations. Hermeneutic relations, by contrast, characterize technology relations through which we are also involved with the world via an artifact, but the artifact is not incorporated and does not become transparent, as it has to be "read." As an example, Ihde shows how when we read the temperature on a thermometer, the instrument reveals to us a certain aspect of the world: "In a hermeneutic relation the world is first transformed into a text, which in turn is read."[120] The thermometer does not enhance my sense of the cold outside, as eyeglasses enhance my vision. Instead, I can read from inside my living room what the outside temperature is. The thermometer enables me to "envision" how cold it is outside precisely without any bodily sensation of warmth or cold. I have to look at the display on the thermometer, and then I have to interpret the numbers that are displayed in order to get a sense of the weather. Ihde discerns here a technological opacity that we do not experience in embodiment relations.

With embodiment relations, Ihde denotes bodily ways of relating to technologies in different use contexts. In a use context, he explains, "I take the technology into my experiencing in a particular way by way of perceiving through such technologies and through the reflexive transformation of my perceptual and body sense."[121] His classic example of embodiment as a kind of body-technology relation is taken from everyday life.[122] In his description of how he puts on eyeglasses, Ihde demonstrates that it is not adequate to say that a technology mediates between us and our environment. Rather, the technology involves a transformation of the body, the world, and the technology itself. It is no coincidence that Ihde chooses eyeglasses as an example with which to explain this concept of embodiment relations, given that eyeglasses embody a key aspect of Ihde's phenomenological account: transparency. He describes how the enhancement of his vision transforms his world from vague to clear; it transforms his body that looks, feels, and moves differently in this world; and last but not least, the eyeglasses themselves are transformed from objects into a part of his body.

Ihde introduces the notion of transparency here to describe what ideally happens when a technology successfully enhances our sense of vision. The "ideal of transparency" postulates a successful embodiment of a technology in terms of its complete incorporation. In the case of eyeglasses, this would mean that I do not see the world better *through* my glasses. Rather, my glasses have become an imperceptible part of my world, of my body; I see with them without seeing them. Ihde's description shows how embodying technologies causes those technologies to become transparent. From this analysis we can infer that the concept of transparency can be applied not only to the technologies in use but also to one's body. As the eyeglasses become a part of my body, I am no longer aware of them, nor am I aware of my body as something that impedes me. I have transformed into a sharp-eyed me, and the glasses are part of that me. Accordingly, Ihde talks about a loss of transparency when this embodied relationship is disturbed. Broken glasses become objects that need to be fixed. Those broken pieces I hold in my hands have lost their transparency in the literal and figurative sense.

This loss of transparency does not apply only to the technology. When I break my glasses, my body does not simply change back into the body I had before ever wearing glasses. Not only the broken glasses but now also my body catches my attention. My potentially sharp-eyed body becomes, without glasses, a visually impaired body. My body loses its transparency when I become aware of it as a not fully functioning body in need of eyeglasses, which suggests that processes of incorporation are to a certain extent nonreversible.

Embodiment relations, according to Ihde, are characterized by the desideratum of "pure transparency" that can never be fulfilled, as technologically enhanced capacities "are always different from my naked capacities."[123] Enhancement also always means a transformation of experiences. Thus, Ihde explains, "the actual, or material, technology always carries with it only a partial or quasi transparency, which is the price for the extension or magnification that technologies give."[124] Nonetheless, it is an extreme form of instrumental transparency that for Ihde is the most characteristic feature of embodiment relations. The more instruments merge with bodies, the more transparent they become. On the other side of the spectrum, we find alterity relations, which denote technologies that do not fall into the category of "hands-on" or "body-on" instruments. These technologies cannot be incorporated but are experienced, Ihde explains, "'as' other to which I relate."[125] The opacity of such technologies is in contrast to the increasing transparency of technologies that relate to bodily perception.

With the concept of "instrumental embodiment" Ihde aims at describing "the co-constitutive relationship between humans and technology in a praxis."[126] His embodied account of optical technologies, for instance, a pair of glasses, as "technics of vision," provides a helpful model to conceptualize microliter technologies in terms of "technics of touch."[127] When I have actively embodied the technics of touch in a molecular laboratory, my pipetting and microcentrifuge-handling body is coconstituted with the world of molecular life. When I am successfully learning how to handle a technology, such as driving a car, playing the piano, hammering, or wearing eyeglasses, my bodily contours are enlarged and begin to coincide with the outlines of the instruments. Transparency articulates that successful incorporation culminates in a loss of awareness of my body and the instrument. Microtechnologies are an example of "the gradual extension of perception into new realms."[128] Phenomenologically speaking, a molecular lifeworld is revealed to us by the embodied practice of micropipetting. Peter-Paul Verbeek emphasizes the importance of this notion of "mutual constitution" in

Ihde's discussion of human-artifact relations.[129] The most defining aspect of embodiment relations in the case of molecular-genetics is that technological artifacts give us access to an imperceptible realm through a "withdrawal" of the tools.[130] The concept of embodied relations describes how routine handling of basic instruments amounts to an experience in which the instruments recede into transparency as they are mastered and incorporated into the bodily schema.

So far, Ihde's model of a "relation of mediation" allows for an illuminating reading of Eppendorf's promotional slogan "In Touch with Life." Ihde accounts for the initially heightened alertness to the tool and one's body in the process of learning, but he contends that "once learned, the embodiment relation can be more precisely described as one in which the technology becomes maximally 'transparent.'"[131] He marks this withdrawal, or effacement, of the instrument-body as the most important characteristic of embodiment relations. The Eppendorf tube in my hand is a means of experiencing in the laboratory rather than an object of experience.[132] Yet, what is problematic is that for Ihde every embodied praxis is fundamentally geared at transparency. Transparency, not praxis, forms the starting point of his thinking about embodiment relations. A closer inspection of another frequently cited example by Ihde exposes this bias toward transparency and shows that his models of embodied perceptions do not stand up to critical scrutiny of different and historically contingent use contexts.

In his pioneering study *Technics and Praxis*, Ihde takes a dentist's probe as his first example of an instrument-embodiment relation because he believes it "to be about as simple as one can imagine."[133] He introduces the notions of "amplification" and "reduction" to describe the hands-on experience when using this precision tool. The dentist's tactile perception is at once impoverished and enhanced when a stainless steel probe is used: imperfections, such as little cracks in a tooth's surface, become amplified, and the dentist "feels" microscopic irregularities with the probe as if it were an extension of his fingertip. Yet, Ihde also observes the loss of a greater richness: wetness and warmth are not sensed by the dentists' "probing hands." Thus, we can speak in the case of instrument-mediated perception of "a reduced experience" in comparison with the "'naked' touch" of a surface.[134] The probe allows the dentist to be "*embodied* at a distance," to get in touch with the tooth through the probe, but this experience, Ihde argues, is never total; there remains always "a sense of *difference*."[135] According to Ihde, it is such a "fringe awareness" in which one remains vaguely aware of using an instrument that distinguishes "in flesh relations" from "instrument-mediated relations."[136] He prefers to describe the dentist-probe relation as "semi-transparent" because the dentist and the probe do not experientially coincide, but "as embodied become a semi-symbiotic unity."[137]

What is conspicuous is that, for Ihde, a certain degree of experiential opacity, that is, semitransparency, arises only in instrument-mediated perceptions. His phenomenological analysis is in this respect at odds with Merleau-Ponty's and Husserl's phenomenological descriptions of the sense of touch. Note that Husserl's famous example of one hand touching the other, which could be described in Ihde's terms as an exemplary "in the flesh experience"[138] or "'naked' touch" perception, cannot be simply reduced to "direct" and fully "transparent" experiences.[139] Instead, the example of a pair of hands touching one another shows, as Jenny Slatman has pointed out, how one always already experiences a sense of difference or otherness with respect to one's own body (see chapter 3).[140]

When Ihde distinguishes between technologically mediated and "in flesh" or "'naked' touch" relations, he seems to imply that "unaided" body perceptions always amount to experiential transparency. My finger's "'naked' touch" and "ordinary in flesh experiences" are, for the sake of his philosophical project, simply understood as "direct" experiences and contrasted to more complex technologically mediated situations that are characterized by an increasing sense of otherness. Embodied perception is at its core understood as transparent, and Ihde attempts to describe how instruments can be incorporated into embodied experience by becoming (at least) partially transparent, too. Ihde's phenomenological analysis appears flawed because it adheres to an unproblematic conception of bodily experience, as if noninstrumental embodiment relations could be described in terms of "pure transparency," and as if opacity arises only when instruments enter into embodiment relations.

Ihde's objective of giving detailed descriptions of the different types of technological relations is compromised by the oversimplified starting point of his analysis. He takes as his reference point the assumed directness of "naked" touch or "in flesh perception." This reference point suggests a context-independent parameter that does not bear up with the idea of praxis, that is, particular historical contingent use contexts, as the starting point of a phenomenological description. A closer inspection of Ihde's example of a dentist's probe substantiates this point of critique. When I visited a dentist surgery after I had been working at the bench in a molecular biology laboratory, my awareness was drawn to similarities between these sites and practices, in which the correct handling of precision tools under strictly sterile regimes is of major importance. During a dental appointment, I asked my dentist about her ordinary working practice and the role that touch played in her daily work.[141] She explained to me that first of all she always puts on a fresh pair of gloves for each patient. The gloves protect both her and her patients from transmissions of infection—microorganisms on her hands could contaminate the interior of her patient's mouth, and vice versa. She also wears glasses to protect herself against contamination with hepatitis B, for instance. All objects that go through her hands or those of her assistant are strictly routed to proceed in one direction through the consulting room. All instruments are first sterilized in an adjoining room and enter her treatment chamber through the right door. After the instruments have been touched and become "contaminated" by practitioners' hands and patients' bodies, they leave the chamber through a second door. When Ihde, in his example, compares "naked" touch experience to instrument-mediated hands-on experiences, he alienates the probe from its use context. In doing so, he alters the situation to fit this example into a phenomenological description geared at transparency. Dentists' hands-on examinations of patient's teeth usually do not entail any "naked" touch experiences. Ihde *makes* from the probe a simple instrument-embodiment example when he isolates the tool from its use context and thus ignores the complex working situation and its experiential effects. His phenomenological description is not rigorous enough to give us insight into dentists' hands-on practices that are marked out by a conscientious awareness of touch.

Ihde's description falls short in articulating the peculiar hands-on/hands-off experiences that come with a sterile regime of touch at work in the dentist's office *and* in molecular life practices. We cannot use Ihde's instrumental transparency model to describe how sterile working procedures oscillate between hands-on and hands-off experiences in which a certain awareness of one's own hands and the instruments one handles remains a crucial aspect of the practice. In Ihde's description of the dentist's probe as an example of an incorporation process, the extension of bodily boundaries and the enhancement of perception are described as an unproblematic and smooth process. As others have pointed out, the phenomenological body in Ihde's theory of perception, and for that matter also in Merleau-Ponty's body philosophy, offers virtually no space for itchy, edgy, weighty, particularly clumsy, passive, or apathetic bodies.[142] Ihde's focus on bodies engaged in knowledge *activities* implicitly favors discussions of active rather than passive bodies, as Andrew Feenberg rightly has pointed out.[143] Ihde's instrument-embodiment model is too much biased toward transparency

230

and, as such, reduces the complexity of particular working situations. His concept of transparency also engages an implicitly ahistorical stance: he uses notions of transparency to describe "basic features" that could apply as much to modern bench scientists as to tool-wielding Neanderthals.

The shortcomings of a model of embodiment geared toward transparency have also been identified in relation to tissue-engineering practices.[144] Mechteld-Hannah Gertrud Derksen and Klasien Horstman point out how Ihde conceptualizes the dysfunction of technology (out-of-focus, dirty, or broken glasses) as a loss of transparency: "If the object is transparent to someone, he or she does not notice it and it is simply part of his or her being-in-the-world. Thus, Ihde's description of embodiment assumes that our normal being-in-the-world implies being directed outward to the world rather than to experiences of our body."[145] Derksen and Horstman's critique of embodiment-as-transparency emphasizes that the outward perspective inherent in Ihde's model allows only for objects, and not bodies, to remain partially opaque. As such, Ihde conflates instrumental and experiential transparency in his descriptions of instrument-embodiment relations. This makes his conceptual framework problematic when we want to describe situations in which (partial) instrumental transparency does not necessarily imply the experiential transparency of one's own body.

Body Awareness and Embodied Technologies

Based on recent empirical studies in psychology, the psychologist Shaun Gallagher develops a more refined vocabulary with which to describe different modes of perceiving our own body. In *How the Body Shapes the Mind*, Gallagher explains that the learning and training process that is accompanied by a special awareness of bodily movements changes during habit or routine formation into a nonreflexive awareness: "My body," he writes, "remains experientially transparent to the agent who is acting."[146] However, it is important that he describes embodied learning in terms of *recessive* reflexive awareness rather than suggesting that embodied learning simply culminates in becoming entirely unconscious of one's body.

To better differentiate between different states of awareness during the process of incorporation, Gallagher discusses several concepts: "body schema," "body image," "body precept," and "proprioception." He shows, for instance, how the notion of proprioception carries slightly different meanings in different disciplines, primarily in the way proprioception is coupled with either awareness or unconsciousness: "Neuroscientists may treat somatic proprioception as an entirely subpersonal, non-conscious function—the unconscious registration in the central nervous system of the body's own limb position. In this sense, it results in information about body posture and limb position, generated in physiological (mechanical) proprioceptors located throughout the body, reaching various parts of the brain, enabling control of movement without the subject being consciously aware of that information."[147] In this latter meaning, the notion comes close to what Merleau-Ponty describes as prereflexive. According to Gallagher, Merleau-Ponty's phenomenological descriptions tend to gloss over the process character of learning experiences. Merleau-Pontian phenomenology lacks the conceptual clarity to account for fluctuations in states of awareness during bodily learning processes. Gallagher proposes an analytical distinction between different concepts, such as body schema and body precept, to account for the dynamics of learning processes involving incorporations. The latter refers to a conscious monitoring of one's bodily movements that, in the case of learning processes, precedes habit formation. It is subsumed under another concept, the body image, that implies awareness of one's own body.[148] The following discussion of hands-on learning and instrument use focuses on Gallagher's notion of body schema that denotes states of receding awareness of one's own body, and on incorporated artifacts.

In their discussion of hands-on training in scientific laboratories, Cyrus Mody and David Kaiser criticized the lack of attention to historical and processual dimensions in accounts of scientific knowledge creation (see chapter 4).[149] Similarly, Gallagher's critique of Merleau-Pontian phenomenology draws attention to the disregard for a historical dimension of knowledge formation as a profound problem in increasingly popular body philosophies. Gallagher identifies an ahistorical stance in phenomenological analyses and highlights instead the process-based character of learning. His analysis exposes how the notion of transparency in phenomenological accounts obscures the complexity of learning processes as relative phenomena. "The schema," Gallagher argues, "works along with a marginal awareness of the body, and Merleau-Ponty often left the relation between the schema and marginal awareness unexplained."[150] The conceptual vocabulary proposed by Gallagher grants (partial) opacity to body awareness. This line of argument is comparable to Kathleen Jordan and Michael Lynch's call for understanding blackboxing of technologies as a process in which instruments migrate from the center of attention to the periphery and become "mere tools" as a reversible process and relative phenomenon.[151] Gallagher's conceptual framework thus allows for the possibility of describing hands-on body-technology experiences and learning processes in a more differentiated and situation-specific manner. However, Gallagher's inward, or recessive reflexive awareness, model of embodiment still presents successful incorporations into one's body schema largely in terms of a self-effacement of one's own body. Both Ihde's and Gallagher's models remain deficient when we want to describe laboratory situations in which instrument use is learned in the context of sterile working principles. These use contexts require a particular continuous awareness of one's own body that cannot be properly articulated with these models.

The phenomenological body that incorporates technologies and instruments is, in fact, a largely ahistorical notion with which it is possible to describe general hands-on aspects of scientific practices and also the constitutive role of the body in all forms of knowing. Yet, this ahistorical body concept is in conflict with the project of historicizing epistemology that explores the peculiarities of hands-on experiences in molecular biology wet labs as specifically different from other contexts. What is needed for a historicized epistemology is, on the one hand, an acknowledgment of the constitutive role of hands-on practices and, on the other hand, an exploration of the role of bodily activities in the creation of scientific knowledge that avoids getting trapped into an *Unterschiebung* of ahistorical body notions (on this point, see the discussion in chapter 3). Phenomenological descriptions of hands-on experiences should respond to the particularities of today's highly artificial technoscientific scenes of knowledge making instead of providing us with descriptions of yet another example of common everyday situations.

The blindness toward experiments, instruments, and hands that Rheinberger, Heidelberger, and Hentschel have diagnosed in the traditional history and philosophy of science can be linked to the dominant rhetorical strategy that scientists employ when they present their scientific results.[152] This strategy has been described as a rhetoric of effortlessness that aims at hiding all the (bodily) labor and exertions that were necessary to attain scientific knowledge.[153] My analysis chimes with this thesis of a modern practice of concealment that presents the products of physical exertions as rationalistic idealities. We can understand Descartes's *Unterschiebung* as such an operation. The rhetoric of effortlessness finds its epistemological complement in hands-off theories of the creation of scientific knowledge and is conspicuously close to narratives of instrumental and experiential transparency both in the rhetoric of scientists themselves and in phenomenological theories of embodiment that obstruct our view of the particularities of technoscientific working situations.

We cannot ignore the obstinate, opaque presence of bodies and recalcitrant benchwork experiences on a theoretical level. Epistemological projects that historicize the creation of knowledge will need rigorous descriptions of site-specific hands-on experiences. In the case of the life sciences, this requires further scrutiny of contamination control technologies and sterile working procedures. To theorize bodywork at contemporary technoscientific sites of knowledge production, we need to go beyond a form of thinking that frames hands-on work in terms of transparency. Our analyses should not begin with a "Cartesian dismissal of the body," or with generic body notions or body phenomenologies, but with a real interest in historically contingent bodies and bodily engagements. Today's scientific sites challenge us, for instance, to think of bodies in their unwieldy, persistent, and contagious presence in order to gain a deeper understanding of hands-on processes of knowledge formation in the life sciences and the particular ways in which these processes bring us "in touch with life."

   ★   ★   ★   ★   ★   ★   ★   ★   ★   ★   ★   ★   ★   ★

Epilogue

I opened this book with a narrative of the Nobel Prize–winning geneticist Barbara McClintock, reflecting on the ways she struggled with her "bodily me." McClintock's story not only alluded to struggles on a practical level but also pointed to unsolved issues on a theoretical level. It describes her as grappling with a Cartesian legacy, one that held the cytogeneticist back from incorporating her body and bodywork into conceptualizations of herself as a scientist. In this epilogue, I return to the ideal of a disembodied knower invoked by McClintock's wish to be free of her body. I discuss visual presentations of life scientists and reflect on the contemporary phenomenon of contamination-controlled spaces that are vital to life science research and its biomedical and pharmaceutical applications. I explore how technoscientific phenomena of cleanroom technologies conspicuously amplify bodily issues we encountered in my discussion of life science laboratory research. I also provide a short ethnographic account of a cleanroom visit and flesh out how problematic views of "disembodied" sciences continue to work through: zooming in on sterile regimes in microbiological cleanrooms makes our need for better and more attentive "body talks" in the sciences of life palpable.

Visualizing Hands-On/Hands-Off Science

In 2011, the German weekly newspaper *Die Zeit* dedicated a special issue of *Zeit Magazin* to the nation's leading scientific researchers under the headings FORSCHER (researcher) and FORSCH (bold, or "do research!")[1] The cover image with the first term features a photograph of an influential German scientist who works in the fields of genetics and cancer research (figure E.1).[2] It shows the geneticist Klaus Rajewsky in his laboratory in a full body suit. A colleague, who is dressed in what looks like a surgical suit, assists him with donning the sterile lab gear. The cover presents Rajewsky in a full-length portrait, standing upright with his hands crossed behind his back and looking straight into the camera. The contrast between his elderly appearance (Rajewksy was then seventy-four) and the futuristic look of the garments emphasizes the impression of his authority in a field that has changed significantly during his lifetime. His female colleague is portrayed in profile, standing right next to him as she extends her gloved hands to adjust the collar of his suit. The portrait illustrates the hierarchical relation between a white male top scientist in a static pose, hands behind his back, and next to him presumably a technician or analyst rendered in an active, assisting pose. Gender issues are obviously at play here, but they are not the topic of this discussion.

The life scientist presents himself as proud and boldly dressed in a full body sterile suit in the presence of an assistant who helps him with the bodily preparations for his work. We can imagine the awkward procedure of donning a sterile suit and the way the suit constrains one's movements. The photo shows at a glance the idealized image of a disembodied knower (with no hands) and at the same time illustrates, in palpable ways, the idea of the body as a "nuisance" that we have to drag with us and that needs special attention because we cannot leave it behind when exploring life at the smallest scale. The image condenses some of the key issues addressed in this book: the more the living cell and its processes become transpar-

236

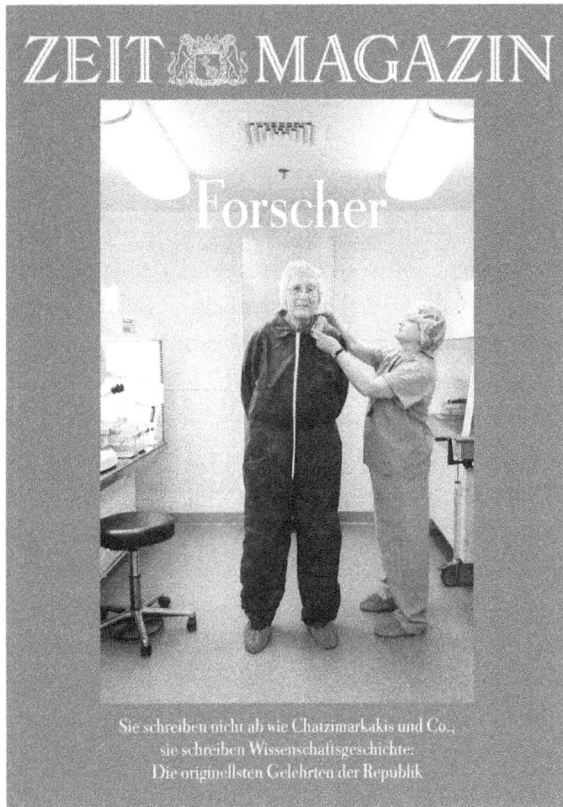

E.1 * Scientist Klaus Rajewsky dons a cleanroom suit with the assistance of an anonymous laboratory staff member. Cover of *Zeit Magazin*, July 7, 2011. Photo @ Andri Pol (https://andripol.com/).

ent and manipulable, the more the researchers' bodies stand out in their opacity and remain resistant to attempts to erase them from the scenes of knowledge on a practical and theoretical level. Life science and biomedical research must be understood within its own tradition that is driven by the desideratum to make bodies transparent and to bring them under the control of the experimenter's hands. In chapter 5, I raised the issue of bodily opacity to offer a counternarrative to persistent ideas of science as a disembodied affair and to question the assumed transparency of technological mediations. In this epilogue I take a closer look at technoscientific contamination-controlled spaces that are designed to deal with the bodily presence of human operators.

Sterile Spaces as Sites of Body Performances

Science and technology studies scholars have written on embodied practices in contamination-controlled technoscientific environments, such as operating theaters, neonatal intensive care units, and material sciences laboratories.[3] The sterile regime in place in an operating room has been described in detail as a ritual that shapes the bodies (of surgeon, nurse, and patient) involved in the surgical procedure.[4] Sterile procedures have been described as rituals that help to establish a separate place: they create a discontinuity between the space of the operating theater and its surroundings.[5] Stefan Hirschauer investigated the operating room not just as a room with special equipment where surgeons encounter patients on whose bodies they operate but as a specific space in which bodies are (co)created.[6] Hirschauer's ethnography makes visible how operating rooms function as spatial configurations from which patients' and surgeons' bodies emerge. Examinations of the procedures and structures at play in the practice of surgery demonstrate that the distinction between sterile and nonsterile areas plays a significant role in establishing active and passive bodies. Hirschauer's observations in "The Manufacture of Bodies in Surgery" challenge us to understand bodily boundaries as flexible and dynamic phenomena that can be shifted and reconfigured in medical practice:

> From the angle of the patient-body, the result of laying connections and of being moved around appears like this: muscles and joints, which move it and hold it together, are now located in the wheels of the stretcher, in the tilting mechanisms at the head and foot of the table, in the straps around the legs, the hinges of the armrest, the lifting mechanism of the jack, pedal-operated by the anaesthetists. The fluids circulating in the body are collected externally or supplied from the outside: urine empties through a catheter into a second bladder under the left arm, the stomach through a tube in the nose into a container, blood is sucked off through a tube and flows into a vessel at the base of the head. . . . The lung of the patient-body is standing diagonally behind its head and breathing for it, sucking and clapping. . . . Technically amplified, the heart of the patient-body is now flashing and bleeping.

238

Bodily boundaries of patients are reconfigured in the encounter with the anesthetist and medical technology. The patient-body and the surgeon-body are not preexistent but context-dependent. The bodily boundaries are redrawn and precarious: "The life of the patient-body seems to be externalized and can be threatened by many errors. Somebody unfamiliar with the new boundaries of the body can endanger its life by turning switches, accidentally pulling cords and tubes off or out, or by getting too close to sheets of linen."[7]

Hirschauer describes "surgical operations as encounters of two disciplined bodies—a parcelled 'patient-body' and an aggregated 'surgeon-body.'"[8] To trace the patient's bodily boundaries is a complex affair; we cannot understand the unconscious living body as an isolated entity, sprawled out on the operation table. The living body's outer limits do not coincide with the patient's skin but are extended and externalized through the interaction with the anesthetist-body and the body of medical devices. The aggregated surgeon's body consists of various members of the operating team. Hierarchically, cooperation guarantees that this body can function as a unit. Sterile regimes play an important role in structuring this body, distributing its parts into different areas around the operating table and within and outside the operating room. Hirschauer's ethnographic account presents surgery as a practice in which bodies are "manufactured" in a coconstructing process.

More recent studies expand on bodily issues in surgery, addressing also the use of virtual technologies in surgical training.[9] Others reflect on experiences of "dis-association" and the development of a "medical gaze" as important concepts in anatomical and medical training; see, for example, Byron Good's anthropological account of medical knowledge practices and Michael Fischer's essay that describes the disruptive impact artworks can have on surgeons who have been trained to develop a detached point of view.[10]

What is particularly interesting is the way in which sterile spaces speak to continuous processes of coconstruction by which all bodies, living and nonliving, are produced through spatialized regimes of cleanliness and touch. Scholars have discussed, for instance, how cleanliness and purity are constant matters of concern in many laboratory sciences. With a case study from material science, Cyrus Mody describes how "keeping things clean (instruments, materials, people) structures the knowledge-making process."[11] Jessica Mesman, focusing on touch and hands-on experiences in neonatal care, has shown how new knowledge is gained when observing and touching children under extreme conditions of sterile separation.[12]

239

Returning now to particular sites of sterile regimes used in life science research, I will take a closer look at isolator technologies that are designed to function without humans operating from the inside. The experimenter's bodily presence manifests itself here as a technoscientific problem. Microbiological cleanrooms thus offer us a thought-provoking imaginary to think about bodies' recalcitrant presence.

Isolators and Biotechnological Cleanrooms

Human contamination: of the many potential sources of contamination in cleanrooms and other manufacturing environments, none is more persistent, pervasive, or pernicious than the human beings who occupy them.

Hugo Huiskamp, cleanroom training presentation

All these clean room procedures are directed at controlling the human body. Within this environment the human body is defined as a virulent contagion. And yet the relationship between a body and a clean room is slightly paradoxical: clean rooms owe their existence to our intellectual ingenuity and physical dexterity, after all, and their purpose is to act as environments for certain specialised forms of human activity, yet they are always on the verge of rejecting our presence. . . . It's hard not to undergo at least a mild degree of self-alienation on hearing our bodies portrayed as great spluttering, oily, flaking parcels of pollutant.

Alex Farquharson, in *Clean Rooms* (exhibition catalog)

The origin of cleanrooms has been traced back to the first hospital theaters in Europe.[13] Today cleanrooms are a worldwide phenomenon and are used in many different fields, such as nanotech, biomedical, material sciences, and space research and in food and pharmaceutical industries. How clean a room has to be depends on the task that is performed in it. Microbiological cleanrooms that are used, for instance, in life science and biomedical research and as pharmaceutical production sites in which aseptically produced injectable medicines are fabricated are much cleaner than operating theaters. This is also true for high-class cleanrooms as they are used in material sciences where research must be conducted in particle-free environments but where bacterial contamination is of no particular concern.[14]

A contemporary isolator is an apparatus that provides a completely sterile enclosed environment featuring a sophisticated ventilation system, including high-efficiency particulate air (HEPA) filters. As the name indicates, this apparatus enables an operator to work with materials in an

240

isolated environment, allowing only indirect access by means of gauntlets. Such isolators may be used, for example, for the breeding of germ-free laboratory animals or for experimental work involving the use of extremely contagious materials. Isolators are also used in pharmaceutical production facilities, where they are generally known as isolator, or barrier, technologies that shield fully automated manufacturing processes from any contact with workers and airborne contamination.[15] An isolator thus either protects the user against contamination with its contents or else protects its contents against contamination by the user and the outer environment.

More particularly, an isolator designates a territory forbidden to human operators; it defines a space in terms of its seclusion from human and other bodies. Exceptions to this definition are isolators used in hospital settings, where they are designed to protect patients suffering from extreme immune deficiency diseases. In this extreme case a patient may be placed in a strictly contamination-controlled and fully isolated environment. One of the most well-known examples of the use of this type of isolator technology is the so-called Bubble Boy, Daniel Vetter (1971–1984), who spent most of his life in a flexible plastic isolator and whose life has since been immortalized in popular culture.

Isolators speak to our imagination as they embody a strong fascination in modern Western thinking for concepts of purity and cleanliness in relation to knowledge making. It is beyond the scope of this book to dive further into the history of isolation technologies, but my fascination for cleanroom technologies was triggered by the first known historical illustrations of experimental setups with glove boxes, which I will briefly describe here.

Traditionally, a glove box is an isolator technology used by chemists for air-sensitive work requiring a separate atmosphere. Glove boxes are also used when working with live pathogens and living organisms if extremely sterile working conditions and a high level of contamination control are required. The first experiments with fully enclosed sterile environments, or so-called isolators, were conducted at the end of the nineteenth century and the beginning of the twentieth century. These experiments were incited by a keen interest in the possibility of "germ-free life." Isolator experiments with animals that were kept in highly contamination-controlled environments gave rise to the research area known as gnotobiology, which is concerned with the study of either germ-free or germ-specified organisms.[16]

The schematic illustration printed in George H. F. Nuttall and Hans Thierfelder's article from 1895 shows an experimental setup with an isolator at the heart of the scene (figure E.2).[17] Inside a large glass container

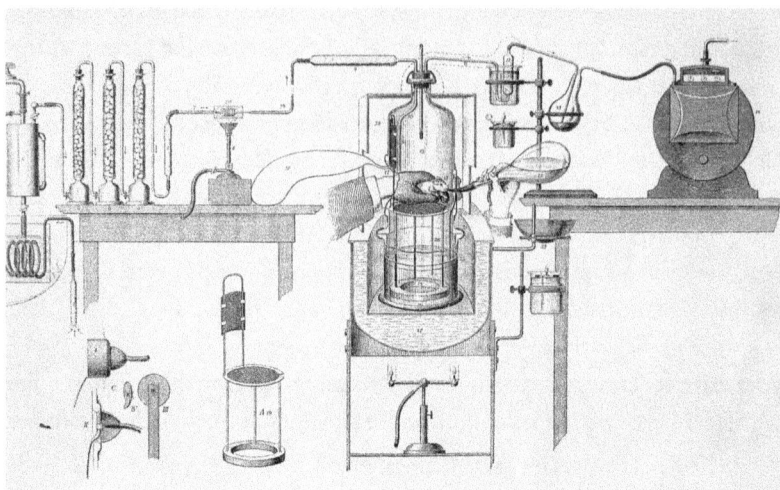

E.2 * Experimental setup with an isolator "for the cultivation of animal life with a digestive tract free of bacteria," documenting the early history of the cultivation of specific pathogen-free animals in modern laboratories. From George H. F. Nuttall and Hans Thierfelder, "Thierisches Leben ohne Bakterien im Verdauungskanal," *Hoppe-Seyler's Zeitschrift für Physiologische Chemie* 21 (1895): table 1.

shaped like a bottle is a mouse that is fed with a mouthpiece inserted via a hole into the inside of the glass cage. The mouthpiece is connected to a flask that contains a fluid medium. Hovering outside of the glass container are a hand that holds the flask and another hand that is inserted into the container and holds the mouse. Both hands extend from cuffed sleeves that are eerily suspended in midair. Only the forearms are visible, and somewhere above the elbows the arms fade away, leaving invisible the imagined adjoining body.

Ernst Küster's article on breeding pathogen-free animals from about twenty years later also provides an image of a fading human operator (figure E.3).[18] In this schematic illustration, the contrast between the gloved hand that extends into the interior of the isolator and the semitranspar-ent contours of a head and a suggested torso is more pronounced. When I came across these two illustrations during my research, I was fascinated by the ambivalent ways in which they seem to manifest a visual rhetoric of self-effacement that aims at turning the investigators' bodies into invisible observers. Yet, at the same time, they represent an experimental technoscientific setup that apparently obstructs such a maneuver.[19] The

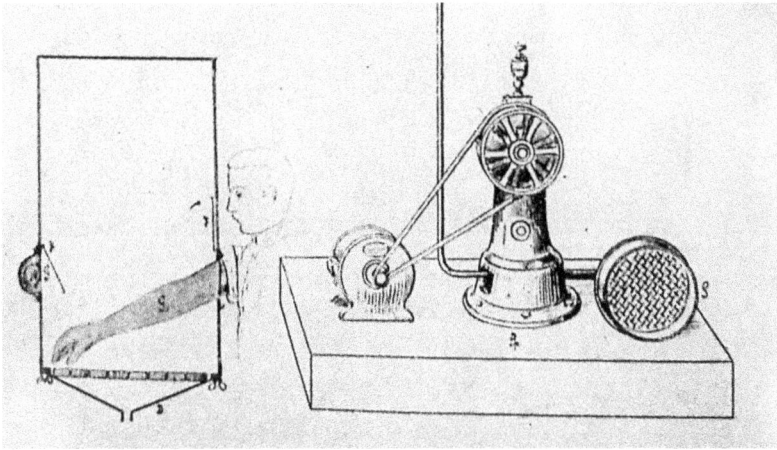

E.3 ★ Detail of "Schematic of the complete setup for the cultivation of pathogen-free animals" ("Schematische Darstellung der Gesamteinrichtung zur Aufzucht keimfreier Tiere"), illustrating the operation of an isolator. From Ernst Küster, "Die keimfreie Züchtung von Säugetieren," in *Handbuch der biochemischen Arbeitsmethoden* (1915), 313.

bodies of researchers resist erasure from scenes of knowledge making: they do not fade fully into transparency but remain (at least partially) visible in their opacity.

These historical illustrations of emerging isolator technologies conspicuously bring into view an epistemic struggle: they visualize the efforts to bring an experimental hands-on praxis in line with a rhetoric of self-effacement that tends to erase scientists' bodily presence from sites of knowledge making. The ambivalent aesthetics of cleanroom environments thus function as a fertile ground for inquiries into questions of embodiment and invite us to explore microbiological laboratories as scenes for and sites of participatory body performances.

More recently, bio artists have explored contamination-controlled spaces as peculiar sites of attention (*Aufmerksamkeitsorte*) and sites of attentive techniques and practices (*Aufmerksamkeitstechniken, Aufmerksamkeitspraktiken*), to use the terminology of the German phenomenologist Bernhard Waldenfels.[20] Artistic research into spatial contexts shaped by contamination and barrier technologies brought the users of these spaces into view as awkward actors who cannot think their bodies away. Bio artists' body performances in cleanroom environments, fume hoods, and grotesque stagings of glove boxes have incited ethical, cultural, and theoretical

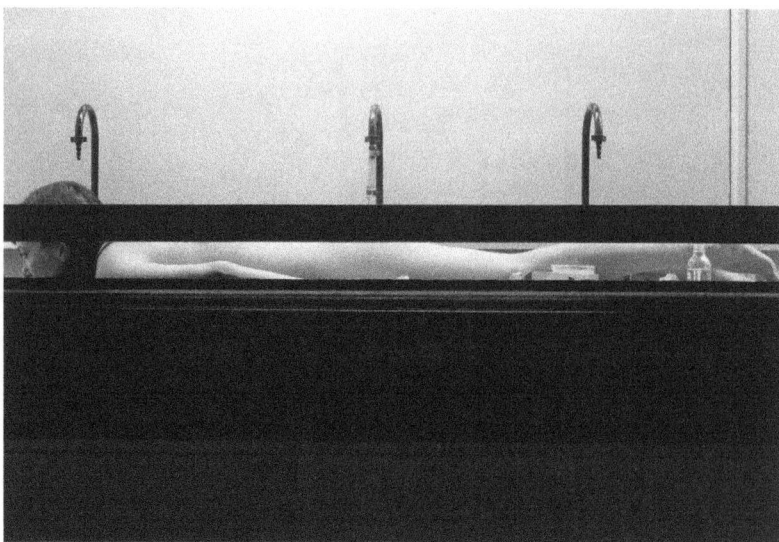

E.4 * Jennifer Willet, *Bioplay*, 2017. Color photograph. Photo credit: Adam Zaretsky.

reflections on how the life sciences bring us in touch with life. The photographic documentation of Jennifer Willet's *Bioplay* performance reflects on the body's precarious presence in sterile laboratory settings (figure E.4).[21]

The one-week performance *The Workhorse Zoo* by Adam Zaretsky in collaboration with Julia Reodica featured a portable cleanroom as part of the 2002 exhibition *Unmediated Vision* at the Salina Art Center in Salina, Kansas.[22] In another setting, Zaretsky turns a glove box into an interactive space, staging a palpable experience of "disembodied life." The crude construction had no technical function, but an aesthetic one. It amplifies the sensory experiences of technoscientific working environments by simulating the bodily constraints of sterile working areas. The workshop participants had to put on gloves that severely constrained their body movements. Hands-on practice was dramatically staged as an awkward attempt to process and manipulate cells from body parts of a freshly slaughtered pig (figures E.5 and E.6).[23]

Oron Catts and Ionat Zurr, founders of the Tissue Culture & Art Project, were among the first artists who exhibited installations featuring tissue-engineered sculptures that "lived" through the time of the exhibition. They also explored, for example, the significance of careful handling and caring hands in life science laboratories (see figures E.7 and E.8).[24] Catts and Zurr

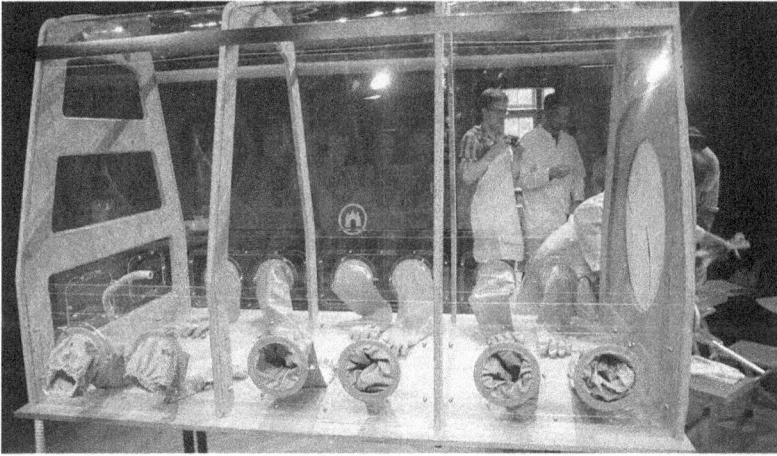

E.5 and E.6 * A glove box constructed for artistic research into sensory dimensions of the "disembodiment of life" and tissue culturing technologies. Installation at a "tissue culture lab" workshop, organized by Adam Zaretsky, as part of his project VivoArts School for Transgenic Aesthetics Ltd. (VASTAL), in collaboration with Oron Catts, at the Waag Society, Amsterdam, September 15, 2009. Photos by the author.

introduced the term *semi-living sculptures* to emphasize the ambivalent status of in vitro life-forms that cannot survive outside the care that is provided by the "artificial womb" of the laboratory.[25] Tissue-engineered "life-forms" are inextricably bound up with a biotech environment in which they must be handled with care under strict sterile regimes. In another article, the bio artists reconceptualized the isolation of cells in terms of a reembodiment

E.7 and E.8 ⋆ (*top*) Tissue Culture & Art Project, *Disembodied Cuisine*, installation at the *L'Art Biotech* exhibition, Nantes, France, 2003. (*bottom*) Tissue-engineered "steaks" cultivated on-site in a bioreactor. Photos courtesy of the artists Ionat Zurr and Oron Catts.

rather than a disembodiment.[26] The aggregated "technoscientific body" that encompasses not only the technoscientific entourage and its life-forms but also the embodied activity of artists/researchers keeps the sculptures "alive" and out of reach of visitors' lethal touch.[27] Bio art performances thus make us aware that the maintenance of living tissue outside of the confines of organic bodies is not only a technological problem but also a question of care.[28]

We can experience in life science's spaces of technoscientific productivity how bodies and their boundaries are continuously reconfigured. Cleanroom environments make the persistent presence of our own bodies palpable; my body functions here as a "transparent source of perception."[29] In addition, it also functions as a strange body of ongoing concern.[30]

Practiographies for Embodied Epistemologies

To further reflect on epistemic implications of cleanroom environments as peculiar sites of body performances, I turn to the work of the medical anthropologist and philosopher Annemarie Mol. With her study on medical practices, Mol does not present a new perspective on objects of knowledge, but she avoids a "perspectivalist" approach altogether: "Instead of being in the point of focus for a variety of eyes, each with their own perspective, they [objects of knowledge] are presented here as objects that are being handled—by hands and knives."[31] Mol mobilizes the concept of performance for research into diseases in medical settings, "for it allows ethnography to engage with the materialities of diseases, with their fleshiness, their physicality. It allows ethnography to shift its attention from the observing eyes of professionals to their hands and instruments; from the group culture of medical specialists, with their shared languages and mutual rivalries, to the material culture of medicine: its hospital buildings, dissection knives, examination tables and artery pulsations. If these shifts are made then perhaps ethnography is no longer the most appropriate term for the discipline. Perhaps it would be better to call it practiography."[32] I suggest here that cleanroom environments call for "practiographies" with which we can explore the aesthetics and epistemic implications of sterile working sites in life sciences research practices. Practiographies of microbiological sites of knowledge making emerge from intimate practical engagements and prompt us to rethink epistemology as a pluralistic project grounded in local hands-on practices. They remind us that our bodies and embodied practices

247

cannot be ignored in reflecting on our understanding of life. The following account of cleanroom experiences puts flesh on my theoretical inquiry into notions of opacity and nuisance as a bodily experience.

Cleanroom Aesthetics

I wondered how it would feel to work in a laboratory setting that required clothing completely designed to prevent bodily contact with the samples. I wanted to know more about these extreme sites at which researchers get "in touch with life" under highly controlled conditions, and where life scientists and other laboratory personnel are doing hands-on work with microbiological media in technoscientific "hands-off environments." I shared this fascination for sterile working sites with a bio artist, Boo Chapple, whom I met during a research stay at the art-science laboratory SymbioticA in Perth, western Australia. Chapple taught a Vivo Arts honors class at Leiden University to which she invited a cleanroom consultant, Hugo Huiskamp, as guest speaker.[33] Huiskamp has specialized in training cleanroom users and invited me to participate in a one-day cleanroom course.[34]

This training for researchers at the Netherlands Organisation for Applied Scientific Research was a general course for all cleanroom users, though most of the participants were researchers in material science–related projects. In this course the bodily presence of the researcher inside a contamination-controlled laboratory setting was of central concern.[35] Bodies and bodily behavior were discussed as the "most important source of contamination," and cleanrooms were described as specific settings in which you have to "turn off" your senses, because the crucial elements—dust particles and microorganisms—cannot be seen, heard, smelled, or felt. The most difficult thing about working in a cleanroom setting is that you have to "act rationally" without relying on your senses, as the cleanroom consultant with years of experience explained to us (see Huiskamp's presentation slides in figure E.9).[36] I was fascinated by this highly "anaesthetic" description of cleanroom environments. I introduce here the notion of "cleanroom aesthetics" to describe the embodied experiences involved in "sensing" life at sites of technoscientific productivity that oscillate between sterile spaces of certainty and "zones of awkward engagement."[37]

Huiskamp brought me in contact with the head of a cleanroom facility at Leiden University Medical Center that complies with the highest biosafety standards in the Netherlands. During an interview, this biological pharmacist explained to me that the cleanroom must comply with the strict safety

248

E.9 ⋆ Cleanroom training presentation slides. Courtesy of Hugo Huiskamp.

standards of the international Good Manufacturing Practice (GMP) guide-lines because it is a production site for medicinal products.[38] The GMP lab is a high-level cleanroom to ensure that cell therapy products do not become polluted. It is an environment with extremely low levels of pollutants in the air, such as dust particles and airborne microbes, to ensure the sterility of the product.[39] The low level of pollution is attained with special HEPA filters and a pressure difference between the "clean corridor" and the inside of the cleanroom. The pharmacist estimates that there are approximately 100,000 times fewer dust particles and bacteria in the air of the cleanroom than in the air we normally breathe in office spaces.

The patients' cells that are processed in this space are not simply research samples; I should see these cell therapy products as medicinal drugs that are not yet registered and are being developed as part of clinical research. Many hospitals now have a GMP facility to develop their own drugs that need to comply with the same strict pharmaceutical regulations, but this cleanroom, which has been in use since 2000, is one of the largest and the best controlled in the Netherlands. All the administration and standardized protocols must ensure the quality and safety of the product. The pharmacist here differenti-ates between innovative and traditional cell therapies, such as bone marrow therapy. The latter are produced in closed systems, meaning the body ma-terials are transferred via tubes into sealed packages and never come into contact with outside air. The T cells, in contrast, are exposed to airborne contamination the moment they are pipetted from one container into an-other (when the cell therapy is successful, the artificially proliferated T cells can inhibit the growth of a tumor or even diminish the malignant swelling).

The T cells that have been isolated from the patient's tumor tissue are proliferated and processed inside the cleanroom and then are reinjected into

249

the patient's body. Generally, medicinal production processes take place in pharmaceutical cleanrooms. Drugs that are destined to be administered to humans are maintained and processed under highly contamination-controlled conditions, but at this facility, clinical research findings can be developed within the academic hospital into an end product for the patient. Usually, it is not easy for an outsider to get permission to visit a cleanroom, so I was very excited when I was allowed to accompany one of the analysts working at this facility on a regular day of work.

Regimes of Touch: Hands-On/Hands-Off

My first visit to a biological cleanroom was a peculiar bodily experience.[40] Entering a space in which I was made conscious of my body as an endangering and endangered entity that needed to be almost totally "sealed" into a special suit induced a strange feeling of uneasiness. At the time of the visit, I was three and a half months pregnant with our second child. When exposing my unborn child to such an environment, I became aware that until then I had naturally conceived of my body as a protective and nourishing environment, not as a source of contamination.

After putting our shoes, clothes, and jewelry into a locker (besides underwear, you were strangely enough allowed to keep on your socks), we changed into a fresh pair of blue suits and slipped into green rubber hospital clogs.[41] I put on a hairnet, taking care that no hair was left uncovered, and a surgical mask that I tied around my head. Next, we had to wash our hands and wrists thoroughly for twenty seconds with Betadine scrub, a microbicidal skin cleanser commonly used for pre- and postoperative washing. Following this, we sprayed our hands with a disinfectant. Just as I finished, my right temple began to itch. The research analyst explained that I was not allowed to touch my face with my freshly washed hands, but that I could touch the outer side of my hairnet, which covered the itching temple.

Putting on the outer white sterile suit demands some handiness because it must be unpacked and unfolded without any part of it touching the floor inside the barrier that divides the lockers' front area from the back area. I learned to climb into the suit while stepping over the barrier. Once on the other side, I had officially entered the clean zone, connecting the "clean corridor" to the GMP lab spaces.[42] The last thing to do before passing through the one-way entrance to the cleanroom facilities was to put on sterile gloves. I learned that preparing hands for working in a cleanroom requires meticulousness and a delicate touch. The entire procedure of unpacking, unfolding,

250

EPILOGUE

and donning sterile gloves depends on touching them only on the inside so that the outside surface of the gloves remains sterile. To concentrate on this task, I had to imagine which of the visible parts formed touchable and untouchable areas: the part that covers the wrists was folded up and partly turned inside out, making it possible to grasp the right glove at its lower end with the left hand and to slide the right hand into the opening without touching the glove's sterile outer surface. Next, I picked up the left glove, now making sure to touch only its sterile outside with my gloved right hand. I slid my gloved right hand under the folded-up part, lifting up the glove so that I could enter its opening with my left hand without touching its outer surface. Then I stretched the gloves' lower part over the cuffs of my sterile suit, thus tightly covering the transitional area between wrist and sleeve. I repeated the last step with my left hand so that the right glove also fit closely over my sterile suit without leaving any bare skin exposed. Finally, we pulled up the hoods of our sterile suits and placed them over the top of the hairnets. The only body parts that were left uncovered were our eyes. The research analyst wore glasses, but as she was allowed to keep her "private" pair on, these qualified—just as my bare eyes—as nonsterile.

After meticulously completing this washing and dressing routine, we were allowed to walk through into a corridor leading to the cleanrooms. We had put all the materials that we would need inside the cleanroom into a service hatch, and we could now collect them from inside the clean zone. We sprayed all the items with disinfectant before putting them back into the trays on the trolley, except for the paperwork and my camera, which, to my surprise, was admitted without further ado. We also repeatedly sprayed our hands with alcohol, and the research analyst explained that we would need to change gloves every hour when working in the cleanroom because they become permeable from the alcohol after an hour. Finally, we entered the heart of the cleanroom facility, the production site, with two laminar flow cabinets on opposite walls. The research analyst turned on one of the laminar flow cabinets (see figures E.10 and E.11). A laminar flow cabinet is an extra-clean workspace inside the cleanroom made of stainless steel, with no gaps or joints where spores might collect. This carefully enclosed bench is designed to prevent contamination of biological samples. Airborne contamination of living cells, or any other particle-sensitive media, is prevented with a HEPA filter through which air is drawn into the cabinet and blown out in a smooth, laminar flow toward the user, thereby preventing particles from the user's body from falling into the cabinet.

251

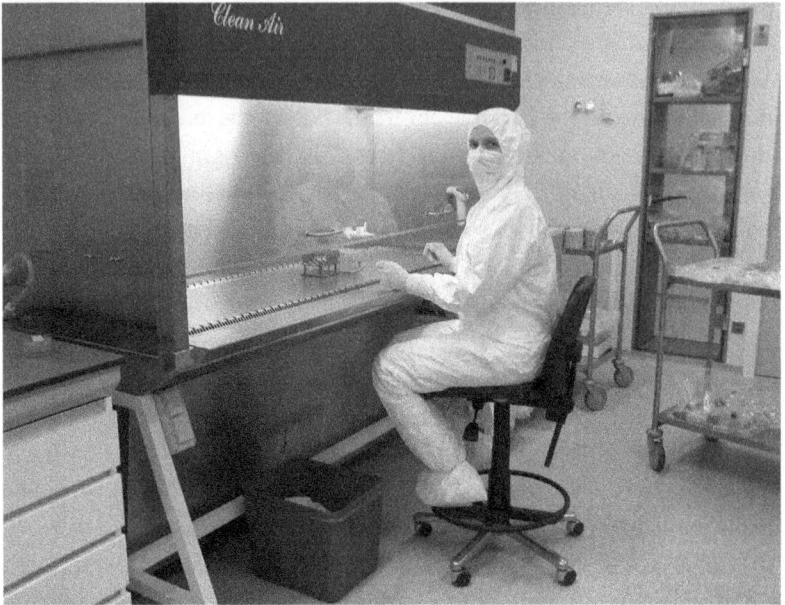
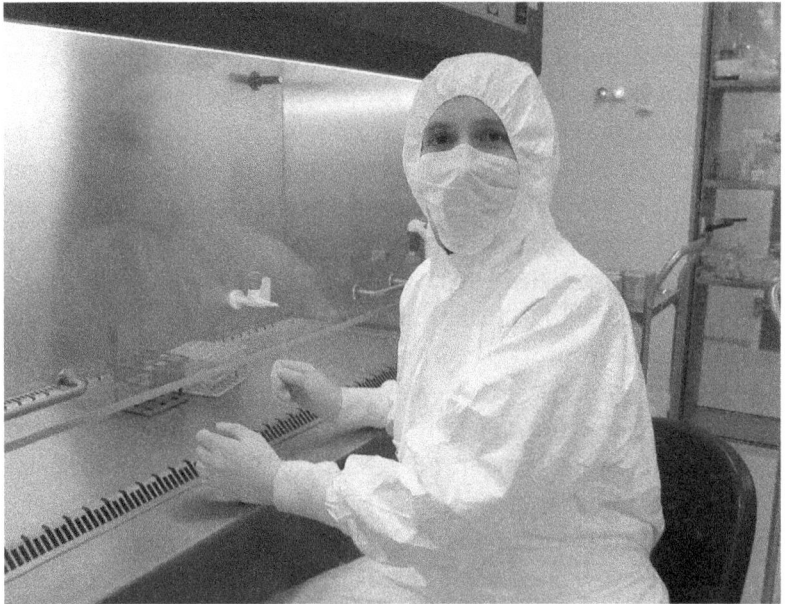

E.10 and E.11 ∗ The author seated at a laminar flow cabinet in the cleanroom at the Interdivisional Good Manufacturing Practice facility, Leiden University Medical Center, which is part of the Department of Clinical Pharmacy and Toxicology. Photo courtesy of the author.

Inside the cleanroom, strict working routines prevail. First, the research analyst has to fill in the standard forms with the codes of all the procedures she is going to perform that day. Then, she prints the labels for monitoring samples, and we stick them onto three petri dishes that are strategically positioned in the cleanroom: one on the bench, one next to the computer, and one inside the laminar flow cabinet that she will use. Next she cleans the inside of the laminar hood with ethanol. In the meantime, I feel my nose tickling, and I need to hold in a sneeze. The hood of my sterile suit tends to drop over my eyes, but the analyst fixes this with a piece of adhesive tape. I am surprised how comfortable the suit feels. I imagined that I would get hot in it and that it would hamper my movements, but this is much less the case than I expected. Much more tiring than moving about in the cleanroom suit is filling in all the paperwork and working in a highly concentrated manner, as the research analyst tells me during our interview: "You have to document and register every step that you do in the cleanroom and write up every action that you perform." In the cleanroom is a separate writing desk. When moving between writing table and working bench, the analysts always have to disinfect their hands. The research analyst explains that you have to be very precise and meticulous, and always have to be on top of everything in the cleanroom. It is quite tiresome because of the hours of monitoring and documentation. I see that the volume of paperwork is enormous, reminding me of Bruno Latour and Steve Woolgar's apt description of the central role of writing in research laboratories.[43]

In the next hour, the research analyst passages one patient's sample of tumor cells in the laminar flow cabinet. Cell passaging is a technique to keep cells alive and growing in culture. The analyst explains later, during an interview, that the cells are passaged at least ten times to make sure the sample contains tumor cells that continue to divide and reproduce, while other cells die after a certain number of passages. I can observe and film her while she performs the protocol, but I am not allowed to handle the cells because this is patient material that must be maintained and processed under strictly standardized conditions according to so-called GMP standards.

The handling of cell samples must comply with strict regulations because the processed cells are meant to become reintegrated into the patient's body. This entails a complex process in which blood cells and tumor cells are cultured first separately and then in one sample to which other substances are added to stimulate the blood cells to develop antibodies to the

253

tumor. The aim of the cell therapy is that the final cell product is injected into the patient's body to "fight" the tumor cells.

The care with which you must handle the samples makes the work in the cleanroom very tiresome, but other aspects also play a role in limiting work shifts in the GMP lab to a few hours (not more than four hours on average). Taking a break to use the bathroom or to eat something is cumbersome. It must be done as specified in the regulations: in the "dirty corridor" you have to take off the outer white sterile suit, which cannot be reused. If you leave the cleanroom, you are allowed to keep on the blue, or green, surgeon's suit that you wear underneath and the hospital clogs, except when using the bathroom or eating something; then you have to discard all lab gear and redo the entire procedure, washing and suiting up again before entering the GMP lab.

The research analyst explains that she works only two days a week in the cleanroom. The protocol for the tumor cell therapy specifies that the cells must be attended to three days a week: one long day on Wednesday, when the research analyst works with her colleague setting up cell samples or adding fresh cells and counting the cells. They have split the other two days between them; on Monday she goes in, and on Friday her colleague goes in. The rest of the week she is mostly busy with all the administrative work pertaining to the quality assessment that comes along with the hands-on work. Her colleague mostly works for the rest of the week downstairs in the research lab. "To work there is different," the analyst explains. "Downstairs you can work on many cell cultures at the same time, so you can work much faster, while you are not allowed to work on more than one patient's sample at a time in the GMP lab."

I am fascinated by how body parts are turned into pharmaceutical products in the GMP lab and then become reincorporated into the body. Downstairs, in the clinical research laboratory, cells are maintained according to the same protocols but under much less strict regulations: because their cultured cells do not go back into the patients' bodies, they do not need to comply with GMP standards.

Mindful, Caring, and Conscientious Hands

I ask the analyst whether it is dangerous to handle the tumor cells. No, she says, you work in a sterile hood, and you do not actually come into contact with the cells; you wear gloves, and you only touch the flasks and tubes. "What you should not do is to suck up the tumor suspension into an injection syringe and then prick your finger and get a big amount of it into your body," she laughs, "but for the rest it is no problem to work with them. You need to work very carefully with your hands to protect your sample. In the laminar flow cabinet you may never work above an open flask, or sample, because you have a constant airflow blowing downward. You always need to work with your hands around your sample to keep it clean."[44]

She adds that you could also work sterile with your bare hands by rubbing them first with alcohol, "but imagine if you once hovered with your hands above your cells then all your cohabiting bacteria will be blown into your sample and you will never get that clean again. Bacteria grow so much faster than cells, and once your sample is infected, it cannot be saved." Handling patients' cells is very responsible work. In the 1990s she had also worked, for example, with bone marrow transplantations in flow cabinets. She continues, "And then you know that the donor had already endured many interventions, and all the hope of the patient and donor is placed on you." "These were the first transplants done," she explains, "and then we also worked very carefully with our hands, and we sometimes took a look to see how the cells got back into the patient via an infusion, and they were so grateful, and we saw them sitting, hand in hand the donor and recipient . . . that was really emotional."

She tells me another anecdote from the GMP lab that made a deep impression on her. A patient's husband had personally handed the tumor tissue from his wife to her. "He really handed it over to me . . . well, I still get goosebumps when I think about it now. And his daughter was there, too, when he gave it to me, and she took a photo. The man overacted a bit, he was very emotional, saying something like 'this is from my dear wife, take good care of it,' then you really . . . I told him that it is not all up to us, sometimes it works well and other times it doesn't. We really do our best . . . you understand?" She further explains that it is not just the cleanroom work but the idea that these people put so much hope in you: "If you don't do it, no one else will do it, and if you do not get it done . . . well you really hope then that it will work out, that the tumor cells will grow, that the blood [cells] also grow and that the patient is still in shape when all is ready and that all will work out somehow . . . but, well, you do not do anything wrong on purpose." She explains that you never know beforehand whether the cells will proliferate in vitro: "You mince the tumor tissue and put it into a flask. You don't know how the cells will do. With some cells you are busy for three months, and it will not work out. Sometimes it needs some time to get going, and once it works it will be fine, but tumors can also grow very slowly."

Besides the laborious preparatory procedure, it is also strenuous to work alone most of the time in the cleanroom. To prevent cross contamination and mixing up of samples, only one pair of hands is allowed to work with the cells of one patient. Sometimes analysts do work in pairs, but then in separate laminar flow cabinets, each with their own sample. Benchwork in the GMP lab of the Department of Clinical Pharmacy and Toxicology is an attentive task that requires mindful hands, conscientious administration, and diligent practitioners. I experienced how working under such strict sterile conditions creates an awareness of our persistent bodily presence; it also made me aware of my body as a source of contamination. Perhaps more important, though, I learned from the analyst that the bodywork she performs in the cleanroom acutely calls for an "aesthetics of care" in handling patients' cells invested with hope.[45] In the aftermath of a COVID-19 pandemic, research into embodied practices of care and contamination control gains new urgency, but that could be the topic of another book.

Conclusion

This book exemplifies the complexity of the notion of hands-on work in epistemic practices that bring us in touch with life. My investigation brought us from seventeenth-century anatomy to contemporary molecular genetics laboratories and microbiological cleanrooms. My explorations of hands-on/hands-off dynamics in experimental wet-lab practices served as a starting point for epistemological reflections. Some final meditations on cleanroom technologies exemplify my thesis that we have to put aesthetic issues in wet-lab practices that bring us in touch with molecular and cellular life at the heart of epistemological investigations. By claiming such a central position for aesthetics—a theory of the cultural, material, and historical conditions of perception—I argue that we have to rethink epistemology as a pluralistic, historicized, and embodied project. My project feeds into a trend that has gained momentum with concepts such as Hans-Jörg Rheinberger's "historical epistemology" and Annemarie Mol's "body multiple."[46] With this book I demonstrated that reconceptualizing and historicizing epistemology as a pluralistic and contextualized project must start with a rethinking of Descartes's modern epistemological project. I have avoided framing such an undertaking as a move away from epistemology as others have done.[47] Instead, I argue that we need to rethink the historical foundations of Cartesian epistemology in the sense of a *répétition génératrice* that places hands-on aesthetics of *doing* experiments in the life sciences at the core of *doing* epistemology.

256

ACKNOWLEDGMENTS

Many people and institutions contributed to the process and the outcome of this project. To all those involved in making the research for this book possible, I am deeply grateful.

My first thanks go to Robert Zwijnenberg and Miriam van Rijsingen, who invited me to join the New Representational Spaces in Arts and Science Interactions research program as a PhD candidate and to become a member of the Arts and Genomics Centre. The research has been generously funded by the Dutch Research Council. The years we spent engaged in energized and fruitful discussions with Danielle Hofmans, Cor van der Weele, Ellen ter Gast, Helen Chandler, and Anne Kienhuis were formative for this book and my development as a scholar working across disciplinary divides in the humanities, sciences, and arts. Special appreciation goes to Robert Zwijnenberg for the many, many hours spent in conversations, exploring ideas and refining concepts that have been invaluable in shaping the direction of this research and encouraged me to develop my own interdisciplinary approach. I feel fortunate for his support and intellectual companionship throughout the duration of this book project that extended far beyond those formative years.

I am grateful to the researchers and institutes that welcomed me into their laboratories, courses, and working spaces. Without their support and collaboration, this book would not have materialized.

At Leiden University, many thanks go to the lecturers of the Life, Science and Technology program and my former "bench partners," in particular Guido Zeegers. I am especially indebted to the intellectual openness, curiosity, and generosity of the researchers at the former Leiden Institute of Chemistry, in particular Huub de Groot and the members of the molecular genetics research laboratory, Riekje Brandsma, Jaap Brouwer, Nora Goosen, Geri Moolenaar, Tineke de Ruijter, and their former doctoral students.

The interdisciplinary nature of this study led me into many unfamiliar places, and I am grateful to those who have introduced me to their fields of specialized expertise. I owe particular thanks to a number of institutions and experts: at Leiden University Medical Center, Loes van Eijk and Jaap Oostendorp; at the Department of Animals in Science and Society, Utrecht University, the lecturers of the course in laboratory animal science; at the Netherlands Vaccine Institute, Piet de With, for generously sharing his knowledge on specific pathogen-free laboratory animals kept in isolators; at TU Delft, the Kavli Institute of Nanoscience, Emil van der Drift for introducing me to nanotech cleanrooms and explaining differences between microbiological and other types of cleanrooms.

Special thanks go to Hugo Huiskamp for generously sharing his expertise during training sessions and extended conversations on microbiological cleanroom users and for providing access to his own archive on cleanroom training materials.

I am grateful to the organizers and participants of the SymbioticA/Arts Catalyst Biotech Art Workshop at the laboratories of Kings College London for introducing me to interdisciplinary research at the intersection of arts and life sciences with thought-provoking discussions and lively debates.

At SymbioticA—An Artistic Laboratory Dedicated to Research and Hands-On Engagement with the Life Sciences, at the University of Western Australia in Perth, I count myself fortunate to have worked with extraordinary bio artists, who urged me to get my hands wet and inspired me with their work. Many of the ideas in this book have in one way or other benefited from engaged discussions, and, in particular, from hands-on training and experiments with Guy Ben-Ary, Oron Catts, Boo Chapple, and Ionat Zurr. Their hospitality and Joan Coakley's organizational support made my research stay Down Under unforgettable. The visit was generously funded with a two-month travel grant by the Dutch Research Council (MCG grant no. 05032502).

I am also indebted to all the artists who inspired me to think deeply about embodied work in life science laboratories, and especially to Boo Chapple, who became a dear friend and whose careful reading and meticulous editing of earlier versions were instrumental in bringing this research project into a written form. Special thanks to all those who kindly provided me with images of their work for this book.

This book has benefited from conversations and critical engagements at various conferences, workshops, and meetings, including invited talks and presentations at the "Visual Cultures in Art and Science: Rethinking Representational Practices in Contemporary Art and Modern Life Sciences" workshop at Berlin-Brandenburg Academy of Sciences and Humanities in Berlin; the "Close Encounters Science Literature Arts" conference of the Society for Science, Literature and the Arts in Amsterdam; the "Perspectives on the Body and Embodiment" conference at University College Dublin; annual meetings of the Society for Social Studies of Science (Cleveland) and the History of Science Society (Cleveland and San Francisco); the "Experimental Crossings Symposium on Interdisciplinarity in Arts Practices" at Zuyd University in Maastricht, the Netherlands; and the colloquium of the Max Planck Group, "Art and Knowledge in Pre-Modern Europe" during an inspiring one-month stay at the Max Plank Institute for the History of Science in Berlin.

Throughout the years of working on this book—and on other projects that alternately provided pleasant diversion and, at times, unwelcome distraction—I found conducive and congenial research and writing environments at several institutions and departments. I feel fortunate to have shared thoughts and working spaces with colleagues at the Faculty of Arts and Social Sciences of Maastricht University, the Leiden Institute for Arts in Society at Leiden University, the Netherlands Graduate School of Science, Technology and Modern Culture (WTMC), the Royal Netherlands Academy of Arts and Sciences, the Vrije Universiteit Amsterdam, and the Faculty of Humanities both at the University of Amsterdam and at Utrecht University.

Over the years, many formal and informal conversations have deepened and broadened my thinking about hands-on knowledge making beyond the scope of this book: special thanks go to Pamela Smith and the members of the Making and Knowing Project at the Department of History of Columbia University in New York, and to Sven Dupré and my former colleagues from the ERC Artechne Project at Utrecht University for providing an inspiring intellectual home, a congenial and supportive atmosphere, and excellent working conditions for several years. Thanks also to Meredith Levin from Columbia University Libraries for helping with translations of essential quotations.

259

Warm thanks go to friends and colleagues for their good company during different writing phases of this project, especially Hieke Huistra and Birgit Reissland for cheering me through the last stages of finalizing the manuscript.

I am grateful to all the interlocutors and readers in and out of academia who commented on ideas, chapters, and drafts for this book.

Many thanks to Joseph Dumit and Michael M. J. Fischer for their feedback which helped me to fully develop my arguments and make my ideas more concise. At Duke University Press, I want to thank Kenneth Wissoker and two anonymous readers who carefully read and reviewed entire drafts for supporting this project. Their perceptive comments and my editors' sound advice have been encouraging throughout, as well as crucial to my completing this monograph in its present state. Many thanks to the editorial team, and, in particular, Kate Mullen and Livia Tenzer, for excellent editorial support in finalizing the manuscript. Special thanks go to Susan Ecklund for meticulous and insightful copyediting; and credit goes to Linda Hallinger for creating the index. Needless to say, the responsibility for any remaining errors in these pages is mine alone.

Finally, I owe an enormous debt to friends and family who helped keep me engaged in this project over the course of many years. Special thanks to my husband, Robert Scheers, who kept encouraging me to write about my own observations and hands-on experiences, and to my father, Guido Boulboullé, for reading many iterations of the entire manuscript.

This book is dedicated to my parents, Anja and Guido, and to Robert, Noor, and Pim, who have been there every step of the way. Thank you for all your love and support.

260

Introduction

Epigraph sources: On p. 8, E. F. Keller, *Feeling for the Organism*, 125. On p. 13, E. F. Keller, *Feeling for the Organism*, 117. On p. 16, Mulligan, "Uplands Farm"; E. F. Keller, *Feeling for the Organism*, 57. On p. 24, Forssberg, "Award Ceremony Speech." On p. 26, Nishikawa, "Japan's Latest Nobel Laureate."

1   The fieldwork for this book was mainly conducted at the Gorlaeus Laboratories, Leiden University, over a period of about a year and a half. It included coursework (552 hours) with practicums selected from the Life Science and Technology curriculum and practical benchwork during an internship at a molecular genetics research laboratory. At Utrecht University, I also participated in a two-week laboratory animal science course (80 hours). The fieldwork included visits and interviews at the isolator facilities for specific pathogen-free animals at the Netherlands Vaccine Institute; at the cleanroom facilities at Kavli Institute for Nanoscience, TU Delft, and at the Leiden University Medical Center; interviews and prolonged conversations with a cleanroom consultant and trainer; and participation in a Cleanroom Behavior Course for microbiological cleanroom users.

My field explorations also included participation in bio art workshops in London, Amsterdam, and Leiden, and a research stay at SymbioticA—An Artistic Laboratory Dedicated to Research and Hands-On Engagement with the Life Sciences, at the University of Western Australia in Perth, in November and December 2006, where artists introduced me to tissue engineering techniques in the research laboratories of the School of Anatomy and Human Biology, and where I worked for the first time with living cells in a flow cabinet.

My field studies benefited greatly from the lively discussions with researchers and bio artists I engaged in as a member of the interdisciplinary research project New Representational Spaces: Investigations of Interactions between and Intersections of Art and Genomics, led by Robert Zwijnenberg and Miriam van Rijsingen.

2   I thank René Peschar for his help with translating the chemical terms of this Dutch protocol. All translations are my own unless otherwise indicated. For information on the translations of works by René Descartes that appear in this book, see the "Note on Descartes's Texts and Their Translation."

3   See, e.g., Walls, "Kayak Games and Hunting Enskilment"; Grasseni, *Skilled Visions*; Grasseni, "Skilled Visions"; Mauss, "Techniques of the Body."

4   On adaptation processes of instruments and experimenters in history of science, see, e.g., Hötteke, "Zur experimentellen Tätigkeit Michael Faradays," 363–64.

5   Oxford English Dictionary, s.v. "pipettor (n.)," July 2023, https://doi.org /10.1093/OED/1788160077.

6   E. F. Keller, *Feeling for the Organism*.

7   See, e.g., the following books by E. F. Keller: *Reflections on Gender and Science*; *Refiguring Life*; *Century of the Gene*; and *Making Sense of Life*.

8   See, e.g., E. F. Keller, *Feeling for the Organism*, 148–49, 197, and on "new biology" esp. 153–70; Landecker, *Culturing Life*; Chadarevian and Kamminga, *Molecularizing Biology and Medicine*; Kay, *Molecular Vision of Life*.

9   E. F. Keller, *Feeling for the Organism*, 36.

10   E. F. Keller, *Feeling for the Organism*, 36.

11   E. F. Keller, *Feeling for the Organism*, 35.

12   Nanney, "Role of the Cytoplasm in Heredity," quoted in E. F. Keller, *Making Sense of Life*, 151.

13   On the fuzzy concept of the gene, see, e.g., Lenoir, "Epistemology Historicized," xviii; on Watson's notion of the "molecular object," see Watson, *Molecular Biology of the Gene*, cited in E. F. Keller, *Feeling for the Organism*, 159; Watson and Crick, "Molecular Structure of Nucleic Acids."

14   Lawler, "Rosalind Franklin Still Doesn't Get the Recognition She Deserves."

15   Gann and Witkowski, "DNA."

16   Pietzsch, "What Is Life?"

17   Max Delbrück, "A Physicist Looks at Biology" [1949], in *Phage and the Origins of Molecular Biology*, ed. John Cairns, Gunther Stent, and James Watson (Cold Spring Harbor, NY: Cold Spring Harbor Laboratory of Quantitative Biology, 1966), 22; cited in E. F. Keller, *Feeling for the Organism*, 165, and in Strauss, "Physicist's Quest in Biology," 647.

18   E. F. Keller, *Feeling for the Organism*, 160.

19   E. F. Keller, *Feeling for the Organism*, 6 and 160.

20   Delbrück, "Experiments with Bacterial Viruses (Bacteriophages)," cited in E. F. Keller, *Feeling for the Organism*, 161.

21   Delbrück, "Experiments with Bacterial Viruses (Bacteriophages)," cited in E. F. Keller, *Feeling for the Organism*, 161.

22   Delbrück, "Experiments with Bacterial Viruses (Bacteriophages)," cited in E. F. Keller, *Feeling for the Organism*, 161.

23   On Delbrück's attitude toward biochemical wet lab work, see, e.g., Strauss, "Physicist's Quest in Biology," 648.

24   Chadarevian and Kamminga, *Molecularizing Biology and Medicine*; Kay, *Molecular Vision of Life*; Rose, *Politics of Life Itself*, 5.

25   E. F. Keller, *Feeling for the Organism*, 94.

26   Watson and Crick, "Molecular Structure of Nucleic Acids."

27   Darlington, *Recent Advances in Cytology*; Kofoid, review of *Recent Advances in Cytology*.

28   E. F. Keller, *Feeling for the Organism*, 90–91.

29   The opening address is reprinted in Darlington and Lewis, *Chromosomes Today*, vol. 1; quoted in E. F. Keller, *Feeling for the Organism*, 91.

30   E. F. Keller, *Feeling for the Organism*, 92.

31   E. F. Keller, *Feeling for the Organism*, 94.

32   E. F. Keller, *Feeling for the Organism*, 147.

33   E. F. Keller, *Feeling for the Organism*, 92.

34   E. F. Keller, *Feeling for the Organism*, esp. 115–17, 198–200.

35   E. F. Keller, *Feeling for the Organism*, 117.

36   E. F. Keller, *Feeling for the Organism*, 200.

37   E. F. Keller, *Feeling for the Organism*, 198.

38   E. F. Keller, *Feeling for the Organism*, 200.

39   E. F. Keller, *Feeling for the Organism*, 101.

40   Witkin, "Remembering Rollin Hotchkiss," 143; on Hotchkiss's work with the bacteriologist René Dubos (1901–1982), see Van Epps, "René Dubos."

41   On the notion of "skilled vision," see Grasseni, *Skilled Visions*; Grasseni, "Skilled Visions." On Hotchkiss's recollections, see his foreword in E. F. Keller, *Feeling for the Organism*, xiv.

42   E. F. Keller, *Feeling for the Organism*, 116–17.

43   E. F. Keller, *Feeling for the Organism*, 113.

44   E. F. Keller, *Feeling for the Organism*, esp. 101 and also 67, 80, 103.

45   On McClintock's staining techniques, see E. F. Keller, *Feeling for the Organism*, 130.

46   E. F. Keller, *Feeling for the Organism*, 40.

47   E. F. Keller, *Feeling for the Organism*, 103.

48   E. F. Keller, *Feeling for the Organism*, 137.

49   E. F. Keller, *Feeling for the Organism*, 69.

50   E. F. Keller, *Feeling for the Organism*, 104.

51   E. F. Keller, *Feeling for the Organism*, 200.

52   E. F. Keller, *Feeling for the Organism*, 80.

53   E. F. Keller, *Feeling for the Organism*, 80.

54   E. F. Keller, *Feeling for the Organism*, 46, 148.

55   Nobel Foundation, "Press Release."

56   Ringertz, "Award Ceremony Speech."

57   Nobel Foundation, "Press Release."

58   Ringertz, "Award Ceremony Speech."

59   Ringertz, "Award Ceremony Speech."

60   "The inheritance of a series of characteristics can *easily be studied simply by examining* the structure, starch content or pigmentation of the individual kernels. Mutations affecting pigmentation are particularly useful, not only because they can be *easily observed*. . . . Another advantage of maize as an experimental system was that individual chromosomes are *easily studied*"; Nobel Foundation, "Press Release" (my emphasis).

61   E. F. Keller, *Feeling for the Organism*, 198.

62   E. F. Keller, *Feeling for the Organism*, 198 (my emphasis).

63   E. F. Keller, *Feeling for the Organism*, 198.

64   E. F. Keller, *Feeling for the Organism*, 198.

65   E. F. Keller, *Feeling for the Organism*, 198.

66   E. F. Keller, *Feeling for the Organism*, 198.

67 E. F. Keller, *Feeling for the Organism*, 80.

68 E. F. Keller, *Feeling for the Organism*, 74.

69 Awiakta, "How the Corn-Mother Became a Teacher of Wisdom," 200; Ducey, "Technologies of Caring Labor."

70 Nobel Foundation, "Press Release."

71 Mulligan, "Uplands Farm."

72 Mulligan, "Uplands Farm."

73 Blackburn, "Elizabeth Blackburn on Barbara McClintock," 0:29–0:38. On Blackburn, see Nobel Foundation, "Elizabeth H. Blackburn."

74 Blackburn, "Elizabeth Blackburn on Barbara McClintock," 1:00–1:05.

75 Nobel Foundation, "Elizabeth H. Blackburn"; Blackburn, "Elizabeth Blackburn on Barbara McClintock," 0:00–1:15.

76 E. F. Keller, *Feeling for the Organism*, 200–201.

77 Such an idealized disembodied and immaterial conception of science has incited many critical responses; see, e.g., Hacking, *Representing and Intervening*; Lawrence and Shapin, *Science Incarnate*; Haraway, *Modest_ Witness@Second_Millennium*; Haraway, *Companion Species Manifesto*; Haraway, *When Species Meet*; Rheinberger, *Epistemology of the Concrete*; Rheinberger, *Toward a History of Epistemic Things*; Barad, *Meeting the Universe Halfway*; Roberts, Schaffer, and Dear, *Mindful Hand*; Daston, *Biographies of Scientific Objects*; Daston and Galison, *Objectivity*. A critical stance toward universalist and reductionist conceptions of science is also reflected in an emerging field focusing on history and formation of knowledge rather than history of science; see, e.g., Daston, "History of Science"; Burke, *What Is the History of Knowledge?* and "Response"; and the peer-reviewed journals KNOW—*A Journal on the Formation of Knowledge* and *Journal for the History of Knowledge*, launched in 2017 and 2019, respectively.

78 E. F. Keller, *Feeling for the Organism*, 147.

79 E. F. Keller, *Feeling for the Organism*, xxii.

80 E. F. Keller, *Feeling for the Organism*, 148, 150.

81 E. F. Keller, *Feeling for the Organism*, 148.

82 Emerson, *Essays and Lectures*, 10, cited in E. F. Keller, *Feeling for the Organism*, 118.

83 E. F. Keller, *Feeling for the Organism*, 118.

84 E. F. Keller, *Feeling for the Organism*, 118.

85 Cerasuolo, *Literature and Artistic Practice*, 69.

86 Not much textual evidence has survived on the training of early Renaissance painters, but we can find in a manuscript copy of Cennino

Cennini's famous *Libro dell'Arte*, dated to the end of the fourteenth century, a description of a painter's extensive apprenticeship; Broecke, *Cennino Cennini's Il Libro dell'Arte*, 138.

87   Zwijnenberg, *Writings and Drawings of Leonardo da Vinci*, 29.

88   "Oder meynen Sie, Prinz, daß Raphael nicht das größte malerische Genie gewesen wäre, wenn er unglücklicher Weise ohne Hände wäre geboren worden?" Lessing, *Emilia Galotti*, 13.

89   Zwijnenberg, *Writings and Drawings of Leonardo da Vinci*, 24, 37. The first decades of the 2000s witnessed a renewed interest in artistic technique and material aspects of artworks and the rise of a new materialism in the humanities; see, for example, two major European research initiatives: Sven Dupré (principal investigator), "ERC ARTECHNE Project," https://artechne.wp.hum.uu.nl; Iris Van der Tuin (chair), "COST Action IS1307 New Materialism: Networking European Scholarship on 'How Matter Comes to Matter,'" https://newmaterialism.eu/.

90   Shapin, *Scientific Revolution*, 4.

91   Lawrence and Shapin, "Introduction: The Body of Knowledge"; Shapin, *Scientific Revolution*, 4; Pickstone, *Ways of Knowing*, 17.

92   Pickstone, *Ways of Knowing*, 17; see also, e.g., Cook, "History of Medicine"; Lawrence and Shapin, *Science Incarnate*; Roberts, Schaffer, and Dear, *Mindful Hand*; P. H. Smith, *Body of the Artisan*; Wolfe and Gal, *Body as Object*.

93   Knorr-Cetina, *Epistemic Cultures*; Myers and Dumit, "Haptics"; Myers, *Rendering Life Molecular*.

94   Mol, *Body Multiple*, 82.

95   For an introductory bibliography, see, e.g., the website of the Embodied Cognition Reading Group, cofounded in 2015 by the author under the auspices of Columbia University's Center for Science and Society, http://blogs.cuit.columbia.edu/embodiedcognition/.

96   Forssberg, "Award Ceremony Speech."

97   Nishikawa, "Japan's Latest Nobel Laureate"; AP News, "The Latest: Discovery on Golf Course Inspired Nobel Winner."

98   See, e.g., Ōmura, "Ivermectin Story."

99   Ōmura, "Philosophy of New Drug Discovery," 259.

100   Ōmura, "Philosophy of New Drug Discovery," 260.

101   Ōmura, "Philosophy of New Drug Discovery," 260.

102   Ōmura, "Philosophy of New Drug Discovery," 259, 268.

103   On the notion of "experimental systems," see Rheinberger, *Toward a History of Epistemic Things*.

104   See AP News, "The Latest: Discovery on Golf Course Inspired Nobel Winner"; Altman, "Nobel Prize in Medicine."

105   E. F. Keller, *Feeling for the Organism*, 189–90; Fink et al., "Transposable Elements (Ty) in Yeast."

106   Alberts, *Molecular Biology of the Cell*. The first edition was published in 1983.

107   On the notion of skilled vision, see Grasseni, *Skilled Visions*. For ethnographies of skilled practices in medical and molecular biology contexts, see, e.g., Prentice, *Bodies in Formation*; Myers, *Rendering Life Molecular*.

108   See, e.g., Fleck, *Genesis and Development of a Scientific Fact*; Hacking, *Representing and Intervening*; Lawrence and Shapin, *Science Incarnate*; Latour, *Science in Action*; Pickering, *Constructing Quarks*; Pickering, *Science as Practice and Culture*; Pickering, *Mangle of Practice*.

109   For these Latin editions, see the entries in the bibliography with "*Renati Descartes*" (of René Descartes) at the start of the title.

110   On the notion of multisited or multilocale ethnographies, see, e.g., Fischer, *Anthropology in the Meantime*, 9. On historical ethnography, see, e.g., Harkness, *Jewel House*, 254–60.

111   Husserl, *Krisis*, 50.

112   See especially Slatman, *Our Strange Body*.

113   On transparency versus opacity in life science research, see also Myers, *Rendering Life Molecular*, 231.

114   Ihde, *Technics and Praxis*; Ihde, *Experimental Phenomenology*; Ihde, *Technology and the Lifeworld*; Ihde, *Instrumental Realism*; Ihde, *Bodies in Technology*; Gallagher, *How the Body Shapes the Mind*.

115   Mol, *Body Multiple*, viii; Mol, "Pathology and the Clinic," 84.

116   For further readings on embodied practices and experimental sciences, see, e.g., Doing, "'Lab Hands' and the 'Scarlet O'"; Hentschel, *Unsichtbare Hände*; Mascia-Lees, *Companion to the Anthropology of the Body*; Mol, *Body Multiple*; Myers and Dumit, "Haptics"; Myers, "Molecular Embodiments"; Myers, *Rendering Life Molecular*; Prentice, *Bodies in Formation*; Polanyi, *Personal Knowledge*; P. H. Smith, *Body of the Artisan*; Wolfe and Gal, *Body as Object*. On anthropological methods for the study of life sciences, biotechnologies, biomedicine, and artistic investigation of emergent life-forms, see esp. Fischer, *Emergent Forms of Life*; Fischer, *Anthropological Futures*; Fischer, "Biopolis"; Fischer, "Science and Technology"; Fischer, *Anthropology in the Meantime*.

Chapter 1. Knowing by *Experience*

1   On the concept and historiography of the seventeenth-century Scientific Revolution, see, e.g., Cohen, *How Modern Science Came into the World*; Teich, *Scientific Revolution Revisited*; Shapin, *Scientific Revolution*. For earlier dating, see, e.g., Neher, *David Gans* (English translation: Neher, *Jewish Thought and the Scientific Revolution*). More recent scholarship criticizes an overtly Eurocentric approach and provides a broader and global perspective; see, e.g., Barrera-Osorio, *Experiencing Nature*; Hasse, *Success and Suppression*; Huff, *Intellectual Curiosity and the Scientific Revolution*; Seed, *Contesting Possession*.

2   I use "new science" and "new scientists," hereafter without quotation marks, to refer to a form of natural inquiry through experiment and to seventeenth-century natural philosophers who understood themselves as adepts of an experimental approach.

3   On premodern natural inquiries, see, e.g., Daston and Park, *Wonders and the Order of Nature*.

4   McAllister, "Virtual Laboratory," 36.

5   Schramm, Schwarte, and Lazardzig, *Collection, Laboratory, Theater*. On theatrical strategies of premodern natural philosophers, see, e.g., Kodera, "Laboratory as Stage."

6   McAllister, "Virtual Laboratory," 38.

7   Schramm, Schwarte, and Lazardzig, *Collection, Laboratory, Theater*, xix.

8   Nelle, "Eucharist and Experiment."

9   Spedding, Ellis, and Heath, "Great Instauration: Plan of the Work," 26, quoted in Nelle, "Eucharist and Experiment," 32. Bacon published *Instauratio magna* (The great instauration) in 1620.

10   McAllister, "Virtual Laboratory," 37.

11   McAllister, "Virtual Laboratory," 37.

12   Nelle, "Eucharist and Experiment," 332.

13   McAllister, "Virtual Laboratory," 44. On the modification of the Scholastic notion of "experience," see also Dear, "Jesuit Mathematical Science." On the distinction between "old empiricism of experience" and "the new empiricism of facts," see Daston and Park, *Wonders and the Order of Nature*, 236–37.

14   Quoted in Drake, *Galileo: Two New Sciences*, 169–70. On a reconstruction of this experiment and discussion about whether Galileo actually performed it, see, e.g., Riess, Heering, and Nawrath, "Reconstructing Galileo's Inclined Plane Experiments."

15   Nelle, "Eucharist and Experiment," 333.

16 McAllister, "Virtual Laboratory," 42.

17 See, e.g., Robert Boyle's (1627–1691) and Thomas Hobbes's (1588–1679) discussion on the leakiness of the air pump and the validity of "elaborate" versus "obvious" experiments in Shapin and Schaffer, *Leviathan and the Air-Pump*, 174–75.

18 Gipper, "Experiment und Öffentlichkeit," 246.

19 Gouk, *Music, Science, and Natural Magic*, 171.

20 Popkin, *Pimlico History of Western Philosophy*, 331–32.

21 Mersenne, *Harmonie universelle*.

22 Gouk, *Music, Science, and Natural Magic*.

23 Klotz, "Vibration und Vernunft," 281.

24 "Il est donc à propos auant que passer outre de sçavoir si le Son, qui est le suiet, ou l'obiet de la Musique & de l'ouye, a un estre reel, & quel il est: car il s'en trouue plusieurs qui croyent que le Son n'est rien, s'il n'est entendu, & que c'est vne simple impression de l'air qui ne doit point estre appellée Son, s'il n'y a quelque oreille qui l'entende & qui la distingue d'avec les autres choses." Mersenne, *Harmonie universelle*, 1; partly cited in Klotz, "Vibration und Vernunft," 381.

25 On the relation between natural philosophy and magic, marvels, and wonders, see, e.g., Daston and Park, *Wonders and the Order of Nature*; Gouk, *Music, Science, and Natural Magic*; Kodera, "Laboratory as Stage."

26 For more on Mersenne's acoustic experiments, see Dear, *Mersenne and the Learning of the Schools*; Gouk, *Music, Science, and Natural Magic*; Klotz, "Vibration und Vernunft"; for more on early modern musical experiments, see, e.g., Cohen, *Quantifying Music*; Coelho, *Music and Science in the Age of Galileo*; Cypess, "Giovanni Battista Della Porta's Experiments"; Drake, "Renaissance Music and Experimental Science"; Gozza, *Number to Sound*; Mancosu, "Acoustics and Optics"; Moyer, *Musica Scientia*; Wardhaugh, *Music, Experiment and Mathematics in England*. **269**

27 Klotz, "Vibration und Vernunft," 285.

28 "Mais les experiences sont si difficiles qu'à moins d'vne chorde de mille pieds on ne peut s'en asseurer; & l'on n'est iamais si certain des endroits où elle reuient à chacque tour, que l'on ne puisse douter si elle n'a point passé outre, & si elle a iustement terminé ses allées & ses venuës aux points que l'on marque." Mersenne, *Harmonie universelle*, 162.

29 Klotz, "Vibration und Vernunft."

30 "De sorte qu'il est tousiours necessaire que la raison supplée quelque chose dans les experiences, qui seules ne peuuent seruir de principes pour les sciences, qui désirent vne parfaite iustesse, que les sens ne peuuent remarquer." Mersenne, *Harmonie universelle*, 162.

31   Klotz, "Vibration und Vernunft," 281.

32   Klotz, "Vibration und Vernunft," 280–81.

33   On "produce," see Klein, "Introduction: Technoscientific Productivity"; Klein, "Styles of Experimentation," 180; on "engender," see Lefèvre, "Science as Labor"; on "enact," see Mol, *Body Multiple*; on "enhance," see Knorr-Cetina, *Epistemic Cultures*.

34   See also Shapin and Schaffer, *Leviathan and the Air-Pump*, on making scientific concepts of imperceptible phenomena into public sensations.

35   See Lefèvre, "Science as Labor."

36   Roberts, Schaffer, and Dear, *Mindful Hand*, xxi.

37   McAllister, " Virtual Laboratory," 44.

38   Nelle, "Eucharist and Experiment," 333.

39   Nelle, "Eucharist and Experiment," 332; Shapin and Schaffer, *Leviathan and the Air-Pump*.

40   Lefèvre, "Science as Labor."

41   Nelle, "Eucharist and Experiment," 332.

42   Nelle, "Eucharist and Experiment," 331.

43   Hausken, introduction to *Thinking Media Aesthetics*, 30. The research field of media aesthetics provides a very useful definition of "aesthetics as a critical reflection on cultural expressions, on technologies of the senses and on the experiences of everyday life" (30).

44   See Lefèvre, "Science as Labor," 214.

45   On use of these metaphors in the media, see, e.g., the opening announcement of the podium discussion "Synthetische Biologie—Leben aus dem Baukasten?" (Synthetic biology—Life out of a construction kit?), Berlin-Brandenburgische Akademie der Wissenschaften, Berlin, February 24, 2010; Lubbadeh, "Synthetische Biologie."

46   Not surprisingly, this area of research attracted the attention of bio artists; see, e.g., an event organized in London in 2011: "Synthesis: Synthetic Biology in Art and Society; A Week-Long Interdisciplinary Exchange Lab and Series of Public Events," accessed May 5, 2019, https://www.artscatalyst.org/synthesis-synthetic-biology-art-society.

47   On the notions of experimentalization and molecularization, see, e.g., Chadarevian and Kamminga, *Molecularizing Biology and Medicine*; Kay, *Molecular Vision of Life*; Rheinberger and Hagner, *Die Experimentalisierung des Lebens*.

48   See also my discussion of the invention of automatic microliter pipettes in chapter 5.

49   See, e.g., Rabinow, *Making PCR*; Landecker, *Culturing Life*.

50   Gipper, "Experiment und Öffentlichkeit."

51   Gipper, "Experiment und Öffentlichkeit," 244n4.

52   Gipper, "Experiment und Öffentlichkeit," 243.

53   Shapin, "Invisible Technician," 556. See also Iliffe, "Technicians," especially
      4n3, with an overview of literature in social studies of science that exam-
      ines "the highly skilled, local and embodied nature of *scientific work*."

54   Shapin, "Invisible Technician," 556.

55   Gouk, *Music, Science, and Natural Magic*, 10–11.

56   On invisible assistants in the history of experimental sciences, see also
      Dupré et al., *Reconstruction, Replication and Re-enactment*, 14; Fors,
      Principe, and Sibum, "From the Library to the Laboratory," 88; Sibum,
      "Experimental History of Science," 81.

57   Shapin, "Invisible Technician," 563.

58   See, e.g., D. Clarke, *Descartes' Philosophy of Science*; Garber, *Descartes
      Embodied*.

59   Gipper, "Experiment und Öffentlichkeit," 242.

60   Here I modified the English translation of Descartes's *Meditations on First
      Philosophy* by John Cottingham, Robert Stoothoff, and Dugald Murdoch
      in Descartes, *The Philosophical Writings*, 1:20, based on retranslating the
      French original, which I consulted in Descartes, *Oeuvres de Descartes*,
      IX:23–24, edited by Charles Adam and Paul Tannery; and in the original
      print edition, Descartes, *Les méditations métaphysiques* (1647), 26–27
      (figures 1.2 and 1.3). For all works by Descartes cited in this book, see
      the first section of the bibliography, "Works by René Descartes."
          Hereafter, translations from Descartes, *The Philosophical Writings*,
      vols. 1 and 2, are cited as CSM 1–2. Translations from Descartes, *The
      Philosophical Writings*, vol. 3, are cited as CSMK. The original texts of
      Descartes's writings are cited from Descartes, *Oeuvres de Descartes*, 11
      vols., as AT I–XI. The first editions of Descartes's works are cited with a
      shortened title and the year of publication. See my "Note on Descartes's
      Texts and Their Translation" at the front of this book.
          For an overview and chronology of Descartes's drafts and the first
      editions of his written works, see, e.g., Descartes, *World and Other Writ-
      ings*, xxx–xxxii.

61   Scholars have situated Descartes's work within a period of major epis-
      temological and methodological transitions; see, e.g., D. Clarke, "Des-
      cartes' Philosophy of Science and the Scientific Revolution," 25.

62   D. Clarke, "Descartes' Philosophy of Science and the Scientific Revolu-
      tion," 258; Garber, *Descartes Embodied*, 2; Popkin, *Pimlico History of
      Western Philosophy*, 337.

63   Gouk, *Music, Science, and Natural Magic*, 10; Clarke, "Descartes' Philosophy of Science and the Scientific Revolution," 258.

64   Gouk, *Music, Science, and Natural Magic*, 171.

65   Gouk, *Music, Science, and Natural Magic*, 175.

66   E.g., Gipper, "Experiment und Öffentlichkeit," 243.

67   See, e.g., Gouk, *Music, Science, and Natural Magic*; Gipper, "Experiment und Öffentlichkeit."

68   Descartes, *Oeuvres de Descartes* (AT I–XI). The extant correspondence comprises 575 letters written by Descartes; see Verbeek and Bos, "Conceiving the Invisible," 163.

69   D. Clarke, *Descartes' Philosophy of Science*, 1, lists as "classics" only *Rules for the Direction of the Mind* and *Discourse on the Method*.

70   Descartes, *Discours de la méthode*, AT VI, 1–78; *Les méditations métaphysiques*, AT IX, 1–72, excluding objections and responses; *Regulae ad directionem ingenii*, AT X, 359–469.

71   D. Clarke, *Descartes' Philosophy of Science*, 1.

72   Garber, *Descartes Embodied*, 85; Hatfield, "The Senses and the Fleshless Eye," 66.

73   D. Clarke, *Descartes' Philosophy of Science*, 2.

74   Garber, *Descartes Embodied*, editor's note.

75   Koyré, *Descartes und die Scholastik*.

76   D. Clarke, "Descartes' Philosophy of Science and the Scientific Revolution," 258.

77   Fancher, *Pioneers of Psychology*, 5; Cook, *Young Descartes*.

78   Garber, *Descartes Embodied*, 1, 9.

79   Snyder, "William Whewell."

80   Gouk, *Music, Science, and Natural Magic*, 10.

81   Musicology encompasses subdisciplines focusing on the embodied performance of musicians. On embodied scientists, see, e.g., Lawrence and Shapin, *Science Incarnate*.

82   The letter has been tentatively attributed to Constantijn Huygens, who had worked with Descartes on optical experiments. Descartes, letter to [Huygens?], June 1645?, AT IV, 222–26; editorial comment, 223. See also Dijksterhuis, *Lenses and Waves*, 56; Dijksterhuis, "Constructive Thinking," 65.

83   "Ce que ie trouue le plus étrange est la conclusion du iugement que vous m'auez enouyé, à sçauoir que ce qui empeschera mes principe d'estre receus dans l'Escole, est qu'ils ne sont pas assez confirmez par l'experience, & que ie n'ay point refuté les raisons des autres. Car i'admire que, nonobstant

que i'aye demontré, en particulier, presque autant d'experiences qu'il y a de lignes en mes écrits, & qu'ayant generalement rendu raison, dans mes Principes, de tous les Phainomenes de la nature, i'aye expliqué, par mesme moyen, toutes les experiences que peuuent estre faites touchant les cors inanimez, & qu'au contraire on n en ait iamais bien expliqué aucune par les principes de la Philosophie vulgaire, ceux qui la suiuent ne laissent pas de m'obiecter le défault d'experiences"; Descartes, letter to [Huygens?], June 1645?, AT IV, 224–25. My translation adapts those in CSMK, 252, and Garber, *Descartes Embodied*, 308n31, and leaves the French term *experience* untranslated. The modern French spelling of *expérience* with an acute accent had not yet been standardized in Descartes's time. In the first French edition of the *Meditations* (*Les méditations métaphysiques*, 1647) and in Adam and Tannery's transcription of Descartes's works in Descartes, *Oeuvre de Descartes* (AT I–XI), the seventeenth-century French term *experience* is spelled without an accent. I put *experience* in italics to distinguish the French from the English term and to indicate that I refer to Descartes's use of this term with its broad connotations of experience and experiment.

84   D. Clarke, *Descartes' Philosophy of Science*, 22.

85   *Experiences* is translated as "observations" in Descartes, letter to [Huygens?], June 1645?, CSMK, 252, and as "experiments" in Garber, *Descartes Embodied*, 308n31. For more on the use of the terms *experience* and *experiment* in Descartes's writings, see, e.g., D. Clarke, *Descartes' Philosophy of Science*, 17–46; Cottingham, *Descartes Dictionary*, 59–60. Cottingham, Stoothoff, and Murdoch provide the following explanation of their choice to translate the French term as "observations": "Fr. *expériences*, a term which Descartes often uses when talking of scientific observations, and which sometimes comes close to meaning 'experiments' in the modern sense (its root being derived from Lat. *experior*, 'to test')"; CSM 1, 143n1. Such an explanation has been problematized by historians and philosophers of science; see, e.g., Daston and Lunbeck, *Histories of Scientific Observation*; Daston and Gallison, *Objectivity*.

86   D. Clarke, *Descartes' Philosophy of Science*, 24.

87   D. Clarke, *Descartes' Philosophy of Science*, 23.

88   D. Clarke, *Descartes' Philosophy of Science*, 23.

89   "Mesme ie remarquois, touchant les experiences, qu'elles sont d'autant plus necessaires, qu'on est plus auancé en connoissance"; Descartes, *Discours de la méthode*, AT VI, 63. In Cottingham, *Descartes Dictionary*, 59, translated as "The further we advance in our knowledge, the greater is the necessity for experiences"; in D. Clarke, "Descartes' Philosophy of Science and the Scientific Revolution," 289, translated as "observations."

273

90 "Car, pour le commencement, il vault mieux ne se seruir que de celles qui se presentent d'elles mesmes a nos sens, & que nous ne sçaurions ignorer, pouruû que nous y facions tant soit peu de reflexion, que d'en chercher de plus rares & estudiées: dont la raison est que ces plus rares trompent souuent, lorsqu'on ne sçait pas encore les causes des plus communes, & que les circonstances dont elles dependent sont quasi tousiours si particulieres & si petites, qu'il est tres malaysé de les remarquer." Descartes, *Discours de la méthode*, AT VI, 63; translation, Cottingham, *Descartes Dictionary*, 60.

91 "& qu'on se serue de plusieurs experiences particulieres"; Descartes, *Discours de la méthode*, AT VI, 64. Cottingham, *Descartes Dictionary*, 60, translates the passage with "experiences" and "special observations"; Descartes, *Discourse on the Method*, CSM 1, 144, translates it as "observations."

92 D. Clarke, *Descartes' Philosophy of Science*, 22, distinguishes between different subcategories of observation: "The first of these sub-categories of observation will be called 'ordinary experience,' while the second is normally called an experiment."

93 "Mais ie voy aussy qu'elles [les experiences] sont telles, & en si grand nombre, que ny mes mains, ny mon reuenu, bien que i'en eusse mille fois plus que ie n'en ay, ne sçauroient suffire pour toutes; en sorte que, selon que i'auray desormais la commodité d'en faire plus ou moins, i'auanceray aussy plus ou moins en la connoissiance de la Nature"; Descartes, *Discours de la méthode*, AT VI, 65. In Descartes, *Discourse on the Method*, CSM 1, 144, "mes mains" is translated as "dexterity." See my discussion of this problematic choice in chapter 2.

94 See the editorial comments by Cottingham, Stoothoff, and Murdoch in Descartes, *Passions of the Soul*, CSM 1, 326; and by Adam and Tannery in Descartes, *Les passions de l'âme*, AT XI, 293. On Picot, see K. Smith, *Descartes Dictionary*.

95 "Mesme à cause que ces experiences sont de deux sortes: les unes faciles, & qui ne dependent que de la reflexion qu'on fait sur les choses qui ses presentent au sens d'elles mesmes; les autres plus rares & difficiles, auxquelles on ne parvient point sans quelque estude & quelque despense." Descartes, *Les passions de l'âme*, AT XI, 319; translation, D. Clarke, *Descartes' Philosophy of Science*, 23.

96 D. Clarke, *Descartes' Philosophy of Science*, 40.

97 D. Clarke, *Descartes' Philosophy of Science*, 272; e.g., Descartes, *Discours de la méthode*, AT VI, 33.

98 D. Clarke, *Descartes' Philosophy of Science*, 272.

99 In the Second Meditation, "the act with which one perceives" is described as "an inspection of the mind that can first be imperfect and confused, as it is before, or clear and distinct as it is now, depending on how my atten-

tion is directed more or less on the things themselves, and on what they consist of." That is, the perceptual act can become clear and distinct as result of a *philosophical introspection*. "Vne inspection de l'esprit, laquelle peut estre imparfaite & confuse, comme elle estoit auparauant, ou bien claire & distincte, comme elle est à present, sélon que mon attention se porte plus ou moins aux choses qui sont en elle, & dont elle est composée"; Descartes, *Les méditations métaphysiques*, AT IX, 25. My translation is adapted from Descartes, *Meditations on First Philosophy*, CSM 2, 21.

100   Garber, *Descartes Embodied*, 5.

101   Garber, *Descartes Embodied*, 4.

102   D. Clarke, *Descartes' Philosophy of Science*, 6.

103   Pickstone, *Ways of Knowing*.

104   For other examples of historically and philosophically informed studies of contemporary scientific practices, see, e.g., Rijcke, "Regarding the Brain"; Rheinberger, *On Historicizing Epistemology*.

105   The first edition was published by Johannes Oporinus in what is today Switzerland (Basel, 1543). On the edition and reception history of Vesalius's *Fabrica*, see Margócsy, Somos, and Joffe, *Fabrica of Andreas Vesalius*.

106   Gaukroger in Descartes, *World and Other Writings*, ix; Bitbol-Hespèriès, "Cartesian Physiology," 369. On Descartes's adolescence, see Cook, *Young Descartes*. For a listing of Descartes's residences during the 1630s in the Low Countries, see Dijksterhuis, "Constructive Thinking," 65.

107   See Berkel, "Dutch Republic"; Davids, "Public Knowledge and Common Secrets"; Davids, "Craft Secrecy in Europe."

108   Bitbol-Hespèriès, "Cartesian Physiology," 354; Gaukroger in Descartes, *World and Other Writings*, xxxi. The treatise was published posthumously in Paris; see Descartes, *L'homme* (1664).

109   Hatfield, "Descartes' Physiology," 336.

110   Gaukroger in Descartes, *World and Other Writings*, ix; Bitbol-Hespèriès, "Cartesian Physiology."

111   Descartes, letter to Mersenne, Amsterdam, December 18, 1629, AT I, 102; mentioned in Bitbol-Hespèriès, "Cartesian Physiology," 369.

112   "I'y trauaille fort lentemant, pource que ie prens beaoucoup plus de plaisir a m'instruire moy-mesme, que non pas a mettre par escrit le peu que ie sçay. I' estudie maintenant en chymie & en anatomie tout ensemble, & apprens tous les iours quelque chose que ie ne trouue pas dedans les liures. . . . Au reste ie passe si doucemant le tans en m'instruisant moy-mesme, que ie ne me mets iamais a escrire en mon traité que par contrainte." Descartes, letter to Mersenne, Amsterdam, April 15, 1630, AT I, 137; translation adapted from CSMK, 21; my emphasis.

113 On the idea of the *Entautorisierung* (deauthorization) of (ancient) textual authorities through a complex and intimately interrelated practice of textual examination (*examinatio*) and hands-on dissections (*autopsia*), see, e.g., De Angelis, *Anthropologien*, esp. 248ff. I return to this point in chapter 2.

114 For further reading on anatomical practices in early modern France, see, e.g., Guerrini, *Courtiers' Anatomists*.

115 CSMK, 21.

116 The full passage reads: "Et celuy dont vous m'écrivez doit avoir l'esprit bien foible, de m'accuser d'aller par les villages, pour voir tuer des pourceaux; car il s'en tue bien plus dans les villes que dans les villages, où ie n'ay iamais esté pour ce sujet. Mais, comme vous m'écrivez, ce n'est pas une crime d'estre curieux de l'Anatomie; & j'ay esté vn hyuer à Amsterdan, que i'allois quasi tous les iours en la maison d'vn boucher, pour luy voir tuer des bestes, & faisois apporter de là en mon logis les parties que ie voulois anatomiser plus à loisir; ce qu i'ay encore fait plusieurs fois en tous les lieux où i'ay esté, & ie ne croy pas qu'aucun homme d'esprit m'en puisse blâmer." Descartes, letter to Mersenne, November 13, 1639, AT II, 621; translation, Bitbol-Hésperiès, "Cartesian Physiology," 355.

117 Descartes stayed in Amsterdam in 1629; see Bitbol-Hésperiès, "Cartesian Physiology," 355.

118 Bitbol-Hésperiès, "Cartesian Physiology," 352.

119 For a detailed discussion of the points on which Descartes disagreed with Harvey about the action of the heart, see Grene, "Heart and Blood."

120 Descartes, letter to Mersenne, Deventer, 1632, AT I, 263; translation, CSMK, 40; partly cited and discussed in Grene, "Heart and Blood," 325 and 328; Petrescu, "Descartes on the Heartbeat," 400n6.

121 Bitbol-Hésperiès, "Cartesian Physiology," 375; Petrescu, "Descartes on the Heartbeat."

122 Cunningham, "Pen and the Sword II," 59.

123 Cunningham, "Pen and the Sword II," 53; on Ruysch, see, e.g., Margóscy, "Philosophy of Wax."

124 Cunningham, "Pen and the Sword I"; Cunningham, "Pen and the Sword II," 60.

125 Cunningham, "Pen and the Sword II," 52.

126 Cunningham, "Pen and the Sword II," 67.

127 Cunningham, "Pen and the Sword I," 645.

128 Cunningham, "Pen and the Sword II," 72.

129 Cunningham, "Pen and the Sword I"; Cunningham, "Pen and the Sword II."

130   On Cartesian physiology, see, e.g., Hatfield, "Descartes' Physiology"; Bitbol-Hespériès, "Cartesian Physiology"; Grene, "Heart and Blood."

131   Cunningham, "Pen and the Sword II," 55.

132   Cunningham, "Pen and the Sword II," 71.

133   Cunningham, "Pen and the Sword I," 635.

134   Francis Glisson, cited in Cunningham, "Pen and the Sword II," 55.

135   Cunningham, "Pen and the Sword II," 56.

136   See Dibon, "Les echanges épistolaires dans l'Europe savante du XVIIe siècle," cited in Miert, "What Was the Republic of Letters?" For more on Descartes's participation in the Republic of Letters, see Miert, *Communicating Observations in Early Modern Letters*; and, in particular, Verbeek and Bos, "Conceiving the Invisible."

137   On the French term *experience*, see note 85.

138   "En effet, i'ay consideré non seulement ce que Vezalius & les autres écriuent de l'Anatomie, mais aussi plusieurs choses plus particulieres que celles qu'ils écriuent, lesquelles i'ay remarquées en faisant moy-mesme la dissection de divers animaux. C'est vn exercice où ie me suis souuent occupé depuis vnze ans, & ie croy qu'il n'y a gueres de Medecin qui y ait regardé de si prés que moy." Descartes, letter to Mersenne, February 20, 1639, AT II, 525.

139   Descartes, letter to Mersenne, February 20, 1639, CSMK, 134.

140   See, e.g., Daston and Gallison, *Objectivity*; Daston and Lunbeck, *Histories of Scientific Observation*.

141   Cunningham, "Pen and the Sword II," 51, 53.

142   "On ne peut douter qu'il n'y ait de la chaleur dans le cœur, car on la peut sentir mesme de la main, quand on ouure le corps de quelque animal viuant." Descartes, *La description du corps humain,* AT XI, 228; translation, Descartes, *Description of the Human Body*, in *World and Other Writings,* 172.

143   Bitbol-Hésperiès, "Cartesian Physiology," 369.

144   On context-sensitive approaches to history of science, see, e.g., the chapter "Does History Have a Future?," in Garber, *Descartes Embodied*, 13–30.

145   Cunningham, "Pen and the Sword II," 55.

146   Descartes, *La description du corps humain*, AT XI, 228. Translation, Descartes, *Description of the Human Body,* in *World and Other Writings,* 172–73.

147   E.g., Descartes, *Discours de la méthode*, AT VI, 50.

148   "Ce qu il auroit encore pù confirmer par vne experience fort apparente, qui est que, si on coupe la pointe du cœur d'vn chien vif, & que

277

par l'incision on mette le doigt dans l'vne de ses concauitez, on sentira manifestement qu'à toutes les fois que le cœur s'accourcira, il pressera le doigt, & qu'il cessera de le presser. a toutes les fois qu'il s'allongera; ce qui semble assurer entierement, que ses concauitez sont plus estroites, lors que le doigt y est plus pressé, que lors qu'il l'est moins. Et toutesfois cela ne prouue autre chose, sinon que les experiences mesme nous donnent souuent occasion de nous tromper, lors que nous n'examinons pas assez toutes les causes qu'elles peuuent auoir." Descartes, *La description du corps humain,* AT XI, 241–42; translation, Descartes, *Description of the Human Body,* in *World and Other Writings,* 180.

149   Descartes, *La description du corps humain,* AT XI, 243. Translation, Descartes, *Description of the Human Body,* in *World and Other Writings,* 181.

150   Bitbol-Hésperiès, "Cartesian Physiology," 374.

151   Vanagt, "Early Modern Medical Thinking on Vision," 578.

152   Descartes, letter to Mersenne April 1, 1640, AT III, 49; translation, CSMK, 146. Hatfield, "Descartes' Physiology," 350n39, provides a discussion with references to mentions in Descartes's writings and English translations. The pineal body is a small, pine cone–shaped gland that is present in vertebrate brains. It is located in the midline of the brain and has been called "the most enigmatic of endocrine organs"; today it is known to produce hormones that have important endocrine functions (e.g., influencing sexual behavior). This tiny organ shrinks during a lifetime— in adult humans it is less than 0.10 centimeters long and weighs only 0.1 gram—which might explain why Descartes had difficulties seeing it during this dissection, or it may have been retained in the *sella turcica* of the skull. Rogers, *Endocrine System,* 86ff.

153   "Mais i'ay bien eu plus de curiosité; car i'ay fait autrefois tuer vne vache, que ie sçauois auoir conceu peu de tems auparavant, expres affin d'en voir le fruit. Et ayant appris, par apres, que les bouchers de ce païs en tuent souuent qui se rencontrent plenes, i'ay fait qu'ils m'ont apporté plus d'vne douzaine de ventres dans lesquels il y auoit de petits veaux, les vns grands comme des fouris, les autres comme des rats, & les autres comme de petits chiens, ou i'ay pu observer beaucoup plus choses qu'en des poulets, a cause que les organes y sont plus grands & plus visibles." Descartes, letter to Mersenne, November 2, 1646, AT IV, 555; mentioned in Bitbol-Héspières, "Cartesian Physiology," 358.

154   See previous discussion with full quotes in notes 112 and 116, referencing Descartes, letter to Mersenne, Amsterdam, April 15, 1630, AT I, 137; and Descartes, letter to Mersenne, November 13, 1639, AT II, 621.

155   Translation, Descartes, *Discourse on the Method,* CSM 1, 136.

156   Descartes, *Discours de la méthode,* AT VI, 50.

157 This observation has been refuted by modern medicine.

158 Cunningham, "Pen and the Sword II," 59.

159 Descartes, *La dioptrique*, AT VI, 79–229; *La description du corps humain*, AT XI, 219–87 (page numbers here refer to the entire treatises; specific passages are discussed above and below). See Hatfield, "Descartes' Physiology," 363n6, for an extensive, though not comprehensive, list with references to Descartes's anatomical studies in his letters and in his unpublished notes in Latin, with references to Descartes, *Excerpta anatomica, varia*, AT XI.

160 On Plempius, see Vanagt, "Early Modern Medical Thinking on Vision."

161 Grene, "Heart and Blood."

162 Descartes, letter to Plempius (in Latin), February 15, 1638, AT I, 526–29. Translation, Grene, "Heart and Blood," 328–30 (without Grene's inserted comments); Grene's translation slightly diverges from Cottingham et al., CSMK, 83, who translate the last sentence as "nothing that we perceive by the senses seems to me more certain than this."

163 Descartes, letter to Plempius, February 15, 1638; translation, Grene, "Heart and Blood," 330.

164 Cunningham, "Pen and the Sword II," 55.

165 Cunningham, "Pen and the Sword II," 70.

166 On the embodied activity of writing in the Renaissance, see, e.g., Goldberg, *Writing Matter*.

167 Cunningham, "Pen and the Sword I," 641.

168 Fernel, *Medicina*; Fernel, *Medicina, physiologiam, pathologiam, methodumque complectens*; Fernel and Plancy, *Universa medicina*.

169 Campbell, "Fernel, Jean François." The Universal Short Title Catalogue (https://www.ustc.ac.uk/) lists forty-one editions of Fernel's *Medicina* published between 1554 and 1679, with some later ones under the title *Universa medicina*, including at least fourteen editions published in hand-sized octavo format (8°). To see this data, search "Fernel Medicina."

170 Hatfield, "Descartes' Physiology," 338; Bitbol-Hespériès, "Cartesian Physiology," 363. On page 362, Bitbol-Hespériès argues that Descartes's indebtedness to Fernel and the Scholastic tradition has been overstated, and she draws attention to Descartes's extensive readings of genuine sources—the same reference books as used by Harvey.

171 Cunningham, "Pen and the Sword I," 649.

172 Cunningham, "Pen and the Sword I," 650.

173 Cunningham, "Pen and the Sword I," 650.

174 Descartes, letter to Mersenne, June 1632, AT I, 254; translation, CSMK, 39.

175 "J'anatomise maintenant les testes de divers animaux, pour expliquer en quoy consistent l'imagination, la memoire &c." Descartes, letter to Mersenne, Deventer, 1632, AT I, 263; CSMK, 40, translates *anatomise* as "dissecting."

176 Cunningham, "Pen and the Sword II," 67.

177 Cunningham, "Pen and the Sword II," 56.

178 Cunningham, "Pen and the Sword II," 56.

179 Cunningham, "Pen and the Sword II," 71.

180 This is, for example, manifest in *Optics*, in the Third Discourse, entitled "Of the Eye" ("De l'oeil"), where Descartes remarks: "I purposely omit many other details which can be observed about this matter, and with which the anatomists swell their books. For I believe that those I have presented here will suffice in order to explain everything relevant to my subject, and that the others which I could add, while in no wise improving your understanding, would only serve to divert your attention." Descartes, *La dioptrique*, AT VI, 108; translation, Descartes, *Optics*, in *Discourse on Method, Optics etc.*, 86. The importance of experiments for Descartes's studies is underscored in Bitbol-Hespériès, "Cartesian Physiology." Grene, "Heart and Blood," and Hatfield, "Descartes' Physiology," discuss Descartes's experimental vivisections as instances of physiology.

181 Cunningham, "Pen and the Sword II," 67.

182 Cunningham, "Pen and the Sword I," 656.

183 Cunningham, "Pen and the Sword I," 648.

184 Cunningham, "Pen and the Sword I," 648.

185 Cunningham, "Pen and the Sword I," 648.

186 Dijksterhuis, "Constructive Thinking," 65.

187 Dijksterhuis, "Constructive Thinking," 66.

188 Dijksterhuis, "Constructive Thinking," 60.

189 Dijksterhuis, "Constructive Thinking," 62.

190 Dijksterhuis, "Constructive Thinking," 62.

191 Dijksterhuis, "Constructive Thinking," 62.

192 Dijksterhuis, "Constructive Thinking," 66.

193 Dijksterhuis, "Constructive Thinking," 60.

194 Descartes, *La dioptrique*, AT VI, 79–228;. Partly translated in Descartes, *Optics*, CSM 1, 166–67. Only the Second Discourse is translated in *Optics*, in Descartes, *World and Other Writings*, 76–85. For a comprehensive German translation of the experiment description, see Leisegang, *Descartes Dioptrik*, 12, 90ff.

195 For further readings on the camera obscura, see, e.g., Hammond, *Camera Obscura*; Lefèvre, "Inside the Camera Obscura."

196 Descartes, *La dioptrique*, AT VI, 115; see also note 200. Translation adapted from Descartes, *Optics*, CSM 1, 166; my modified translation keeps closer to the French source text.

197 "Pour ce qui est de couper l'œil d'vn beuf en sorte qu'on y puisse voir le mesme qu'en la chamber obscure, comme i'ay escrit en la Dioptrique, ie vous assure que i'en ay fait l'experience, & quoy que c'ait esté sans beaucoup de soins ny de precautions, elle n'a pas laissé pour cela de reussir; mais ie vous diray comment. Ie pris l'œil d'vn vieux boeuf (ce qu'il faut observer, car celuy des ieunes veaux n'est pas transparent), & ayant choisi la moitié d'vne coquille d'œuf, qui estoit telle que cet œil pouuoit aysement estre mis & aiusté dedans sans changer sa figure, ie couppay en rond avec des ciseaux fort tranchans & vn peu esmoussez a la pointe les deux peaux, *corneam* & *vueam*, sans offencer la troisiesme, *retinam*. Et la piece ronde que ie couppay n'estoit qu'enuiron de la grandeur d'vn sous, & elle auoit le nerf optique pour centre. Puis, quand elle fut ainfy coupée tout autour, sans que ie l'eusse encore ostée de sa place, ie ne fis que tirer le nerf optique, & elle suiuit auec la *retinam*, qui se rompit sans que l'humeur vitrée fust aucunement offensée, si bien que l'ayant couuerte de ma coquille d'œuf, ie vis derriere ce que ie voulois; car la coquille d'œuf estoit assez transparente pour cet effet. Et ie l'ay monstré a d'autres depuis en mesme sorte, mesme sans coquille d'œuf, auec vn papier derriere. Il est vray que l'œil est suiet a se rider vn peu au deuant, & ainsy a rendre l'image moins parfaite; mais on y peut obuier en le pressant vn peu aux costez auec les doigts, ou aussy en prenant l'œil d'vn bœuf fort fraischement tué & le tenant tousiours dans l'eau, si tost qu il est tiré de la teste, mesme pendant qu'on en couppe les peaux, iusques a ce qu'il soit aiusté dans la coquille. Voila pour vostre Ire lettre." Descartes, letter to Mersenne, March, 31, 1638, AT II, 86–88.

198 "Mais vous en pourrés eftre encores plus certain, si, prenant l'œil d' vn homme fraischement mort, ou, au defaut, celuy d'vn bœuf ou de quelqu'autre gros animal, vous coupés dextrement vers le fonds les trois peaux qui l'enuelopent, en sorte qu'vne grande partie de l'humeur M, qui y est, demeure découuerte, sans qu'il y ait rien d'elle pour cela qui se re- spende; puis, l'ayant recouuerte de quelque cors blanc, qui soit si delié que le iour passe au trauers, comme, par exemple, d'vn morceau de papier ou de la coquille a vn œuf, RST, que vous mettiés cet œil dans le trou d'vne fenestre fait exprés, comme Z, en sorte qu'il ait le devant, BCD, tourné vers quelque lieu où il y ait diuers obiets, comme V, X,Y, esclairés par le soleil; & le derriere, où est le cors blanc RST, vers le dedans de la chambre, P, où vous serés, & en laquelle il ne doit entrer aucune lumiere, que celle qui pourra penetrer au trauers de cet œil, dont vous sçaués que toutes les

parties, depuis C iusques a S, sont transparentes." Descartes, *La dioptrique*, AT VI, 115; translation adapted and modified from Descartes, *Optics*, CSM 1, 166; and Descartes, *Optics*, in *Discourse on Method, Optics etc.*, 91–93.

199   The long-neglected genre of recipe literature has received much attention in history of science scholarship since the turn of the millennium. See, e.g., Eamon, *Science and the Secrets of Nature*; Leong and Rankin, *Secrets and Knowledge in Medicine and Science*; Leong, *Recipes and Everyday Knowledge*; Making and Knowing Project et al., *Secrets of Craft and Nature*.

200   "Car, cela fait, si vous regardés sur ce cors blanc RST, vous y verrés, non peutestre sans admiration & plaisir, vne peinture, qui representera fort naiuement en perspectiue tous les objets qui seront au dehors vers VXY, au moins si vous faites en sorte que cet œil [for the illustration see figure 1.5] retiene sa figure naturelle, proportionée a la distance de ces obiets." Descartes, *La dioptrique*, AT VI, 115–17; translation adapted and modified from Descartes, *Optics*, CSM 1, 166–67; and from Descartes, *Optics*, in *Discourse on Method, Optics etc.*, 91–93.

201   "Car, pour peu que vous le pressiés plus ou moins que de raison, cete peinture en deuiendra moins distincte. Et il est a remarquer qu'on doit le presser vn peu dauantage, & rendre sa figure vn peu plus longue, lors que les obiets sont fort proches, que lors qu'ils sont plus esloignés." Descartes, *La dioptrique*, AT VI, 117; translation adapted and modified from Descartes, *Optics*, CSM 1, 167; and Descartes, *Optics*, in *Discourse on Method, Optics etc.*, 93.

202   Dijksterhuis, "Constructive Thinking," 66.

203   See Descartes, letter to Mersenne, November, 13, AT II, 621.

204   "S'il estoit possible de couper l'oeil par la moitié, sans que les liquers don't il est rempli s'escoulassent, ni qu'aucune de ses parties changeast de place, et que le plan de la section passast iustement par le milieu de la prunelle, il paroistroit tel qu'il est representé en cete figure." Descartes, *La dioptrique*, AT VI, 105–6; translation, Descartes, *Optics*, in *Discourse on Method, Optics etc.*, 84.

205   "ABCB est vne peau assés dure & espaisse, qui compose comme vn vaze rond dans lequel toutes ses parties interieures sont contenues. DEF est vne autre peau deliée, qui est tendue ainsi qu'vne tapisserie au dedans de la precedente. ZH est le nerf nommé optique, qui est composé d'vn grand nombre de petits filets, dont les extremités s'estendent en tout l'espace GHI, où, se meslant avec vne infinité de petites veines & arteres, elles composent vne espece de chair extremement tendre & delicate, laquelle est comme vne troisiesme peau, qui couure tous les fons de la seconde. K, L, M sont trois sortes de glaires ou humeurs fort transparentes, qui remplissent tout l'espace contenu au dedans de ces peaux, & ont chacune

la figure, en laquelle vous la voyés icy representée." Descartes, *La dioptrique*, AT VI, 106; translation slightly modified from Descartes, *Optics*, in *Discourse on Method, Optics etc.*, 84.

Chapter 2. Descartes's Manual Meditations

Epigraph source: On p. 81, Hagen, *Meditation Now or Never*, 10.

1 Descartes, letter to Princess Elisabeth of Bohemia, June 28, 1643, AT III, 692–93; translation adapted from Shapiro, *Correspondence*, 70, my emphasis. Descartes uses the expression "& que i'ay donné tout le reste de mon temps au relasche des sens," translated by Lisa Shapiro as "and that I give all the rest of my time to relaxing the senses" and as "to the relaxation of the senses" by Cottingham et al., Descartes, *Correspondence*, CSMK, 227. I translate the French word "relasche," literally loosening or relaxing, in the sense of "giving free rein to the senses," because Descartes uses the same word in this sense in a passage in the Second Meditation, which I will discuss further on. Note also that the French "l'entendement seul," here translated as "the understanding alone," is less precisely given as "the intellect alone" in the widely cited CSMK translation.

2 The term *meditator* is used in Rorty, "The Structure of Descartes' Meditations," 4–5.

3 This term is used by the French Cartesian Pierre Petit (1598–1667) in a letter to the diplomat Pierre Chanut (1601–1667) in which he describes Blaise Pascal's experiments on a vacuum; quoted in Gipper, "Experiment und Öffentlichkeit," 247.

4 Descartes, *Les méditations métaphysiques* (1647), 26–27; Descartes, *Les méditations métaphysiques*, AT IX, 23–24; Descartes, *Meditations on First Philosophy*, CSM 2, 20.

5 The term *wax argument* appears in handbooks and online resources, e.g., Hatfield, *Routledge Philosophy Guidebook to Descartes*, 125–31; "Wax Argument," *Wikipedia*, accessed April 25, 2024, https://en.wikipedia.org/wiki/Wax_argument.

6 On the Latin phrasing of this Aristotelian concept, see Cranefield, "On the Origin of the Phrase, *Nihil est in intellectu quod non prius fuerit in sensu*," 77–80.

7 Descartes, *Les méditations métaphysiques*, AT IX, 23–24; translation, Descartes, *Meditations on First Philosophy*, CSM 2, 20.

8 E.g., Rorty, *Essays on Descartes' "Meditations."*

9 "Mais ie voy bien ce que c'est: mon esprit se plaist de s'égarer, & ne se peut encore contenir dans les iustes bornes de la vérité. Relachons-luy

donc encore vne fois la bride, afin que, venant cy-apres à la retirer douce-ment & à propos, nous le puissions plus facilement régler & conduire"; Descartes, *Les méditations métaphysique*, AT IX, 23. My translation stays as close as possible to the French wording and deviates from Descartes, *Meditations on First Philosophy*, CSM 2, 20 (cited above, note 7).

10 Descartes, *Les méditations métaphysiques* (1647), 26 (figure 1.2.); Des-cartes, *Les méditations métaphysiques*, AT IX, 23–24. I adapted and modi-fied the English translation in Descartes, *Meditations on First Philosophy*, CSM 2, 20.

11 "How-to" literature, technical writings, and early modern recipe texts have received much attention from historians of science; for further read-ings, see, e.g., DiMeo and Pennell, *Reading and Writing Recipe Books*; Leong, *Recipes and Everyday Knowledge*; Leong and Rankin, *Secrets and Knowledge in Medicine and Science*; Long, *Artisans/Practitioners and the Rise of the New Sciences*; P. H. Smith and Schmidt, *Making Knowledge in Early Modern Europe*; P. H. Smith, Meyers, and Cook, *Ways of Making and Knowing*; P. H. Smith, *From Lived Experience to the Written Word*; Hagendijk, "Reworking Recipes." See also these collaborative research projects: The Making and Knowing Project (2014–20), Columbia Uni-versity, New York, https://www.makingandknowing.org/; ARTECHNE— Technique in the Arts, 1550–1950 (2015–20), Utrecht University and University of Amsterdam, https://artechne.wp.hum.uu.nl. See also the blog posts of the Recipes Project, https://recipes.hypotheses.org/.

12 Descartes, *Discours de la méthode*, AT VI, 47; Descartes, *Description du corps humain*, AT XI, 228.

13 Descartes, *Les méditations métaphysiques* (1647), 26 (figure 1.2.); Des-cartes, *Les méditations métaphysiques*, AT IX, 23–24. I adapted and modi-fied the English translation in Descartes, *Meditations on First Philosophy*, CSM 2, 20.

14 Descartes, letter to Mersenne, November 13, 1639 , AT II, 621; cited and quoted in translation in Bitbol-Hespériès, "Cartesian Physiology," 355.

15 On the use of tables and the rise of experimental sciences in the seven-teenth century, see, e.g., Daniel Garber's discussion of Francis Bacon in Garber, *Descartes Embodied*, 301–3.

16 On the genre of the recipe and the experimental essay, see, e.g., DiMeo, "Communicating Medical Recipes"; V. Keller, "'Everything Depends upon the Trial.'"

17 My argument chimes with Garber's observation that "one can see in the *Discourse* [*on Method*] a clear anticipation of an important later liter-ary form, the grant application"; Garber, *Descartes Embodied*, 85. Other scholars have traced the rhetoric of the scientific periodical back to the

experimental essays of Robert Boyle, an early fellow of the Royal Society, e.g., Paradis, "Montaigne, Boyle, and the Essay of Experience," cited in V. Keller, "'Everything Depends upon the Trial.'"

18   The term *protocol* originates in the sixteenth century in trade documents, where it was used for an original note or minute of a transaction. In the seventeenth century it became a formal expression in diplomacy indicating the original draft or record of a diplomatic document. In the nineteenth century, *protocol* was used for the etiquette of precedence; only in the nineteenth century do we find the earliest use of the term *prothocoll* from the French *prothocole* (modern French *protocole*) in scientific contexts. See "protocol," in *The Concise Oxford Dictionary of English Etymology*, edited by T. F. Hoad (Oxford: Oxford University Press, 2003).

19   "Ad lectorem candidum, et magneticae philosophiae studiosum. . . . Qui eadē experiri voluerit, non oscitāter & ineptè, sed prudenter, artificiosè & appositè corpora tractet; ne ille (cùm res non successerit) inscius nostras arguat inuētiones: nihil enim in istis libris depromptum, quod non exploratum, sępissiméq; actū & transactum apud nos fuerit"; Gilbert, *De magnete*, fol. iir, iiv. Translation by S. P. Thompson (1901) in Gilbert, *On the Magnet*, fols. iir and iiv. Garber, *Descartes Embodied*, 299, cites the less precise yet widespread translation by P. Fleury Mottelay, first published in New York in 1893 and reprinted as Dover editions in the twentieth century.

20   See my discussion of Descartes's use of the term *experience* in chapter 1.

21   On Descartes's rejection of spiritual and ancient authorities, see, e.g., Garber, *Descartes Embodied*, 315; Rorty, *Essays on Descartes' "Meditations,"* ix.

22   P. H. Smith, *Body of the Artisan*; P. H. Smith and Schmidt, *Making Knowledge in Early Modern Europe*; P. Smith, Meyers, and Cook, *Ways of Making and Knowing*.

23   Zilsel, "Genesis of the Concept of Scientific Progress"; Long, *Artisans/ Practitioners and the Rise of the New Sciences, 1400–1600.*

24   P. H. Smith, *Body of the Artisan*.

25   P. H. Smith, *Body of the Artisan*, 8.

26   Rankin, "How to Cure the Golden Vein," 113–37; Cook, "Preservation of Specimens," 302–29.

27   Cook, "Preservation of Specimens," 302.

28   Descartes, *Discours de la méthode*, AT VI, 77–78; translation, Descartes, *Discourse on the Method*, CSM 1, 151.

29   Grene, "Heart and Blood," 333.

30   Nelle, "Eucharist and Experiment," 332.

31   Shapin and Schaffer, *Leviathan and the Air-Pump*.

32   Nelle, "Eucharist and Experiment," 328.

33   Amélie Oksenberg Rorty, e.g., proposes "to read the *Meditations* as a
      work within the traditional meditational form, rather than a treatise
      composed of a series of arguments"; Rorty, "The Structure of Descartes'
      Meditations," 2.

34   Hatfield, "The Senses and the Fleshless Eye"; Rorty, "Descartes on Think-
      ing with the Body"; Sepper, "Texture of Thought."

35   Hatfield, *Routledge Philosophy Guidebook to Descartes*, 41.

36   Hatfield, *Routledge Philosophy Guidebook to Descartes*, 38.

37   On "thinning" of knowledge, see, e.g., V. Keller, "'Everything Depends
      upon the Trial'"; Porter, "Thin Description."

38   E.g., Hatfield, "The Senses and the Fleshless Eye"; Rorty, "Descartes on
      Thinking with the Body."

39   Rorty, *Essays on Descartes' "Meditations,"* ix.

40   Hatfield, "The Senses and the Fleshless Eye," 47.

41   Hatfield, "The Senses and the Fleshless Eye," 47.

42   Hatfield, "The Senses and the Fleshless Eye," 48.

43   Hatfield, "The Senses and the Fleshless Eye," 73n6. On Descartes's school
      education, see also Cook, *Young Descartes.*

44   Hatfield, "The Senses and the Fleshless Eye," 47.

45   Hatfield, "The Senses and the Fleshless Eye," 69.

46   Hatfield, "The Senses and the Fleshless Eye," 47.

47   Rorty, *Essays on Descartes' "Meditations,"* xi.

48   Hatfield, "The Senses and the Fleshless Eye," 69.

49   Hatfield, "The Senses and the Fleshless Eye," 49–51. Subsequent page
      numbers to this essay are provided parenthetically in the text.

50   Sepper, "Texture of Thought," 737.

51   Williams, "Introductory Essay on Descartes' *Meditations,*" 246.

52   Hatfield, "The Senses and the Fleshless Eye," 69.

53   See note 3 on Pierre Petit's use of this term in the seventeenth century.

54   On recipe literature, see note 11.

55   Shapin, "'The Mind Is Its Own Place,'" 191.

56   Hatfield, *Routledge Philosophy Guidebook to Descartes,* xv, with cross-
      references to Hatfield, "Senses and the Fleshless Eye," and Hatfield, "Reason,
      Nature, and God in Descartes."

57   McAllister, "Virtual Laboratory," 53. On the history of laboratories as
      spaces of natural inquiries, see, e.g., Klein, "Laboratory Challenge"; Shapin,
      "House of Experiment in Seventeenth-Century England"; Schramm,

Schwarte, and Lazardzig, *Collection, Laboratory, Theater*; P. H. Smith, "Laboratories."

58 Ophir and Shapin, "The Place of Knowledge", 13–14.

59 McAllister, "Virtual Laboratory," 53.

60 Garber, *Descartes Embodied*, 320.

61 Dijksterhuis, "Constructive Thinking," 65.

62 See chapter 1, note 116.

63 On the reception of Cartesianism in French salon culture, see, e.g., Gipper, "Experiment und Öffentlichkeit"; Ranea, "A 'Science for *Honnêtes hommes*."

64 Descartes, letter to Plempius, February 15, 1638, AT I, 526; translated from Latin into English by Grene, "Heart and Blood," 327, with a detailed discussion of how Descartes conducted this vivisection to refute Galen's theory of the movement of the blood. Cottingham et al., CSMK, 81, translate the passage as "an utterly decisive experiment, which I was interested to observe, . . . and which I performed again today in the course of writing this letter," thus rephrasing the epistemically important phrase "most certain experiment" (*certissimo experimento*) and substituting "observing" and "performing" for seeing. For reflections on how Descartes's thinking was shaped in the act of writing, see Nancy, "Dum Scribo"; more generally on embodied aspects of early modern writing practices, see, e.g., Goldberg, *Writing Matter*.

65 "Et pource que ie ne me fie gueres aux experiences que ie n'ay point faites moy-mefme, i'ay fait faire vn tuyau de douze pieds pour ce fuiet; mais i'ay si peu de mains, & les artifans sont si mal ce qu'on leur commande, que ie n'en ay pû apprendre autre chose"; Descartes, letter to [Huygens], February 18 or 19, 1643, AT III, 617. According to AT, the letter might refer to a note addressed to "M. de Zuylichem" that was posted from the Dutch city of Endegeest. Shortened and translated as "I have little trust in experiments which I have not performed myself," in D. Clarke, "Descartes' Philosophy of Science and the Scientific Revolution," 285n22.

66 Shapin and Schaffer, *Leviathan and the Air-Pump*, Garber, *Descartes Embodied*, esp. chap. 14.

67 The following exposition draws on Garber, *Descartes Embodied*, 296–328.

68 Garber, *Descartes Embodied*, 316–17

69 Garber, *Descartes Embodied*, 318–19

70 Garber, *Descartes Embodied*, 325–26

71 Garber, *Descartes Embodied*, 314–15

72 Shapin, "Invisible Technician," 561.

73 "Mais ie voy aussy qu'elles sont telles, & en si grand nombre, que ny mes mains, ny mon revenu, bien que i'en eusse mille fois plus que ie n'en ay, ne

287

sçauroient suffire pour toutes; en sorte que, selon que i'auray desormais la commodité d'en faire plus ou moins, i'auanceray aussy plus ou moins en la connaissance de la Nature." Descartes, *Discours de la méthode*, AT VI, 65; translation, Descartes, *Discourse on the Method*, CSM 1, 144.

74   On Descartes's collaborative experimental work in optics and anatomy, see Dijksterhuis, "Constructive Thinking"; Bitbol-Hespériès, "Cartesian Physiology," 374.

75   Gipper, "Experiment und Öffentlichkeit."

76   Burnyeat, "Idealism and Greek Philosophy," 35 blanc (my emphasis).

77   Burnyeat, "Idealism and Greek Philosophy," 38.

78   Burnyeat, "Idealism and Greek Philosophy," 33.

79   Hatfield, "The Senses and the Fleshless Eye," 72.

80   P. H. Smith, *Body of the Artisan*, 113.

81   P. H. Smith, *Body of the Artisan*, 20.

82   Shapin, "Invisible Technician," 561.

83   Cook, "Medicine," 3:411.

84   Descartes, letter to [Huygens], February 18 or 19, 1643, AT III, 617. On this letter, see note 65.

85   P. H. Smith, *Body of the Artisan*, 20.

86   Mody and Kaiser, "Scientific Training," 377–402.

87   P. H. Smith, *Body of the Artisan*, 239.

88   See the discussion in chapter 1 on Cunningham, "Pen and the Sword I"; Cunningham, "Pen and the Sword II."

89   See also Pamela Smith on "models of the new active gathering of knowledge"; P. H. Smith, *Body of the Artisan*, 20.

90   The complex notion of *scientiae* and debates on the (dis)continuities between Renaissance epistemology and epistemologies associated with the new science of the seventeenth century have been discussed in classic and more recent standard works; see, e.g., Jardine, "Problems of Knowledge and Action"; Demeter, Láng, and Schmal, "Scientia in the Renaissance."

91   Nelle, "Eucharist and Experiment."

Chapter 3. Making Modern Epistemology

Epigraph source: On p. 112, Didi-Huberman, "Before the Image," 38.

1   The *Meditations* were first published in Latin in 1641 but were soon translated into French (1647) with Descartes's approval; Descartes, *Renati Des-Cartes Meditationes de prima philosophia* (1641); Descartes, *Les*

288

méditations métaphysiques (1647); Descartes, *Les méditations métaphysiques*, AT IX, 1–72, objections and replies, 73–245; English translation, Descartes, *Meditations on First Philosophy*, CSM 2, 1–62, objections and replies, 63–397. I use the italicized term *experience* to refer to the French word, spelled without accent in the first authorized French edition (1647) and its historical meaning as "experience" and "experiment"; see chapter 1, notes 83 and 85.

2   Rabinow, "Epochs, Presents, Events," 31.

3   Husserl, *Crisis of European Sciences*; Husserl, *Die Krisis der europäischen Wissenschaften.*

4   Vogt, "Ancient Skepticism."

5   Sandkühler, Pätzold, and Egenbogen, "Phänomenologie," 1015.

6   Husserl, *Cartesianische Meditationen und Pariser Vorträge* (1929); first published in French in 1931; Husserl, *Cartesian Meditations.*

7   "Vn certain mauuais genie, non moins rusé & trompeur que puissant, qui a employé toute son industrie à me tromper." Descartes, *Les méditations métaphysiques*, AT IX, 17.

8   Descartes, *Les méditations métaphysiques*, AT IX, 18–19; translation, Descartes, *Meditations on First Philosophy*, CSM 2, 16.

9   Husserl, *Ideen zu einer reinen Phänomenologie und phänomenologischen Philosophie: Erstes Buch*, 66: "Davon sehen wir ab, uns interessiert nicht jede analytische Komponente des Zweifelversuchs, daher auch nicht seine exakte und vollzureichende Analyse."

10 Husserl, *Ideen zu einer reinen Phänomenologie und phänomenologischen Philosophie: Erstes Buch*, 64: "Die Herausstellung einer absoluten zweifellosen Seinssphäre."

11 For the following paragraph, see Husserl, *Ideen zu einer reinen Phänomenologie und phänomenologischen Philosophie: Erstes Buch*, 65.

12 Husserl, *Ideen zu einer reinen Phänomenologie und phänomenologischen Philosophie: Erstes Buch*, 65.

13 For the wax passage, see Descartes, *Les méditations métaphysiques* (1647), 26 (figure 1.2); Descartes, *Les méditations métaphysiques*, AT IX, 23–24; Descartes, *Meditations*, CSM 1, 20.

14 For a critical analysis of a thinking in origins, see Derrida, *Edmund Husserl's "Origin of Geometry."*

15 Slatman, *L'expression au-delà de la représentation*, 5.

16 Merleau-Ponty, "Philosopher and His Shadow," 160. Original quotation: "Je größer das Denkwerk eines Denkers ist, das sich keineswegs mit dem Umfang und der Anzahl seiner Schriften deckt, um so reicher ist das in diesem Denkwerk Ungedachte, das heißt jenes, was erst und allein

durch dieses Denkwerk als das Noch-nicht-Gedachte heraufkommt";
Heidegger, *Der Satz vom Grund*, 123–24.

17 On Heidegger's notion of "the Unthought" as the beginning of a think-ing process that points beyond the work of a great philosopher, see also Samuel Ijsseling, "Das Ende der Philosophie," 287.

18 On the concept of displaced resemblance, see Didi-Huberman, "Before the Image, before Time," 33. On experimental protocol as a term that only in the nineteenth century became used in the sense of a record of a scientific observation, see note 18 in chapter 2.

19 *Oxford English Dictionary*, s.v. "anachronism (n.)," July 2023, https://doi .org/10.1093/OED/6842940724.

20 Didi-Huberman, "Before the Image," 35.

21 Didi-Huberman, "Before the Image," 37.

22 Didi-Huberman, "Before the Image"; Didi-Huberman, *Fra Angelico*.

23 Didi-Huberman, "Before the Image," 37.

24 Didi-Huberman, "Before the Image," 42.

25 Didi-Huberman, "Before the Image," 35.

26 Didi-Huberman, "Before the Image," 37.

27 Didi-Huberman, "Before the Image," 37.

28 Garber, *Descartes Embodied*, 5; Didi-Huberman, "Before the Image," 36.

29 Garber, *Descartes Embodied*, 5.

30 On the historiography of Descartes's search for a "universal method," see, e.g., Brissey, "Rule VIII of Descartes' *Regulae ad directionem ingenii*."

31 See, e.g., Daston and Galison, *Objectivity*; Daston and Lunbeck, *Histories of Scientific Observation*.

32 See, e.g., Rheinberger, *On Historicizing Epistemology*; and the *Journal of the History of Knowledge,* especially the introduction to the first issue, Dupré and Somsen, "What Is the History of Knowledge?"

33 Didi-Huberman, "Before the Image," 41

34 Didi-Huberman, "Before the Image," 41.

35 Didi-Huberman, "Before the Image," 41; Didi-Huberman, *Fra Angelico*.

36 Didi-Huberman, "Before the Image," 40.

37 Robins and Trigger, "Recent Phase of Aboriginal Occupation."

38 Robins and Trigger, "Recent Phase of Aboriginal Occupation," 39.

39 Robins and Trigger, "Recent Phase of Aboriginal Occupation."

40 Robins and Trigger, "Recent Phase of Aboriginal Occupation," 41.

41   Binford, "Smudge Pits and Hide Smoking," 10, cited in Robins and Trig-
     ger, "Recent Phase of Aboriginal Occupation," 41.

42   See, e.g., the entry "Wax Argument" on Wikipedia. "The famous
     wax thought experiment" is mentioned in Newman, "Descartes'
     Epistemology."

43   Newman, "Descartes' Epistemology."

44   The wax passage precedes the first mention of the concepts "clear and
     distinct" in the Second Meditation where Descartes speaks about an in-
     spection of the mind: "mais seulement une inspection de l'esprit, laquelle
     peut estre imparfaite & confuse, comme elle estoit auparauant, ou bien
     claire & distincte, comme elle est à present"; Descartes, *Les méditations
     métaphysiques*, AT IX, 25. The "idées claires & distinctes" appear in the
     Third Meditation; Descartes, *Les méditations métaphysiques*, AT IX, 35.

45   Newman, "Descartes' Epistemology," uses this phrasing in his discus-
     sion of the wax passage: Descartes, *Meditationes de prima philosophia*,
     AT VII, 30–31; Descartes, *Les méditations métaphysiques*, AT IX, 23–24;
     Descartes, *Meditations on First Philosophy*, CSM 2, 20.

46   See MIT's mission statement, accessed May 2, 2022, https://www.mit.edu
     /about/mission-statement/.

47   See, e.g., Sharon Traweek's now classical study on tinkering with detec-
     tors by experimental high-energy physicists; Traweek, *Beamtimes and
     Lifetimes*.

48   Shapin, *Scientific Revolution*, 2 (my emphasis).

49   Lawrence and Shapin, *Science Incarnate*, 4.

50   See, e.g., Newen, Bruin, and Gallagher, *Oxford Handbook of 4E Cogni-
     tion*; Ward and Stapleton, "Es Are Good."

51   Slatman, *Our Strange Body*, 42.

52   P. H. Smith, *Body of the Artisan*; Cook, "Medicine."

53   Slatman, *Our Strange Body*, 40.

54   Slatman refers to a famous passage in the second book of Edmund
     Husserl's *Ideas Pertaining to a Pure Phenomenology and to a Phenom-
     enological Philosophy*, in which Husserl describes the phenomenon of
     two hands touching and being touched; Husserl, *Ideen zu einer reinen
     Phänomenologie und phänomenologischen Philosophie, Zweites Buch*,
     §36. For the following exposition, see Slatman, *Our Strange Body*, 70,
     74–76.

55   In Dutch, *lijf* and *lichaam*.

56   Slatman, *Our Strange Body*, 75

57   Slatman, *Our Strange Body*, 74.

58 Slatman, *Our Strange Body*, 75.

59 Slatman, *Our Strange Body*, 75.

60 Slatman, *Our Strange Body*, 75–76.

61 Taylor, *Sources of the Self*, 146.

62 On objectifying laboratory animals in scientific practices, see Michael Lynch's ethnographic study "Sacrifice and the Transformation of the Animal Body."

63 Taylor, *Sources of the Self*, 145.

64 Taylor, *Sources of the Self*, 146.

65 Taylor, *Sources of the Self*, 146.

66 Slatman, *Our Strange Bodies*, 40–41.

67 On strangeness as constitutive experience for living beings, see Slatman, "Sense of Life."

68 Husserl, *Die Krisis der europäischen Wissenschaften*, 49; translation, Husserl, *Crisis of European Sciences*, 50;. I cite from the latter in the following discussion.

69 On Husserl's critique of Kant's geometric conception of space as a priori knowledge, see also Boulboullé, "Die verführerische Transparenz des geometrischen Raums."

70 Husserl, *Crisis of European Sciences*, 48–49.

71 Taylor, *Sources of the Self*, 149.

72 On the process of purification and a retrospective conception of modern science, see Hans-Jörg Rheinberger's discussion of Gaston Bachelard's work, Rheinberger, *Epistemology of the Concrete*, esp. 21, 31–32.

73 On traditions of concealment in modern theoretical discourses on science and their roots in Baroque aesthetic theory, see McAllister, "Die Rhetorik der Mühelosigkeit."

74 Rheinberger, *On Historicizing Epistemology*, 63.

75 Zilsel, "Genesis of the Concept of Scientific Progress"; P. H. Smith, *Body of the Artisan*.

76 Husserl, *Crisis of European Sciences*, 48.

77 See, e.g., Derrida, *Edmund Husserl's "Origin of Geometry."*

78 Rheinberger, *Epistemology of the Concrete*, 25–36.

79 Rheinberger, *On Historicizing Epistemology*, 90.

80 Rheinberger, *Epistemology of the Concrete*, 27.

81 The need for reflections on STS foundations is discussed in Kontopodis, Niewöhner, and Beck, "Investigating Emerging Biomedical Practices."

82   On tacit knowledge, see Polanyi, *Personal Knowledge*.

83   Latour and Woolgar, *Laboratory Life*; Knorr-Cetina, *Epistemic Cultures*; Ihde, *Technics and Praxis*.

84   Latour and Woolgar, *Laboratory Life*; Knorr-Cetina, *Epistemic Cultures*.

Chapter 4. Revisiting Laboratory Cultures

Epigraph source: On p. 164, E. F. Keller, "The Biological Gaze," 108.

1   Rheinberger, *On Historicizing Epistemology*.

2   Rheinberger, *On Historicizing Epistemology*, 3.

3   Rheinberger, *On Historicizing Epistemology*, 2 (my emphasis). Though Rheinberger acknowledges here the importance of Ludwik Fleck's work, he glosses over the central importance of the pragmatists and logical empiricists in Europe and the United States for an empirical understanding of epistemology, including the members and associates of the Vienna Circle, chaired by Moritz Schlick, whose focus on verification and testing in scientific methodology reflects their belief in the importance of the empirical and the retractability of concepts, and the school of the American pragmatists (among which Charles Sanders Peirce, William James, and John Dewey). I thank my first reviewer for pointing this out.

4   Rheinberger, *On Historicizing Epistemology*, 3 (my emphasis).

5   Rheinberger, *On Historicizing Epistemology*, 90.

6   E.g., P. H. Smith, *Body of the Artisan*; P. H. Smith, "Laboratories"; Cook, "Medicine."

7   P. H. Smith, *Body of the Artisan*, 240.

8   Knorr-Cetina, *Epistemic Cultures*; Schatzki, Knorr-Cetina, and von Savigni, *Practice Turn in Contemporary Theory*. Ludwik Fleck's studies of scientific communities were influential here; see Rheinberger, *On Historicizing Epistemology*, 57–64; Fleck, *Genesis and Development of a Scientific Fact*.

9   Rheinberger, *On Historicizing Epistemology*, 90.

10   For an introduction to science and technology studies and a short history on its discipline formation, see, e.g., Biagioli, *Science Studies Reader*; *Handbook of Science and Technology Studies* (Hackett et al., 3rd ed., and Felt et al., 4th ed.). Michael Fischer offers cultural genealogies of science studies with special attention to the feminist tradition that has been highly sensitive to body issues in Fischer, *Anthropological Futures*, 50–113.

11   Kontopodis, Niewöhner, and Beck, "Investigating Emerging Biomedical Practices," 602.

293

12   Literature on ethnographic research methods is abundant; see, e.g., Given, *Sage Encyclopedia of Qualitative Research Methods*.

13   Doing, "Give Me a Laboratory," 280; this article appears in Hackett et al., *Handbook of Science and Technology Studies*, 279–95.

14   Doing, "Give Me a Laboratory."

15   Doing, "Give Me a Laboratory," 12. Another pioneering study worth mentioning that is not further treated here is Traweek, *Beamtimes and Lifetimes*.

16   Latour, "Visualisation and Cognition," 3.

17   Latour and Woolgar, *Laboratory Life*, 274.

18   Latour and Woolgar, *Laboratory Life*, 274.

19   Latour and Woolgar, *Laboratory Life*, 280.

20   Latour and Woolgar, *Laboratory Life*, 280.

21   Latour and Woolgar, *Laboratory Life*, 280.

22   Knorr-Cetina, *Epistemic Cultures*, 96.

23   Annemarie Mol introduces this ethnographic tool to raise epistemological questions in Mol, *Body Multiple*, viii.

24   On fieldwork, see the introduction, note 1. My field research comprised life science and technology coursework with full participation in selected courses (including experimental work, writing reports, and taking exams). After that, I conducted research during a two-month internship in a molecular genetics research laboratory that specialized in working with in vivo yeast models and visits to a research lab that specialized in working with bacteria and in vitro models. During the course of my internship, the yeast group consisted of a scientist, an analyst, a PhD student, and a bachelor's student conducting a three-month research internship in this laboratory. My research group shared a lab space with another technician, and occasionally an international postdoc working at one of the benches. After this period, I paid a few more irregular visits to the lab and kept in touch sporadically with one of the researchers who supervised my internship and who remained an important contact for occasional conversations on my observations.

25   Knorr-Cetina, *Epistemic Cultures*, 106.

26   Knorr-Cetina, *Epistemic Cultures*, 106.

27   Knorr-Cetina, *Epistemic Cultures*, 106.

28   Knorr-Cetina, *Epistemic Cultures*, 106.

29   Latour and Woolgar, *Laboratory Life*, 45.

30   Daston, *Wunder, Beweise und Tatsachen*; Daston, *Eine kurze Geschichte der wissenschaftlichen Aufmerksamkeit*.

31 "Die kognitieven Leidenschaften: Staunen und Neugier im Europa der frühen Neuzeit," in Daston, *Wunder, Beweise und Tatsachen*, 77–97.

32 Daston, *Wunder, Beweise und Tatsachen*, 79.

33 Daston, *Wunder, Beweise und Tatsachen*, 79.

34 Daston, *Wunder, Beweise und Tatsachen*, 89.

35 Daston, *Wunder, Beweise und Tatsachen*, 89–90.

36 Daston, *Eine kurze Geschichte der wissenschaftlichen Aufmerksamkeit*, 25.

37 Robert Hooke, "General Scheme," in *The Posthumous Works of Robert Hooke* [1705], 61–62, cited in Daston and Park, *Wonders and the Order of Nature*, 315–16. See also Daston, *Eine kurze Geschichte der wissenschaftlichen Aufmerksamkeit*, 25–26.

38 On context and background of "science wars," see, e.g., Fischer, *Emergent Forms of Life*, 5–6.

39 Latour, "Insiders and Outsiders in the Sociology of Science," 206.

40 Latour, "Insiders and Outsiders in the Sociology of Science," 206.

41 Latour, "Insiders and Outsiders in the Sociology of Science," 201.

42 Latour, "Insiders and Outsiders in the Sociology of Science," 206.

43 Gellner, *Thought and Change*, 105–13; see also Marcus and Fischer's discussion of defamiliarization and epistemological critique in *Anthropology as Cultural Critique*, 141–57.

44 See also Daston and Lunbeck, *Histories of Scientific Observation*.

45 Latour and Woolgar, *Laboratory Life*, 45–53.

46 Latour, "Visualisation and Cognition," 3.

47 Latour and Woolgar, *Laboratory Life*, 105.

48 Latour and Woolgar, *Laboratory Life*, 69.

49 Latour and Woolgar, *Laboratory Life*, 69–71.

50 Latour discusses body questions and an embodied approach to learning in later work; see, e.g., Latour, "How to Talk about the Body?"

51 Latour and Woolgar, *Laboratory Life*, 27.

52 C. P. Snow, "The Two Cultures and the Scientific Revolution: The Rede Lecture 1959," in Snow, *Two Cultures*.

53 Latour and Woolgar, *Laboratory Life*, 27, 39: "We attempt to capitalise on the experiences of observation of a laboratory in situ: by being close to localised scientific practices the observer has a preferential situation from which to understand how scientists themselves produce order."

54 Latour and Woolgar, *Laboratory Life*, 29, 33.

55 On Cartesian geometrization, see, e.g., Maull, "Cartesian Optics."

56 Pickering, "Space," 2.

57 On the notion of situated knowledge, see, e.g., Haraway, "Situated Knowledges."

58 E.g., on the construction of bodies in medical spaces and sterile regimes in operating theaters, see Hirschauer, "Manufacture of Bodies in Surgery." Another debate concerns the use of virtual bodies in medical schools and the use of virtual models in protein modeling; e.g., Prentice, *Bodies in Formation*; Myers, "Molecular Embodiments"; Myers, *Rendering Life Molecular*.

59 It exceeds the scope of this book to give an overview of the growing body of literature on embodied cognition. For early introductions to the field, see, e.g., Blum, Krois, and Rheinberger, *Verkörperungen*; Shapiro, *Routledge Handbook of Embodied Cognition*; Shapiro and Spaulding, "Embodied Cognition."

60 Merleau-Ponty, *Phenomenology of Perception*.

61 Blum, Krois, and Rheinberger, *Verkörperungen*, 5.

62 E.g., Latour, *Pasteurization of France*.

63 Latour and Woolgar, *Laboratory Life*, 245.

64 Latour and Woolgar, *Laboratory Life*, 245 (my emphasis).

65 On "invisible technicians" teaching in the basements, see also White, *Idea Factory*.

66 Shapin, "Invisible Technician," 554.

67 Latour and Woolgar, *Laboratory Life*, 245.

68 For details on the field study, see the materials and methods section in Latour and Woolgar, *Laboratory Life*, 39–40.

69 Latour and Woolgar, *Laboratory Life*, 29.

70 Latour and Woolgar, *Laboratory Life*, 277.

71 Latour and Woolgar, *Laboratory Life*, 282.

72 Latour and Woolgar, *Laboratory Life*, 283.

73 Latour and Woolgar, *Laboratory Life*, 41.

74 On the notion of "strange body," see Slatman, *Our Strange Body*. The original Dutch title translates literally as strange or foreign (*vreemd*) body (*lichaam*) (Slatman, *Vreemd lichaam*).

75 Lynch, *Art and Artifact in Laboratory Science*, 2.

76 On a "rhetoric of authentication," see Hirschauer, "Putting Things into Words," 414.

77 Slatman, *Our Strange Body*.

78 Latour draws attention to the materiality of instruments and how they become part of embodied work in the lab and field and to the role of eyes, hands, and noses in knowledge practices in later work; Latour, *Pandora's Hope*; Latour, "Visualisation and Cognition"; Latour, "How to Talk about the Body?"

79 This section, from page 164 to 165, is based on Landecker, "New Times for Biology"; Landecker, *Culturing Life*, 28–67 (chapter 1, "Autonomy").

80 Landecker, "New Times for Biology," 687.

81 E. F. Keller, "Beyond the Gene," 293.

82 E. F. Keller, "Beyond the Gene," 293.

83 Landecker, "New Times for Biology," 670.

84 Landecker, "New Times for Biology," 670.

85 See, e.g., E. F. Keller, *Refiguring Life*; E. F. Keller, "Rethinking the Meaning of Biological Information"; Kay, *Who Wrote the Book of Life?*; Rheinberger, "Gene Concepts."

86 SymbioticA is described as an "artistic laboratory dedicated to the research, learning, critique and hands-on engagement with the life sciences," https://www.symbiotica.uwa.edu.au/, accessed 26 April 2024. Guy Ben-Ary, artist in residence, and I performed and documented the umbilical cord experiment in December 2006.

87 Knorr-Cetina, *Epistemic Cultures*.

88 Landecker, *Culturing Life*, 10–11.

89 E. F. Keller, "Biological Gaze," 108.

90 Rheinberger, "Die Evidenz des Präparates," 4.

91 On embodied experiences of learning in (bio)medical contexts, see, e.g., Good, *Medicine, Rationality, and Experience*; Prentice, *Bodies in Formation*.

92 See chapter 1 and P. H. Smith, *Body of the Artisan*; Cook, "Medicine." **297**

93 E. F. Keller, "Biological Gaze," 113.

94 E. F. Keller, "Biological Gaze," 109.

95 E. F. Keller, "Biological Gaze," 107.

96 Slatman, "Sense of Life," 305.

97 E. F. Keller, "Biological Gaze," 110–13; Hacking, *Representing and Intervening*, 189–90.

98 Knorr-Cetina, *Epistemic Cultures*; Latour and Woolgar, *Laboratory Life*.

99 Merz, review of *Epistemic Cultures*, 123.

100 Knorr-Cetina, *Epistemic Cultures*, 84.

101 Shapin, "Invisible Technician," 563.

102 Knorr-Cetina, *Epistemic Cultures*, 37.

103 P. H. Smith, *Body of the Artisan*.

104 Knorr-Cetina, *Epistemic Cultures*, 94.

105 Knorr-Cetina, *Epistemic Cultures*, 85–86.

106 Knorr-Cetina, *Epistemic Cultures*, 155. On issues of sterility and contamination in biomedical and science studies, see also, e.g., Fox, "Space, Sterility and Surgery"; Mody, "A Little Dirt Never Hurt Anyone"; Hirschauer, "Manufacture of Bodies in Surgery"; and the discussion of surgical practices in relation to molecular biological practices in Knorr-Cetina *Epistemic Cultures*, 219.

107 See, e.g., Knorr-Cetina, *Epistemic Cultures*, 144, 155.

108 Knorr-Cetina, *Epistemic Cultures*, 86.

109 Knorr-Cetina, *Epistemic Cultures*, 94.

110 Knorr-Cetina, *Epistemic Cultures*, 95.

111 Knorr-Cetina, *Epistemic Cultures*, 95.

112 McAllister, "Virtual Laboratory."

113 Knorr-Cetina, *Epistemic Cultures*, 95.

114 Knorr-Cetina, *Epistemic Cultures*, 96–99.

115 Knorr-Cetina, *Epistemic Cultures*, 97.

116 Knorr-Cetina, *Epistemic Cultures*, 97.

117 Knorr-Cetina, *Epistemic Cultures*, 99–100.

118 E.g., Dreyfus, *What Computers Can't Do*; Dreyfus, *What Computers Still Can't Do*.

119 Knorr-Cetina, *Epistemic Cultures,* 86; on object-oriented epistemic practices, see also Knorr-Cetina, "Objectual Practice."

120 Knorr-Cetina, *Epistemic Cultures*, 98.

121 Knorr-Cetina, *Epistemic Cultures*, 99.

122 Knorr-Cetina, *Epistemic Cultures*, 98.

123 Knorr-Cetina, *Epistemic Cultures*, 99.

124 Knorr-Cetina, *Epistemic Cultures*, 99–100.

125 Knorr-Cetina, *Epistemic Cultures*, 99.

126 Knorr-Cetina, *Epistemic Cultures*, 99.

127 Knorr-Cetina, *Epistemic Cultures*, 99.

128 Shapin, "Invisible Technician."

129 Shapin, "Invisible Technician," 562.

130 Shapin, "Invisible Technician," 563.

131 Shapin, "Invisible Technician," 563.

132 Knorr-Cetina, *Epistemic Cultures*, 32.

133 Knorr-Cetina, *Epistemic Cultures*, 221, 223.

134 Knorr-Cetina, *Epistemic Cultures*, 225.

135 Mody and Kaiser, "Scientific Training," 385.

136 Mody and Kaiser, "Scientific Training," 384.

137 Knorr-Cetina, *Epistemic Cultures*, 221.

138 Knorr-Cetina, *Epistemic Cultures*, 228.

139 Under the supervision of an analyst/technician and senior researcher, I became, in the molecular genetics laboratory, a committed, enthusiastic, and conscientious bench worker with—to my surprise—unexpectedly good motor skills. Because I was quick on the uptake with debates on research questions, the supervising researcher appeared at some point to consider the possibility of letting me do a PhD at her lab. However, she did not pursue this idea when I pointed out that I lacked a sound theoretical foundation, with no undergraduate or master's training in biology and with noticeably rusty mathematical skills.

140 Mody and Kaiser, "Scientific Training," 377.

141 Mody and Kaiser, "Scientific Training," 378.

142 Mody and Kaiser, "Scientific Training," 378.

143 Doing, "'Lab Hands' and the 'Scarlet O.'"

144 Mody, "Sounds of Science," 176.

145 Mody, "Sounds of Science," 175.

146 Mody and Kaiser, "Scientific Training."

147 Hentschel, *Unsichtbare Hände*.

148 Mody and Kaiser, "Scientific Training," 378.

149 Mody and Kaiser, "Scientific Training," 378.

150 Mody and Kaiser, "Scientific Training," 390.

151 Mody and Kaiser, "Scientific Training," 381, 384.

152 H. M. Collins, *Changing Order*, 13–16. More generally on tacit knowledge, see also H. M. Collins, *Tacit and Explicit Knowledge*.

153 Mody and Kaiser, "Scientific Training," 377.

154 Mody and Kaiser, "Scientific Training," 391.

155 Latour and Woolgar, *Laboratory Life*, 245.

156 Mody and Kaiser, "Scientific Training," 389.

157 Mody and Kaiser, "Scientific Training," 389.

158 See, e.g., Lynch, *Art and Artifact in Laboratory Science*, 2.

Chapter 5. In Touch with Life

Epigraph texts and sources: On p. 189, "Was brauche ich zum Arbeiten? Einen Arbeitsplatz in einem Labor mit einer Gehnemigung für gentechnische Versuche, drei Pipetten, mit denen man Volumina zwischen 0 und 1000 μl pipettieren kann, der Rest ist Luxus. Hat man zumindest den Eindruck"; "Der Molekularbiologe, auch Molli genannt, hantiert die meiste Zeit mit winzigen Mengen zumeist klarer, farbloser Lösungen"; in Mühlhardt, *Der Experimentator Molekularbiologie/Genomics*, 7, 1. On p. 216, Barker, *At the Bench*, 187.

1   Kay, *Molecular Vision of Life*, 5.

2   Rheinberger, *Toward a History of Epistemic Things*.

3   Kay, *Molecular Vision of Life*, 4.

4   Hacking, *Representing and Intervening*; E. F. Keller, "Biological Gaze." On the role of preparation technologies in the biological sciences, see also Rheinberger, *Epistemology of the Concrete*, 233–43. Bruno Latour provided palpable descriptions of scientific practices in other disciplines such as soil sampling; see his conceptualization of these manipulations in terms of "circulating references" in Latour, *Pandora's Hope*, 24–77.

5   "Een goed experiment kan niet zonder een goede pipet, behandel de pipetten dus met respect!" Vijgenboom et al., "Handleiding Introductie & Biochemie Practicum I," 1.

6   Online resources for the proper use of micropipettes are abundant; see, e.g., "Using Micropipettes Correctly" by Clare M. O'Connor in the Biology Library, a collaborative open-access pedagogical initiative, hosted by LibreTexts Projects, that offers a "Living Library" with continuously updated online textbook materials, accessed April 27, 2024, https://bio .libretexts.org/Bookshelves/Cell_and_Molecular_Biology/Book%3A _Investigations_in_Molecular_Cell_Biology_(O%27Connor)/02%3A _Mastering_the_micropipette/2.01%3A_Using_micropipettes_correctly.

7   See "Experiment 1. IJken van een pipet" (figure I.1), a first-year biochemistry practicum for life science and technology students, from Vijgenboom et al., "Handleiding Introductie & Biochemie Practicum I."

8   Rheinberger, *Epistemology of the Concrete*, 217.

9   Heidelberger, "Experiment und Instrument," 386.

10 Heidelberger, "Experiment und Instrument," 386–87.

11 Rheinberger, *Toward a History of Epistemic Things*.

12 Rheinberger, *Epistemology of the Concrete*, xi.

13 See, e.g., Lynch, *Art and Artifact in Laboratory Science*; Rabinow, *Making PCR*; Rheinberger, *Toward a History of Epistemic Things*; Landecker,

Culturing Life; Myers, *Rendering Life Molecular;* Biagioli, *Science Studies Reader;* Jasanoff et al., *Handbook of Science and Technology Studies* (see also Hackett et al., 3rd ed., and Felt et al., 4th ed.).

14   Rheinberger, *Epistemology of the Concrete,* 217.

15   E.g., Rheinberger, *Epistemology of the Concrete*; Daston, *Biographies of Scientific Objects*; Daston and Lunbeck, *Histories of Scientific Observation*; Daston and Gallison, *Objectivity.*

16   Rose, "Politics of Life Itself," 13–14; Shostak, "Emergence of Toxicogenomics"; Chadarevian and Kamminga, *Molecularizing Biology and Medicine*; Kay, *Molecular Vision of Life.*

17   Pfeiffer, "Die 'Marburg Pipette,'" 51.

18   Pfeiffer, "Die 'Marburg Pipette.'" On the Marburg pipette as an icon of modern biological technology and biomedical research, see Klingenberg, "When a Common Problem Meets an Ingenious Mind"; on it being a "mainstay in laboratory work," see Lähteenmäki, Hodgson, and Michael, "Tips for Your Pipette," 1030; on it being "a lifelong friend" to bench workers, see Martin, "Art of the Pipet."

19   Interview with Günther Bechtler and Wilhelm Bergmann on July 28, 1999, in Pfeiffer, "Die 'Marburg Pipette,'" 134–62.

20   Pfeiffer, "Die 'Marburg Pipette,'" 89.

21   Eppendorf AG, headquartered in Hamburg, Germany; see the corporate website, Eppendorf, accessed April 27, 2024, https://corporate.eppendorf .com/.

22   Pfeiffer, "Die 'Marburg Pipette,'" 152.

23   Chadarevian and Kamminga, *Molecularizing Biology and Medicine,* 7.

24   See note 15.

25   See Mody and Kaiser, "Scientific Training"; Doing, "'Lab Hands' and the 'Scarlet O.'"

26   See, e.g., Ihde, *Instrumental Realism*; P.-P. Verbeek, *What Things Do.*

27   The invention of the first adjustable automatic pipette has also been claimed by an American company that brought the Gilson pipette onto the market in the early 1970s; see Klingenberg, "When a Common Problem Meets an Ingenious Mind"; Pfeiffer, "Die 'Marburg Pipette,'" 86.

28   The following historical summary is based on Pfeiffer, "Die 'Marburg Pipette.'"

29   A detailed account of these experiments is presented in Pfeiffer, "Die 'Marburg Pipette,'" 44–45.

30   Pfeiffer presents an overview of pipettes used in the twentieth century before and after the invention of the piston strike pipette (*Kohlenhubpipette*); Pfeiffer, "Die 'Marburg Pipette,'" 6–12.

31 "Glasblasen, also Pipettenspitzen zu machen, gehörte zu einer der Haupt-
    beschäftigungen von uns. Wir haben Glasrohre gekauft. Das konnte man
    bei der Firma Kroge. Dann haben wir die über den Bunsenbrenner gezo-
    gen und so fein ausgezogen, wie wir sie haben wollten. Ja, wir mussten
    alles machen: Geräte, bis auf den Photometer, die konnten wir so kriegen,
    aber sonstige Geräte (zum Beispiel Destillierer, Apparate) musste man
    sich alles selbst zusammenbauen." Interview with Hans-Jürgen Hohorst
    on June 11, 1999, in Pfeiffer, "Die 'Marburg Pipette,'" 104.

32 "Wenn man nicht aufpaßte, hatte man das ganze Zeug im Mund. Man
    musste ja immer saugen. Und man musste diesen Widerstand erfühlen.
    Das war alles andere als ideal." Interview with Martin Klingenberg on
    June 28, 1999, in Pfeiffer, "Die 'Marburg Pipette,'" 122.

33 On the important role of Louis Pasteur's research on fermentation and
    first experiments with vaccination and Latour's attempt to rewrite his-
    tories of scientific discoveries as events that cannot be understood as
    the deeds of a genius with a brilliant idea or theory, but as occurrences
    unfolding within a network of different material entities and (social)
    actors, see Latour, *Pasteurization of France*.

34 "Der ein phänomenaler Bastler war, aber keine akademische Ausbildung
    hatte." Interview with Hohorst on June 11, 1999, in Pfeiffer, "Die 'Marburg
    Pipette,'" 107.

35 Pfeiffer, "Die 'Marburg Pipette.'"

36 Doing, "'Lab Hands' and the 'Scarlet O.'"

37 Doing, "'Lab Hands' and the 'Scarlet O.'"

38 Doing, "'Lab Hands' and the 'Scarlet O,'" 308.

39 "Ja. Da musste man auch mal mit ins Labor kommen, muß man heute
    auch noch. Bei Professor Bücher war das noch etwas Besonderes. Nun
    hing er auch sehr an der Werkstatt, kam oft und hat was gefragt. Er war
    ja handwerklich auch recht geschickt. Wir mussten dann hochkommen
    und zuschauen, wofür es dann gebraucht wurde. Das war schon immer
    interessant. Er war aber auch sehr ungeduldig. Alles musste immer sehr
    schnell gehen (lacht). Er war halt ständig in der Werkstatt und hat ge-
    schaut wie weit man ist. Aber ich glaube, so war er nicht nur mit der
    Werkstatt, sondern mit allen." Interview with Willi Bender on June 29,
    1999, in Pfeiffer, "Die 'Marburg Pipette,'" 132.

40 "In Marburg hatte er eine Versuchswerkstatt zur Verfügung, die war für
    die Zeit ungewöhnlich gut. Das hatte Professor Bücher gewollt. Er hatte
    dieses Institut nur übernommen unter der Auflage, dass er eine nagelneue
    Versuchswerkstatt bekomme, die top sein musste. Und Schnitger, wie er da
    als junger Mann hinkam, hat diese Versuchswerkstatt zwar genutzt, aber
    sich sofort eine eigene nebenan im Keller gebaut. Da gab es einen Meister,

einen klasse ausgebildeten Feinmechaniker-Meister, der im Wesentlichen dem Klingenberg und dem Schnitger zuarbeiten sollte. Schnitger hat ihm so richtig das Wasser abgegraben. Er sagte: 'Was du da so machst, ist ja alles ein guter Wille, aber so richtig habt ihr ja nischt drauf.' Und dann hat er darauf bestanden, dass sie ihm in so einem Kellerraum, der übrigens zugemauerte Fenster hatte—das haben Sie sicher schon gehört: er wollte unabhängig von der Tageszeit sein. Er hat in diesem zugemauerten Raum, der auch kühl war und eben nicht bestimmt war durch Tageszeiten oder Winter und Sommer, sich Maschinen hingestellt—übrigens eine kleine 'Hommel'-Maschine: das ist eine ganz besondere Bearbeitungsmaschine, die kannte er, weiß der Teufel woher, die hatte man früher in den U-Booten; diese hatte er sich gekauft oder bestellt und hat mit ihr er dann selbst solche Sachen gebaut." Interview with Wilhelm Bergmann on July 28, 1999, in Pfeiffer, "Die 'Marburg Pipette,'" 136–37.

41   P. H. Smith, *Body of the Artisan*; Roberts, Schaffer, and Dear, *Mindful Hand.*

42   Pfeiffer, "Die 'Marburg Pipette,'" 46.

43   Pfeiffer, "Die 'Marburg Pipette,'" 44–45.

44   A detailed description of the working mechanism and patent specifications, including a reprint of the patent letter, can be found in Pfeiffer, "Die 'Marburg Pipette,'" 55.

45   Pfeiffer provides a precise description of the first fifty models built by Schnitger and Helmut Funke, head of the experimental workshop of the Institute of Physiological Chemistry in Marburg; Pfeiffer, "Die 'Marburg Pipette,'" 46–47. An example of this model belongs to the collection of the Deutsche Museum in Munich, Germany.

46   Pfeiffer, "Die 'Marburg Pipette,'" 86; see also Rabinow, *Making* PCR; Landecker, *Culturing Life.*

47   Landecker, *Culturing Life*, 1.

48   Etzkowitz and Leydesdorff, "Theme Paper Triple Helix I"; on the concept of the triple helix, see also, e.g., Etzkowitz and Leydesdorff, "Dynamics of Innovation."

49   Rheinberger, *Epistemology of the Concrete*, 25–36.

50   Rheinberger, "Cytoplasmic Particles," 273.

51   Rheinberger, "Cytoplasmic Particles," 272.

52   A. E. Clarke and Fujimura, *Right Tools for the Job.*

53   Ravetz, *Scientific Knowledge and Its Social Problems*, 93, cited in A. E. Clarke and Fujimura, *Right Tools for the Job*, 15.

54   Rheinberger, *Toward a History of Epistemic Things*, 29.

55   A. E. Clarke and Fujimura, *Right Tools for the Job*, 14; Schaffer, "Glass Works."

56   Callon and Latour, "Unscrewing the Big Leviathan," 285.

57   Jordan and Lynch, "Sociology of a Genetic Engineering Technique."

58   Latour, *Pandora's Hope*, 304.

59   Latour, *Pandora's Hope*, 304.

60   On conceptual reflections, see, e.g., Beurton, Falk, and Rheinberger, *Concept of the Gene in Development and Evolution*.

61   Rheinberger, *Epistemology of the Concrete*, 217.

62   Rheinberger, *Epistemology of the Concrete*, 217.

63   Rheinberger, *Epistemology of the Concrete*, 217.

64   Rheinberger, *Epistemology of the Concrete*, 219.

65   Rheinberger, *Epistemology of the Concrete*, 218.

66   Rheinberger, *Epistemology of the Concrete*, 218.

67   Rheinberger, *Epistemology of the Concrete*, 218.

68   Rheinberger, *Epistemology of the Concrete*, 218–19.

69   *Say It Isn't So: Naturwissenschaften im Visier der Kunst*, exhibition, Museum für moderne Kunst, Bremen, May 12 to September 16, 2007; see also Friese et al., *Say It Isn't So*, 235–40.

70   Ophthalmology is a "branch of medicine that is concerned with the study and treatment of diseases and disorders of the eye"; *Oxford English Dictionary*, s.v. "ophthalmology (n.)," July 2023, https://doi.org/10.1093/OED/1655844763.

71   Herwig Turk, *agents,* photo series in collaboration with Paulo Pereira, 2007, accessed April 26, 2024, https://www.herwigturk.net/de/ausgewaehlte-arbeiten/agents#fotos; Herwig Turk, *hands on (vers.3),* video work, 2014, accessed July 26, 2023, https://www.herwigturk.net/en/selected-works/hands-on#videos.

72   Reichle, "Art of Making Science"; Reichle, "Unter Beobachtung"; Hentschel, *Unsichtbare Hände*.

73   Hentschel, *Unsichtbare Hände*, 13.

74   Polanyi, *Personal Knowledge*.

75   Reichle, "Art of Making Science"; Reichle, "Unter Beobachtung"; Herwig Turk, conversation with the author, October 8, 2008.

76   Reichle, "Art of Making Science," 17.

77   On this point, see also Reichle, "Taube Bilder und Sehende Hände," 180.

78   Meinel, *Instrument—Experiment*.

79   Pereira in Turk and Pereira, *Blindspot*, 29.

80   On "unreflective action," see, e.g., Rietveld, "Situated Normativity"; Rietveld, "McDowell and Dreyfus on Unreflective Action."

81   Shusterman, *Body Consciousness*, 3.

82   Fischer, *Emergent Forms of Life*, 102–5.

83   Fischer, *Emergent Forms of Life*, 105.

84   Prentice, *Bodies in Formation*, 111.

85   For Ancari's video work, see "Yuri Ancari—Da Vinci—Inside Palais de Tokyo—2014," YouTube, https://youtu.be/4cBoqgJfyXw. For the 2013 Venice Biennale, *Il Palazzo Enciclopedico (The Encyclopedic Palace)*, curated by Massimiliano Gioni, see https://www.labiennale.org/en/il -palazzo-enciclopedico.

86   See "Making Clinical Sense: A Comparative Study of How Doctors Learn in Digital Times," accessed August 3, 2022, https://www .makingclinicalsense.com/.

87   A. E. Clarke and Fujimura, *Right Tools for the Job*, 3.

88   A. E. Clarke and Fujimura, *Right Tools for the Job*, 6.

89   A. E. Clarke and Fujimura, *Right Tools for the Job*, 5.

90   A. E. Clarke and Fujimura, *Right Tools for the Job*, 5.

91   A. E. Clarke and Fujimura, *Right Tools for the Job*, 6.

92   Knorr-Cetina, *Epistemic Cultures*.

93   See the introduction, note 1.

94   E.g., Hirschauer, "Manufacture of Bodies in Surgery"; Mol, *Body Multiple*; Myers, "Molecular Embodiments"; Myers, *Rendering Life Molecular*; Mesman, *Uncertainty in Medical Innovation*; Doing, "'Lab Hands' and the 'Scarlet O'"; Mody, "A Little Dirt Never Hurt Anyone"; Mody, "Sounds of Science"; Landecker, *Culturing Life*; Prentice, *Bodies in Formation*.

95   E.g., Ihde, *Technics and Praxis*; Ihde, *Experimental Phenomenology*.

96   Rheinberger, "Cytoplasmic Particles," 276.

97   For the following section, see Rheinberger, *Toward a History of Epistemic Things*, 26–28.

98   Rheinberger, *Toward a History of Epistemic Things*, 28.

99   Rheinberger, *Toward a History of Epistemic Things*, 28.

100   Rheinberger, *Toward a History of Epistemic Things*, 28.

101   Rheinberger, *Toward a History of Epistemic Things*, 28.

102   Rheinberger, *Historische Epistemologie zur Einführung*, 96; Rheinberger, *On Historicizing Epistemology*, 63.

103   Gene technology practicum led by Dr. J. A. Brandsma, April 11 to May 2, 2006; for more information on my fieldwork in a molecular genetics research laboratory, see the introduction, note 1, and chapter 4, note 24.

104 A. E. Clarke and Fujimura, *Right Tools for the Job*.

105 Ihde, *Instrumental Realism*.

106 Eppendorf AG webinar by Dr. Hanaë Henke and Dr. Jessica Wagener, "Protect Your Cells with Proper Pipetting—How Liquid Handling Influences Your Cell Culture Work," November 2, 2017, https://handling -solutions.eppendorf.com/fileadmin/Community/Liquid_Handling /countonit/Biosafety_at_the_highest_level/Webinar_Webinar -pipetting-in-Cell-Culture_Protect-Your-Cells.pdf

107 Brey, "Technology and Embodiment in Ihde and Merleau-Ponty," 45.

108 Ihde, *Technics and Praxis*.

109 Merleau-Ponty, *Phenomenology of Perception*.

110 Slatman, *Our Strange Body*, 68.

111 Merleau-Ponty, *Phenomenology of Perception*, 160.

112 Merleau-Ponty, *Phenomenology of Perception*, 166.

113 Merleau-Ponty, *Phenomenology of Perception*, 166.

114 Slatman, *Our Strange Body*, 68.

115 Shusterman, *Body Consciousness*.

116 On "experiential transparency," see Gallagher, *How the Body Shapes the Mind*; on "instrumental transparency," see Ihde, *Instrumental Realism*.

117 Shusterman, *Body Consciousness*, 3.

118 Ihde, *Instrumental Realism*, 29.

119 Ihde, *Instrumental Realism*, 79.

120 Ihde, *Technology and the Lifeworld*, 92.

121 Ihde, *Technology and the Lifeworld*, 72.

122 Ihde, *Technology and the Lifeworld*, 73–74.

123 Ihde, *Instrumental Realism*, 75.

124 Ihde, *Instrumental Realism*, 75.

125 Ihde, *Technology and the Lifeworld*, 107.

126 Olesen, "Technological Mediation," 242.

127 Ihde, *Technology and the Lifeworld*, 73.

128 Ihde, *Technology and the Lifeworld*, 75.

129 P.-P. Verbeek, "Don Ihde," 130.

130 Ihde, *Technology and the Lifeworld*, 40.

131 Ihde, *Technology and the Lifeworld*, 73.

132 P.-P. Verbeek, "Don Ihde," 124.

133 Ihde, *Technics and Praxis*, 18.

134 Ihde, *Experimental Phenomenology*, 141.

135 Ihde, *Technics and Praxis*, 19.

136 Ihde, *Technics and Praxis*, 20.

137 Ihde, *Technics and Praxis*, 19.

138 Ihde, *Technics and Praxis*, 20.

139 See also Slatman, "On the (Im)Possibility of Immediate Bodily Experience."

140 Slatman, *Our Strange Body*, 68–76.

141 Personal conversation with my dentist in Amsterdam, April 2, 2012.

142 I thank Sheena Hyland for bringing Merleau-Ponty's ignorance of itchy bodies to my attention at the UCD Body Conference: Perspectives on the Body and Embodiment in Dublin. See Hyland, "Merleau-Ponty, Beckett and the Body."

143 Feenberg, "Active and Passive Bodies: Comments on Don Ihde's Bodies in Technology"; Feenberg, "Active and Passive Bodies: Don Ihde's Phenomenology of the Body."

144 Derksen and Horstman, "Engineering Flesh."

145 Derksen and Horstman, "Engineering Flesh," 271–72.

146 Gallagher, *How the Body Shapes the Mind*, 73.

147 Gallagher, *How the Body Shapes the Mind*, 6–7.

148 Gallagher, *How the Body Shapes the Mind*.

149 Mody and Kaiser, "Scientific Training."

150 Gallagher, *How the Body Shapes the Mind*, 20.

151 Jordan and Lynch, "Sociology of a Genetic Engineering Technique."

152 Rheinberger, *Toward a History of Epistemic Things*; Rheinberger, *Epistemology of the Concrete*; Rheinberger, *On Historicizing Epistemology*; Heidelberger, "Experiment und Instrument"; Hentschel, *Unsichtbare Hände*.

153 McAllister, "Die Rhetorik der Mühelosigkeit."

Epilogue

Epigraph Sources: On p. 240, Cleanroom training presentation by Hugo Huiskamp, guest speaker at Boo Chapple's Vivo Arts Course, the Arts and Genomics Centre at Leiden University, session on sterility and cleanrooms (June 5, 2009); Alex Farquharson, quoted in La Frenais and Daily, *Clean Rooms*, n.p. The catalog accompanied a show curated by Arts Catalyst in association with Gallery Oldham, Oldham, Greater

Manchester, UK, October 5 through November 11, 2022, featuring art-works by Critical Art Ensemble, Gina Czarnecki, and Neal White.

1   *Zeit Magazin*, July 14, 2011. In German, *forsch* can be an adjective that alludes to a "bold" attitude, or an imperative, "(do) research!" The second cover image shows a man in a suit who looks successful and "bold." For the cover images, see "Ace Covers 2011 / Zeit Magazin," accessed July 15, 2022, https://coverjunkie.com/cover-categories/best-of-the-rest/forsch/.

2   For an online version of the magazine's interview with Rajewsky, see "Bisher war es aufregend, jetzt wird es dramatisch," *Zeit Online*, July 14, 2022, http://www.zeit.de/2011/29/Gelehrte-Rajewsky-Humanbiologe.

3   On operating theaters, see Hirschauer, "Manufacture of Bodies in Surgery"; on neonatal intensive care units, see Mesman, *Uncertainty in Medical Innovation*; on material sciences laboratories, see Mody, "A Little Dirt Never Hurt Anyone."

4   Katz, "Ritual in the Operating Room"; Hirschauer, "Manufacture of Bodies in Surgery."

5   Katz, "Ritual in the Operating Room," 344–45.

6   Hirschauer, "Manufacture of Bodies in Surgery."

7   Hirschauer, "Manufacture of Bodies in Surgery," 291.

8   Hirschauer, "Manufacture of Bodies in Surgery," 279.

9   See Fischer, *Emergent Forms of Life*, 316–17; Prentice, *Bodies in Formation*. See also Yuri Ancarani's video work *Da Vinci* (2012), documenting surgeons' work and training with a surgical robot, https://www.yuriancarani.com/works/da-vinci/.

10 Good, *Medicine, Rationality, and Experience*; Fischer, *Emergent Forms of Life*, 101–5.

11 Mody, "A Little Dirt Never Hurt Anyone," 7.

12 Mesman, *Uncertainty in Medical Innovation*.

13 Whyte, *Cleanroom Design*; Whyte, *Cleanroom Technology*.

14 These rooms are often controlled for smaller particles than are microbiological cleanrooms because microorganisms do not attach to particles smaller than 5 micrometers.

15 Whyte, *Cleanroom Design*, 17.

16 "Germfree life, biological condition characterized by the complete absence of living microorganisms. Gnotobiology comprises the study of germfree plants and animals, as well as living things in which specific microorganisms, added by experimental methods, are known to be present"; Britannica, "Germfree Life." On the history of gnotobiological

experiments, see, e.g., Dolan, "100 Years Plus of Gnotobiology"; Basic and Bleich, "Gnotobiotics."

17   Nuttall and Thierfelder, "Thierisches Leben ohne Bakterien im Verdauungskanal."

18   Küster, "Die keimfreie Züchtung von Säugetieren."

19   On a rhetoric of self-effacement in scientific writing, see also McAllister, "Die Rhetorik der Mühelosigkeit."

20   Waldenfels, *Phänomenologie der Aufmerksamkeit.*

21   On Jennifer Willet, see, e.g., Willet, "Bodies in Biotechnology"; Willet, "(RE)Embodying Biotechnology." On Willet's *Bioplay*, a clandestine performance in a fume hood and its photographic documentation in the laboratory, see, e.g., Zwijnenberg, "BIOPLAY."

22   See "Workhorse Zoo," accessed March 22, 2024, http://emutagen.com/whzoogl.html.

23   The workshop was organized in collaboration with the Australian artist Oron Catts. It was part of Zaretsky's artist-in-residency at the Waag Society in Amsterdam, where the artist established a temporary research and education institute, VivoArts School for Transgenic Aesthetics Ltd. (VASTAL). The public event was announced as the VASTAL Tissue Culture Lab and took place on September 15, 2009, at the Waag Society, Amsterdam. See see Waag Society and VASTAL, *Glove Box Performance*, accessed July 17, 2022, https://vimeo.com/8968488.

24   The artists Oron Catts and Ionat Zurr have specialized since the 1990s in the practice of tissue engineering under the name the Tissue Culture & Art Project (TC&A); see "The Tissue Culture & Art Project," accessed July 15, 2022, https://tcaproject.net/.

25   See publications by TC&A at http://tcaproject.org/publications, and especially Zurr, "Growing Semi-Living Art"; Catts, Zurr, and Ary, "Tissue Culture & Art(ificial) Wombs." On the history of the art of tissue culturing, see, e.g., Landecker, *Culturing Life.*   309

26   Catts and Zurr, "Towards a New Class of Being."

27   Zurr and Catts, "Ethical Claims of Bio-Art."

28   This topic was addressed at "The Aesthetics of Care?," a symposium that was part of the exhibition *Biofeel*, Biennale of Electronic Arts Perth in 2002.

29   Shusterman, *Body Consciousness*, 3.

30   On the notion of "strange bodies," see also Slatman, *Our Strange Body.*

31   Mol, "Pathology and the Clinic," 82.

32   Mol, "Pathology and the Clinic," 84.

33   Chapple was invited by the Arts and Genomics Centre at Leiden University, where she taught the third edition of this course in 2009; earlier editions were taught by bio artists Adam Zaretsky and Jennifer Willet.

34   I participated in the VCCN Cleanroom Gedrags Cursus at the Netherlands Organisation for Applied Scientific Research, known as TNO (https://www.tno.nl/en), in Delft, July 9, 2009, and had several conversations and an interview with Hugo Huiskamp, the cleanroom consultant and course instructor.

35   See ISO 14644-5:2004 Annex C on "Personnel," in International Organization for Standardization, *Cleanrooms and Associated Controlled Environments, Part 5*. This publication sets the international standards for operating cleanrooms; first published in 2004 by the International Organization for Standardization (ISO) in Geneva, it is periodically updated.

36   Huiskamp also gave this presentation in Boo Chapple's Vivo Arts Course during a session on sterility and cleanrooms on June 5, 2009. I quote here also from syllabi that were handed out to participants of the VCCN Cleanroom Gedrags Cursus.

37   Kontopodis, Niewöhner, and Beck, "Investigating Emerging Biomedical Practices."

38   Standards for GMP are set by the European Union; see European Commission, *Eudralex*, Vol. 4, *Good Manufacturing Practice (GMP) Guidelines,* https://health.ec.europa.eu/medicinal-products/eudralex/eudralex-volume-4_en.

39   The ISO classifies cleanrooms according to the number and size of particles in the air. In ISO 14644-1:2015, a cleanroom is defined as follows: "A room in which the concentration of airborne particles is controlled, and which is constructed and used in a manner to minimize the introduction, generation, and retention of particles inside the room and in which other relevant parameters, for example, temperature, humidity, and pressure are controlled as necessary." International Organization for Standardization, *Cleanrooms and Associated Controlled Environments, Part 1*, clause 2.1.1. More generally on cleanroom design and technology, see the classic texts Whyte, *Cleanroom Design*; Whyte, *Cleanroom Technology*; Ramstorp, *Introduction to Contamination Control*; Carlberg, *Cleanroom Microbiology*.

40   Cleanroom visit and interview with cleanroom user (research analyst), Interdivisional Good Manufacturing Practice facility, Department of Clinical Pharmacy and Toxicology, Leiden University Medical Center, October 5, 2009.

310

41   The research analyst never puts on necklaces when she has a cleanroom working day scheduled, but she wears her plain golden wedding ring, since it will be fully covered by gloves and has no sharp edges that could damage the gloves.

42   The GMP guidelines are used, for example, in the pharmaceutical industry to ensure quality of medicinal production processes.

43   Latour and Woolgar, *Laboratory Life*.

44   All quotes are taken from my English translations from the anonymized interview I conducted in Dutch with the analyst of the Department of Clinical Pharmacy and Toxicology, Leiden University Medical Center, October 5, 2009.

45   See note 28 on the symposium "Aesthetics of Care?"

46   Rheinberger, *On Historicizing Epistemology*; Mol, *The Body Multiple*.

47   Latour and Woolgar, *Laboratory Life*.

Works by René Descartes

Historical Editions

Descartes, René. *Discours de la méthode pour bien conduire sa raison et chercher la verité dans les sciences. Plus la dioptrique, les meteores et la géometrie, qui sont des essais de cette méthode.* Leiden: L'imprimerie de Jan Maire, 1637. Reproduction, https://gallica.bnf.fr/ark:/12148/btv1b86069594.

Descartes, René. *L'homme de René Descartes et un traitté de la formation du foetus; avec les remarques de Louis de La Forge, . . . sur le traitté de l'homme de René Descartes et sur les figures par luy inventées.* Paris: Charles Angot, 1664. Reproduction, http://gallica.bnf.fr/ark:/12148/bpt6k574850.

Descartes, René. *Les méditations métaphysiques de René Des-Cartes touchant la première philosophie . . . traduites du latin de l'auteur par M. le D. D. L. N. S., et les objections faites contre ces méditations par diverses personnes très doctes, avec les réponses de l'auteur, traduites par M. C. L. R.* Paris: J. Camusat et P. le Petit, 1647. Reproduction, https://gallica.bnf.fr/ark:/12148/btv1b86015099.

Descartes, René. *Les passions de l'ame.* Paris: Henry Le Gras, 1649. Reproduction, https://gallica.bnf.fr/ark:/12148/btv1b8601505n.

Descartes, René. *R. Des-Cartes Regulae ad directionem ingenii ut et inquisitio veritatis per lumen naturale.* Amsterdam: Pieter Blaeu, 1701. Reproduction, https://dspace.library.uu.nl/handle/1874/10227.

Descartes, René. *Renati Des-Cartes Meditationes de prima philosophia, in qua Dei existentia et animae immortalitas demonstratur.* Paris: M. Soly, 1641. Reproduction, https://gallica.bnf.fr/ark:/12148/btv1b86002964.

Descartes, René. *Renati Des-cartes Meditationes de prima philosophia: in quibus de prima philosophia, in quibus Dei existentia, et animae humanae a corpore distinctio, demonstrantur.* 2a ed. septimis objectionibus antehac non visis aucta. Amsterdam: Ludovicus Elzevirius, 1642.

Modern Edition

Descartes, René. *Oeuvres de Descartes.* 12 vols. New ed. Edited by Charles Adam and Paul Tannery. New ed. Paris: Vrin, 1996. [Abbreviated as AT I–XII]

Translations into English

Descartes, René. *Descartes: Selected Philosophical Writings.* Translated by John Cottingham, Robert Stoothoff, and Dugald Murdoch. Cambridge: Cambridge University Press, 1988.

Descartes, René. *Discourse on Method, Optics, Geometry, and Meteorology.* Translated and edited by Paul J. Olscamp. Indianapolis, IN: Bobbs-Merrill, 1976.

Descartes, René. *The Philosophical Writings of Descartes.* Vols. 1 and 2. Translated by John Cottingham, Robert Stoothoff, and Dugald Murdoch. Cambridge: Cambridge University Press, 1985. [Abbreviated as CSM 1–2]

Descartes, René. *The Philosophical Writings of Descartes.* Vol. 3, *The Correspondence.* Translated by John Cottingham, Robert Stoothoff, Dugald Murdoch, and Anthony Kenny. Cambridge: Cambridge University Press, 1991. [Abbreviated as CSMK]

Descartes, René. *The World and Other Writings.* Translated and edited by Stephen Gaukroger. Cambridge: Cambridge University Press, 1998.

314

Other Works Cited

Alberts, Bruce. *Molecular Biology of the Cell.* 6th ed. New York: Garland Science, 2015.

Altman, Lawrence K. "Nobel Prize in Medicine Awarded to 3 Scientists for Parasite-Fighting Therapies." *New York Times,* October 5, 2015, Science. https://www.nytimes.com/2015/10/06/science/william-c-campbell-satoshi-omura-youyou-tu-nobel-prize-physiology-medicine.html.

Antoine-Mahut, Delphine, and Stephen Gaukroger, eds. *Descartes' Treatise on Man and Its Reception.* New York: Springer, 2016.

AP News. "The Latest: Discovery on Golf Course Inspired Nobel Winner." October 5, 2015. https://apnews.com/fd1c045359424d1cbd77462d03a9491c.

Awiakta, Marilou. "How the Corn-Mother Became a Teacher of Wisdom: A Story in Counterpoint—Two Mind-Sets, Two Languages." In *Transformations: Feminist Pathways to Global Change*, edited by Torry D. Dickinson and Robert K. Schaeffer, 195–203. Boulder, CO: Paradigm, 2008.

Bacon, Francis, Lisa Jardine, and Michael Silverthorne. *The New Organon*. Cambridge: Cambridge University Press, 2000.

Barad, Karen. *Meeting the Universe Halfway: Quantum Physics and the Entanglement of Matter and Meaning*. Durham, NC: Duke University Press, 2007.

Barker, Kathy. *At the Bench: A Laboratory Navigator*. Updated edition. Cold Spring Harbor, NY: Cold Spring Harbor Laboratory Press, 2005.

Barrera-Osorio, Antonio. *Experiencing Nature: The Spanish American Empire and the Early Scientific Revolution*. Austin: University of Texas Press, 2006.

Basic, Marijana, and André Bleich. "Gnotobiotics: Past, Present and Future." *Laboratory Animals* 53, no. 3 (2019): 232–43. https://doi.org/10.1177/0023677219836715.

Berkel, Klaas van. "The Dutch Republic: Laboratory of the Scientific Revolution." BMGN—*Low Countries Historical Review* 125, nos. 2–3 (January 1, 2010): 81–105. https://doi.org/10.18352/bmgn-lchr.7116.

Bertoloni Meli, Domenico. *Thinking with Objects: The Transformation of Mechanics in the Seventeenth Century*. Baltimore: Johns Hopkins University Press, 2006.

Beurton, Peter J., Raphael Falk, and Hans-Jörg Rheinberger, eds. *The Concept of the Gene in Development and Evolution: Historical and Epistemological Perspectives*. Cambridge: Cambridge University Press, 2000.

Biagioli, Mario, ed. *The Science Studies Reader*. New York: Routledge, 1999.

Binford, Lewis R. "Smudge Pits and Hide Smoking: The Use of Analogy in Archaeological Reasoning." *American Antiquity* 32, no. 1 (1967): 1–12. https://doi.org/10.2307/278774.

Bitbol-Hésperiès, Annie. "Cartesian Physiology." In *Descartes' Natural Philosophy*, edited by Stephen Gaukroger, John Andrew Schuster, and John Sutton, 349–82. London: Routledge, 2000.

Blackburn, Elizabeth H. "Elizabeth Blackburn on Barbara McClintock." Video recording. Oral History Collection, CSHL Digital Archives, June 2000. http://library.cshl.edu/oralhistory/interview/cshl/barbara-mcclintock/mcclintock/.

Blum, Andreas L., John Michael Krois, and Hans-Jörg Rheinberger, eds. *Verkörperungen*. Berlin: Akademie Verlag, 2012. https://doi.org/10.1524/9783050062549.

Boulboullé, Jenny. "Die verführerische Transparenz des geometrischen Raums: Der Raum als erkenntnistheoretisches Problem bei Kant und Husserl." Master's thesis, University of Amsterdam, 2004.

Brey, Philip. "Technology and Embodiment in Ihde and Merleau-Ponty." In *Metaphysics, Epistemology, and Technology*, edited by Carl Mitcham, 45–58. New York: JAI Press, 2000.

Brissey, Patrick. "Rule VIII of Descartes' *Regulae ad directionem ingenii*." *Journal of Early Modern Studies* 3, no. 2 (2014): 9–31. https://doi.org/10.5840/jems20143212.

Britannica, The Editors of Encyclopaedia. "Germfree Life." *Encyclopedia Britannica*, November 21, 2018. https://www.britannica.com/science/germfree-life.

Broecke, Lara. *Cennino Cennini's* Il Libro dell'Arte: *A New English Translation and Commentary with Italian Transcription*. London: Archetype, 2015.

Burke, Peter. "Response." *Journal for the History of Knowledge* 1, no. 1 (2020): 1–7. https://doi.org/10.5334/jhk.27.

Burke, Peter. *What Is the History of Knowledge?* Cambridge: Polity Press, 2016.

Burnyeat, Myles. F. "Idealism and Greek Philosophy: What Descartes Saw and Berkeley Missed." *Philosophical Review* 91, no. 1 (January 1982): 3–40.

Callon, Michel, and Bruno Latour. "Unscrewing the Big Leviathan: How Actors Macro-Structure Reality and How Sociologists Help Them to Do So." In *Advances in Social Theory and Methodology: Toward an Integration of Micro- and Macro-Sociologies*, edited by K. Knorr-Cetina and A. V. Cicourel, 277–303. Boston: Routledge and Kegan Paul, 1981.

Campbell, Gordon. "Fernel, Jean François." In *The Oxford Dictionary of the Renaissance*, 277. Oxford: Oxford University Press, 2005.

Carlberg, David M. *Cleanroom Microbiology for the Non-Microbiologist*. 2nd ed. Boca Raton, FL: CRC Press, 2005.

Catts, Oron, and Ionat Zurr. "Towards a New Class of Being—The Extended Body." *Artnodes* no. 6 (November 2006). https://doi.org/10.7238/a.v0i6.755.

Catts, Oron, Ionat Zurr, and Ben Guy Ary. "Tissue Culture & Art(ificial) Wombs: An Installation of Semi-Living Worry Dolls inside a Bioreactor." In *Next Sex: Sex in the Age of Its Procreative Superfluousness*, edited by Gerfried Stocker and Christine Schöpf, 225–55. Vienna: Springer, 2000.

Cerasuolo, Angela. *Literature and Artistic Practice in Sixteenth-Century Italy*. Translated by Helen Glanville. Leiden: Brill, 2016.

Chadarevian, Soraya de, and Harmke Kamminga, eds. *Molecularizing Biology and Medicine: New Practices and Alliances, 1910s–1970s*. Amsterdam: Harwood Academic Publishers, 1998.

Clarke, Adele E., and Joan H. Fujimura, eds. *The Right Tools for the Job: At Work in Twentieth-Century Life Sciences*. Princeton, NJ: Princeton University Press, 1992. https://doi.org/10.1515/9781400863136.

Clarke, Desmond. *Descartes' Philosophy of Science*. Manchester: Manchester University Press, 1982.

Clarke, Desmond. "Descartes' Philosophy of Science and the Scientific Revolution." In *The Cambridge Companion to Descartes*, edited by John Cottingham, 258–85. Cambridge: Cambridge University Press, 1992. https://doi.org/10.1017/CCOL0521366232.010.

Coelho, Victor, ed. *Music and Science in the Age of Galileo*. Dordrecht: Kluwer Academic, 1992.

Cohen, H. Floris. *How Modern Science Came into the World: Four Civilizations, One 17th-Century Breakthrough*. Amsterdam: Amsterdam University Press, 2010.

Cohen, H. Floris. *Quantifying Music: The Science of Music at the First Stage of Scientific Revolution 1580–1650*. Dordrecht: Springer, 1984.

Cold Spring Harbor Laboratory. "Uplands Farm." Accessed April 30, 2019. https://www.cshl.edu/research/plant-biology/uplands-farm/.

Collins, Harry. M. *Changing Order: Replication and Induction in Scientific Practice*. Chicago: University of Chicago Press, 1992.

Collins, Harry. M. *Tacit and Explicit Knowledge*. Chicago: University of Chicago Press, 2012.

Collins, James B., and Karen L. Taylor, eds. *Early Modern Europe: Issues and Interpretations*. Malden, MA: Blackwell, 2006.

Cook, Harold J. "The History of Medicine and the Scientific Revolution." *Isis* 102, no. 1 (2011): 102–8. https://doi.org/10.1086/658659.

Cook, Harold J. "Medicine." In *The Cambridge History of Science*. Vol. 3, *Early Modern Science*, edited by Katharine Park and Lorraine Daston, 407–34. Cambridge: Cambridge University Press, 2006.

Cook, Harold J. "The Preservation of Specimens and Takeoff in Anatomical Knowledge in the Early Modern Period." In *Ways of Making and Knowing: The Material Culture of Empirical Knowledge*, edited by Pamela H. Smith, Amy R. W. Meyers, and Harold J. Cook, 302–29. Ann Arbor: University of Michigan Press, 2014.

Cook, Harold J. *The Young Descartes: Nobility, Rumor, and War*. Chicago: University of Chicago Press, 2018.

Cottingham, John, ed. *The Cambridge Companion to Descartes*. Cambridge: Cambridge University Press, 1992.

Cottingham, John. *A Descartes Dictionary*. Oxford: Blackwell Reference, 1993.

Cranefield, Paul F. "On the Origin of the Phrase, *Nihil est in intellectu quod non prius fuerit in sensu*." *Journal of the History of Medicine and Allied Sciences* 25, no. 1 (1970): 77–80.

Cunningham, Andrew. *The Anatomist Anatomis'd: An Experimental Discipline in Enlightenment Europe*. Farnham, UK: Ashgate, 2010.

Cunningham, Andrew. "The Pen and the Sword: Recovering the Disciplinary Identity of Physiology and Anatomy before 1800 I: Old Physiology—The Pen." *Studies in History and Philosophy of Biological and Biomedical Sciences* 33, no. 4 (2002): 631–65.

Cunningham, Andrew. "The Pen and the Sword: Recovering the Disciplinary Identity of Physiology and Anatomy before 1800: II: Old Anatomy—The Sword." *Studies in History and Philosophy of Biological and Biomedical Sciences* 34, no. 1 (2003): 51–76. https://doi.org/10.1016/S1369-8486(02)00069-9.

Cypess, Rebecca. "Giovanni Battista Della Porta's Experiments with Musical Instruments." *Journal of Musicological Research* 35, no. 3 (July 2, 2016): 159–75. https://doi.org/10.1080/01411896.2016.1180946.

317

Darlington, C. D. *Recent Advances in Cytology*. 2nd ed. London: Churchill, 1937.

Darlington, C. D., and K. R. Lewis, eds. *Chromosomes Today*. Vol. 1, *Proceedings of the First Oxford Chromosome Conference, July 28–31, 1964*. Edinburgh: Oliver and Boyd, 1966.

Daston, Lorraine, ed. *Biographies of Scientific Objects*. Chicago: University of Chicago Press, 2000.

Daston, Lorraine. *Eine kurze Geschichte der wissenschaftlichen Aufmerksamkeit*. Band 7. Munich: Carl Friedrich von Siemens Stiftung, 2001.

Daston, Lorraine. "The History of Science and the History of Knowledge." *KNOW: A Journal on the Formation of Knowledge* 1, no. 1 (March 1, 2017): 131–54. https://doi.org/10.1086/691678.

Daston, Lorraine. *Wunder, Beweise und Tatsachen: Zur Geschichte der Rationalität*. Frankfurt: Fischer-Taschenbuch-Verlag, 2001.

Daston, Lorraine, and Peter Galison. *Objectivity*. New York: Zone Books, 2007.

Daston, Lorraine, and Elizabeth Lunbeck. *Histories of Scientific Observation*. Chicago: University of Chicago Press, 2011.

Daston, Lorraine, and Katharine Park. *Wonders and the Order of Nature, 1150–1750*. New York: Zone Books, 1998.

Davids, Karel. "Craft Secrecy in Europe in the Early Modern Period: A Comparative View." *Early Science and Medicine* 10, no. 3 (January 1, 2005): 341–48. https://doi.org/10.1163/1573382054615398.

Davids, Karel. "Public Knowledge and Common Secrets: Secrecy and Its Limits in the Early-Modern Netherlands." *Early Science and Medicine* 10, no. 3 (January 1, 2005): 411–27. https://doi.org/10.1163/1573382054615424.

De Angelis, Simone. *Anthropologien: Genese und Konfiguration Einer "Wissenschaft vom Menschen" in der Frühen Neuzeit*. Berlin: Walter de Gruyter, 2010.

Dear, Peter. "Jesuit Mathematical Science and the Reconstitution of Experience in the Early Seventeenth Century." *Studies in History and Philosophy of Science Part A* 18, no. 2 (June 1987): 133–75.

Dear, Peter. *Mersenne and the Learning of the Schools*. Ithaca, NY: Cornell University Press, 1988.

Delbrück, Max. "Experiments with Bacterial Viruses (Bacteriophages) [lecture delivered on January 17, 1946]." *Harvey Lectures* 41 (1946): 161–87.

Demeter, Tamás, Benedek Láng, and Dániel Schmal. "Scientia in the Renaissance, Concept Of." In *Encyclopedia of Renaissance Philosophy*, edited by Marco Sgarbi, 1–15. Cham: Springer International Publishing, 2019.

Derksen, Mechteld-Hanna Gertrud, and Klasien Horstman. "Engineering Flesh: Towards an Ethics of Lived Integrity." *Medicine, Health Care and Philosophy* 11, no. 3 (September 2008): 269–83.

Derrida, Jacques. *Edmund Husserl's "Origin of Geometry": An Introduction*. Lincoln: University of Nebraska Press, 1989.

Dibon, Paul. "Les echanges épistolaires dans l'Europe savante du XVIIe siècle." *Revue de Synthèse* 81–82 (1976): 31–50.

Dickinson, Torry D., and Robert Schaeffer, eds. *Transformations: Feminist Pathways to Global Change: An Analytical Anthology*. Boulder, CO: Paradigm, 2008.

Didi-Huberman, Georges. "Before the Image, before Time: The Sovereignty of Anachronism." In *Compelling Visuality: The Work of Art in and out of History*, edited by Robert Zwijnenberg and Claire Farago, 31–44. Minneapolis: University of Minnesota Press, 2003.

Didi-Huberman, Georges. *Fra Angelico: Dissemblance and Figuration*. Chicago: University of Chicago Press, 1995.

Dijksterhuis, Fokko Jan. "Constructive Thinking: A Case for Dioptrics." In *The Mindful Hand: Inquiry and Invention from the Late Renaissance to Early Industrialisation*, edited by Lissa Roberts, Simon Schaffer, and Peter Dear, 59–82. Amsterdam: Koninkliijke Nederlandse Akademie van Wetenschappen, 2007.

Dijksterhuis, Fokko Jan. *Lenses and Waves: Christiaan Huygens and the Mathematical Science of Optics in the Seventeenth Century*. Vol. 9, *Archimedes*. Dordrecht: Springer Netherlands, 2005.

DiMeo, Michelle. "Communicating Medical Recipes: Robert Boyle's Genre and Rhetorical Strategies for Print." In *The Palgrave Handbook of Early Modern Literature and Science*, edited by Howard Marchitello and Evelyn Tribble, 209–28. London: Palgrave Macmillan, 2017. https://doi.org/10.1057/978-1-137-46361-6_10.

DiMeo, Michelle, and Sara Pennell, eds. *Reading and Writing Recipe Books, 1550–1800*. Manchester: Manchester University Press, 2013.

Dobre, Mihnea, and Tammy Nyden, eds.. *Cartesian Empiricisms*. Dordrecht: Springer, 2013.

Doing, Park. "Give Me a Laboratory and I Will Raise a Discipline: The Past, Present, and Future Politics of Laboratory Studies in STS." In *The Handbook of Science and Technology Studies*, 3rd ed., edited by Edward J. Hackett, Olga Amsterdamska, Michael Lynch, and Judy Wajcman, 279–95. Cambridge, MA: MIT Press, 2008.

Doing, Park. "'Lab Hands' and the 'Scarlet O': Epistemic Politics and (Scientific) Labor." *Social Studies of Science* 34, no. 3 (June 2004): 299–323. https://doi.org/10.1177/0306312704043677.

Dolan, K. P. "100 Years Plus of Gnotobiology." *Animal Technology and Welfare* 3, no. 2 (2004): 59–78.

Drake, Stillman. *Galileo Galilei: Two New Sciences, Including Centers of Gravity and Force of Percussion*. 2nd ed. Toronto: Wall and Emerson, 2000.

Drake, Stillman. "Renaissance Music and Experimental Science." *Journal of the History of Ideas* 31, no. 4 (1970): 483–500. https://doi.org/10.2307/2708256.

Dreyfus, Hubert L. *What Computers Can't Do: The Limits of Artificial Intelligence*. Rev. ed. New York: Harper and Row, 1979.

Dreyfus, Hubert L. *What Computers Still Can't Do: A Critique of Artificial Reason*. Cambridge, MA: MIT Press, 1992.

Ducey, Ariel. "Technologies of Caring Labor: From Objects to Affect." In *Intimate Labors: Cultures, Technologies, and the Politics of Care*, edited by Eileen Boris, 18–32. Stanford, CA: Stanford University Press, 2010.

Dupré, Sven, Anna Harris, Julia Kursell, Patricia Lulof, and Maartje Stols-Witlox, eds. *Reconstruction, Replication and Re-Enactment in the Humanities and Social Sciences*. Amsterdam: Amsterdam University Press, 2020. https://doi.org/10.5117/9789463728003.

Dupré, Sven, and Geert Somsen. "What Is the History of Knowledge?" *Journal for the History of Knowledge* 1, no. 1 (July 15, 2020). https://doi.org/10.5334/jhk.29.

Eamon, William. *Science and the Secrets of Nature: Books of Secrets in Medieval and Early Modern Culture*. Princeton, NJ: Princeton University Press, 1994.

Ebenstein, Joanna, Colin Dickey, and Chiara Ambrosio, eds. *The Morbid Anatomy Anthology*. Brooklyn, NY: Morbid Anatomy Press, 2015.

Emerson, Ralph Waldo. *Essays and Lectures*. Edited by Joel Porte. New York: Library of America, 1983.

Etzkowitz, Henry, and Loet Leydesdorff. "The Dynamics of Innovation: From National Systems and 'Mode 2' to a Triple Helix of University-Industry-Government Relations." *Research Policy* 29, no. 2 (February 1, 2000): 109–23. https://doi.org/10.1016/S0048-7333(99)00055-4.

Etzkowitz, Henry, and Loet Leydesdorff. "Theme Paper Triple Helix I: The Triple Helix-University-Industry-Government Relations: A Laboratory for Knowledge Based Economic Development." 1995. https://www.leydesdorff.net/th1/index.htm.

European Commission. *Eudralex*. Vol. 4, *Good Manufacturing Practice (GMP) Guidelines*. Brussels: European Commission, 1989–. https://health.ec.europa.eu/medicinal-products/eudralex/eudralex-volume-4_en.

Fancher, Raymond E. *Pioneers of Psychology*. 3rd ed. New York: Norton, 1996.

Fancher, Raymond E., and Alexandra Rutherford. *Pioneers of Psychology*. 5th ed. New York: W. W. Norton, 2017.

Feenberg, Andrew. "Active and Passive Bodies: Comments on Don Ihde's Bodies in Technology." *Techné: Research in Philosophy and Technology* 7, no. 2 (2003): 125–30. https://doi.org/10.5840.techne2003725.

Feenberg, Andrew. "Active and Passive Bodies: Don Ihde's Phenomenology of the Body." In *Postphenomenology: A Critical Companion to Ihde*, edited by Evan Selinger, 189–96. Albany: State University of New York Press, 2006.

Felt, Ulrike, Rayvon Fouché, Clark A. Miller, and Laurel Smith-Doerr, eds. *The Handbook of Science and Technology Studies*. 4th ed. Cambridge, MA: MIT Press, 2017.

Fernel, Jean. *Medicina*. Venice: Apud Balthassarem Constantinum, 1555. Reproduction, https://edit16.iccu.sbn.it/titolo/CNCE018761.

Fernel, Jean. *Medicina: Ad Henricum II Galliarum regem christianissimum.* Paris: Apud Andream Wechelum, 1554. Reproduction: https://gallica.bnf.fr/ark:/12148/bpt6k542831.

Fernel, Jean. *Medicina, physiologiam, pathologiam, methodumque complectens.* Venice: Bosellus, 1555. Reproduction, https://www.digitale-sammlungen.de/view/bsb10191161?page=2%2C3.

Fernel, Jean, and Guillaume Plancy. *Universa medicina, tribus et viginti libris absoluta: Ab ipso quidem authore ante obitum diligenter recognita, et quatuor libris nunquam ante editis ad praxim tamen perquam necessariis aucta.* 2 vols. Paris: Apud A. Wechelum, 1567. https://babel.hathitrust.org/cgi/pt?id=ucm.5316546508&view=1up&seq=5.

Fink, G., P. Farabaugh, G. Roeder, and D. Chaleff. "Transposable Elements (Ty) in Yeast." *Cold Spring Harbor Symposia on Quantitative Biology* 45 (1980): 575–80.

Fischer, Michael M. J. *Anthropological Futures.* Durham, NC: Duke University Press, 2009.

Fischer, Michael M. J. *Anthropology in the Meantime: Experimental Ethnography, Theory, and Method for the Twenty-First Century.* Durham, NC: Duke University Press, 2018.

Fischer, Michael M. J. "Biopolis: Asian Science in the Global Circuitry." *Science, Technology and Society* 18, no. 3 (2013): 379–404. https://doi.org/10.1177/0971721813498500.

Fischer, Michael M. J. *Emergent Forms of Life and the Anthropological Voice.* Durham, NC: Duke University Press, 2003.

Fischer, Michael M. J. "Science and Technology, Anthropology Of." In *International Encyclopedia of the Social and Behavioral Sciences*, 182–85. Amsterdam: Elsevier, 2015. https://doi.org/10.1016/B978-0-08-097086-8.12143-9.

Fleck, Ludwik. *Genesis and Development of a Scientific Fact.* Translated by Thaddeus J. Trenn and Frederick Bradley. Chicago: University of Chicago Press, 1935.

Fors, Hjalmar, Lawrence M. Principe, and Heinz Otto Sibum. "From the Library to the Laboratory and Back Again: Experiment as a Tool for Historians of Science." *Ambix* 63, no. 2 (April 2, 2016): 85–97. https://doi.org/10.1080/00026980.2016.1213009.

Forssberg, Hans. "Award Ceremony Speech: The Nobel Prize in Physiology or Medicine 2015." The Nobel Prize. Accessed May 1, 2019. https://www.nobelprize.org/prizes/medicine/2015/ceremony-speech/.

Fox, Nick J. "Space, Sterility and Surgery: Circuits of Hygiene in the Operating Theatre." *Social Science and Medicine* 45, no. 5 (September 1997): 649–57. https://doi.org/10.1016/S0277-9536(96)00381–4.

Friese, Peter, Guido Boulboullé, and Susanne Witzgall, eds. *Say It Isn't So.* Exhibition catalog. Heidelberg: Museum für moderne Kunst, 2007.

Gallagher, Shaun. *How the Body Shapes the Mind.* Oxford: Clarendon Press, 2005.

Gann, Alexander, and Jan A. Witkowski. "DNA: Archives Reveal Nobel Nominations." *Nature* 496 (2013): 434. https://doi.org/10.1038/496434a.

Garber, Daniel. *Descartes Embodied: Reading Cartesian Philosophy through Cartesian Science.* Cambridge: Cambridge University Press, 2001.

Garber, Daniel. Review of *Cartesian Empiricisms*, edited by Mihnea Dobre and Tammy Nyden. HOPOS: *The Journal of the International Society for the History of Philosophy of Science* 5, no. 2 (2015): 374–77. https://doi.org/10.1086/682722.

Gaukroger, Stephen, John Andrew Schuster, and John Sutton, eds. *Descartes' Natural Philosophy.* London: Routledge, 2000.

Gellner, Ernest. *Thought and Change.* London: Weidenfeld and Nicolson, 1964.

Gilbert, William. *De magnete, magneticisque corporibus, et de magno magnete tellure, physiologia nova.* London: Petrus Short, 1600.

Gilbert, William. *On the Magnet, Magnetick Bodies Also, and on the Great Magnet the Earth; a New Physiology, Demonstrated by Many Arguments & Experiments.* Translated and edited by Sylvanus P. Thompson. London: Chiswick Press, 1901. http://archive.org/details/b28038009.

Gipper, Andreas. "Experiment und Öffentlichkeit: Cartesianismus und Salonkultur im französischen 17. Jahrhundert." In *Spektakuläre Experimente: Praktiken der Evidenzproduktion im 17. Jahrhundert*, edited by Helmar Schramm, Ludger Schwarte, and Jan Lazardzig, 242–59. Berlin: Walter de Gruyter, 2006. https://doi.org/10.1515/9783110201970.242.

Given, Lisa M. *The Sage Encyclopedia of Qualitative Research Methods.* Los Angeles: Sage, 2008.

Goldberg, Jonathan. *Writing Matter: From the Hands of the English Renaissance.* Stanford, CA: Stanford University Press, 1990.

Good, Byron. *Medicine, Rationality, and Experience: An Anthropological Perspective.* Cambridge: Cambridge University Press, 1994.

Gouk, Penelope. *Music, Science, and Natural Magic in Seventeenth-Century England.* New Haven, CT: Yale University Press, 1999.

Gozza, Paolo. *Number to Sound: The Musical Way to the Scientific Revolution.* Dordrecht: Springer Science and Business Media, 2013.

Grasseni, Cristina. *Skilled Visions: Between Apprenticeship and Standards.* New York: Berghahn Books, 2007.

Grasseni, Cristina. "Skilled Visions: Toward an Ecology of Visual Inscriptions." In *Made to Be Seen: Perspectives on the History of Visual Anthropology*, edited by Marcus Banks and Jay Ruby, 19–44. Chicago: University of Chicago Press, 2011.

Grene, Marjorie. "The Heart and Blood: Descartes, Plemp, and Harvey." In *Essays on the Philosophy and Science of René Descartes*, edited by Stephen Voss, 324–36. New York: Oxford University Press, 1993.

Guerrini, Anita. *The Courtiers' Anatomists: Animals and Humans in Louis XIV's Paris.* Chicago: University of Chicago Press, 2015.

Hackett, Edward J., Olga Amsterdamska, Michael Lynch, and Judy Wajcman, eds. *The Handbook of Science and Technology Studies.* 3rd ed. Cambridge, MA: MIT Press, 2008.

Hacking, Ian. *Representing and Intervening: Introductory Topics in the Philosophy of Natural Science.* Cambridge: Cambridge University Press, 1983.

Hagen, Steve. *Meditation Now or Never.* New York: HarperOne, 2007.

Hagendijk, Thijs. "Reworking Recipes: Reading and Writing Practical Texts in the Early Modern Arts." PhD diss., Utrecht University, 2020. https://dspace .library.uu.nl/handle/1874/397503.

Hammond, John H. *The Camera Obscura: A Chronicle.* Bristol: Hilger, 1981.

Haraway, Donna J. *The Companion Species Manifesto: Dogs, People, and Significant Otherness.* Chicago: Prickly Paradigm Press, 2012.

Haraway, Donna J. *Modest_Witness@Second_Millennium. FemaleMan_Meets_ OncoMouse: Feminism and Technoscience.* New York: Routledge, 1997.

Haraway, Donna J. "Situated Knowledges: The Science Question in Feminism and the Privilege of Partial Perspective." In *Simians, Cyborgs, and Women: The Reinvention of Nature,* edited by Donna J. Haraway, 183–202. New York: Routledge, 1991.

Haraway, Donna J. *When Species Meet.* Minneapolis: University of Minnesota Press, 2008.

Harkness, Deborah E. *The Jewel House: Elizabethan London and the Scientific Revolution.* New Haven, CT: Yale University Press, 2007.

Hasse, Dag Nikolaus. *Success and Suppression: Arabic Sciences and Philosophy in the Renaissance.* Cambridge, MA: Harvard University Press, 2016.

Hatfield, Gary. "Descartes' Physiology and Its Relation to His Psychology." In *The Cambridge Companion to Descartes,* edited by John Cottingham, 335–70. Cambridge: Cambridge University Press, 1992. https://doi.org/10 .1017/CCOL0521366232.012.

Hatfield, Gary. "Reason, Nature, and God in Descartes." In *Essays on the Philosophy and Science of René Descartes,* edited by Stephen Voss, 259–87. New York: Oxford University Press, 1993.

Hatfield, Gary C. *Routledge Philosophy Guidebook to Descartes and the Meditations.* London: Routledge, 2003..

Hatfield, Gary. "The Senses and the Fleshless Eye: The Meditations as Cognitive Exercises." In *Essays on Descartes' Meditations,* edited by Amélie Oksenberg Rorty, 45–81. Berkeley: University of California Press, 1986.

Hausken, Liv. "Introduction." In *Thinking Media Aesthetics: Media Studies, Film Studies and the Arts,* edited by Liv Hausken, 29–50. Frankfurt: PL Academic Research, 2013.

Heidegger, Martin. *Der Satz vom Grund.* Pfullingen: Neske, 1957.

Heidelberger, Michael. "Experiment und Instrument." In *Spektakuläre Experimente: Praktiken der Evidenzproduktion im 17. Jahrhundert,* edited by Helmar Schramm, Ludger Schwarte, and Jan Lazardzig, 378–97. Berlin: Walter de Gruyter, 2006.

Hentschel, Klaus, ed. *Unsichtbare Hände: Zur Rolle von Laborassistenten, Mechanikern, Zeichnern u. a. Amanuenses in der physikalischen Forschungs- und*

Entwicklungsarbeit. Diepholz: Verlag für Geschichte der Naturwissenschaften und der Technik, 2008.

Hirschauer, Stefan. "The Manufacture of Bodies in Surgery." *Social Studies of Science* 21, no. 2 (1991): 279–319.

Hirschauer, Stefan. "Putting Things into Words: Ethnographic Description and the Silence of the Social." *Human Studies* 29, no. 4 (March 1, 2007): 413–41. https://doi.org/10.1007/s10746-007-9041-1.

Hooke, Robert. *The Posthumous Works of Robert Hooke*. Edited by Richard Waller. New York: Johnson Reprint, 1969 [1705].

Hötteke, Dietmar. "Zur experimentellen Tätigkeit Michael Faradays." In *Instrument—Experiment: Historische Studien*, edited by Christoph Meinel, 360–68. Berlin: Verlag für Geschichte der Naturwissenschaften und der Technik, 2000.

Huff, Toby E. *Intellectual Curiosity and the Scientific Revolution: A Global Perspective*. Cambridge: Cambridge University Press, 2010.

Husserl, Edmund. *Cartesianische Meditationen und Pariser Vorträge*. Edited by Stephan Strasser. The Hague: Martinus Nijhoff, 1950.

Husserl, Edmund. *Cartesian Meditations: An Introduction to Phenomenology*. Dordrecht: Kluwer Academic Publishers, 1982.

Husserl, Edmund. *The Crisis of European Sciences and Transcendental Phenomenology: An Introduction to Phenomenological Philosophy*. Evanston, IL: Northwestern University Press, 1970.

Husserl, Edmund. *Die Krisis der europäischen Wissenschaften und die transzendentale Phänomenologie*. Edited by Walter Biemel. The Hague: Nijhoff, 1954.

Husserl, Edmund. *Ideen zu einer reinen Phänomenologie und phänomenologischen Philosophie: Erstes Buch*. Edited by Walter Biemel and Marly Biemel. The Hague: Martinus Nijhoff, 1950.

Husserl, Edmund. *Ideen zu einer reinen Phänomenologie und phänomenologischen Philosophie: Zweites Buch*. Edited by Walter Biemel and Marly Biemel. The Hague: Martinus Nijhoff, 1952.

Hutchins, Barnaby, Christoffer Basse Eriksen, and Charles Wolfe. "The Embodied Descartes: Contemporary Readings of *L'Homme*." In *Descartes' Treatise on Man and Its Reception*, edited by Delphine Antoine-Mahut and Stephen Gaukroger, 287–304. Cham: Springer, 2016.

Hyland, Sheena. "Merleau-Ponty, Beckett and the Body: A Phenomenology of Aches and Pains." Paper presented at "UCD Body Conference: Perspectives on the Body and Embodiment," University College Dublin, Ireland, 2007.

Ihde, Don. *Bodies in Technology*. Minneapolis: University of Minnesota Press, 2002.

Ihde, Don. *Experimental Phenomenology: An Introduction*. Albany: State University of New York Press, 1986.

Ihde, Don. *Instrumental Realism: The Interface between Philosophy of Science and Philosophy of Technology*. Bloomington: Indiana University Press, 1991.

Ihde, Don. *Technics and Praxis*. Dordrecht: Reidel, 1979.

Ihde, Don. *Technology and the Lifeworld: From Garden to Earth*. Bloomington: Indiana University Press, 1990.

Iliffe, Rob. "Technicians." *Notes and Records of the Royal Society* 62, no. 1 (March 20, 2008): 3–16. https://doi.org/10.1098/rsnr.2007.0053.

Ijsseling, Samuel. "Das Ende der Philosophie als Anfang des Denkens." In *Heidegger et l'idée de la phénoménologie*, edited by Franco Volpi et al., 285–99. Dordrecht: Kluwer Academic, 1988.

International Organization for Standardization (ISO). *Cleanrooms and Associated Controlled Environments, Part 1: Classification of Air Cleanliness by Particle Concentration*. 2nd ed. Geneva: ISO, 2015.

International Organization for Standardization (ISO). *Cleanrooms and Associated Controlled Environments, Part 5: Operations*. Geneva: ISO, 2004–8.

Jardine, Nicholas. "Problems of Knowledge and Action: Epistemology of the Sciences." In *The Cambridge History of Renaissance Philosophy*, edited by C. B. Schmitt, Eckhard Kessler, Jill Kraye, and Quentin Skinner, 685–712. Cambridge: Cambridge University Press, 1988. https://doi.org/10.1017/CHOL9780521251044.021.

Jasanoff, Sheila, Gerald Markle, James Petersen, and Trevor Pinch, eds. *Handbook of Science and Technology Studies*. Thousand Oaks, CA: Sage, 1995.

Jordan, Kathleen, and Michael Lynch. "The Sociology of a Genetic Engineering Technique: Ritual and Rationality in the Performance of the Plasmid Prep." In *The Right Tools for the Job: At Work in Twentieth-Century Life Sciences*, edited by Adele E. Clarke and Joan H. Fujimura, 77–114. Princeton, NJ: Princeton University Press, 1992.

Katz, Pearl. "Ritual in the Operating Room." *Ethnology* 20, no. 4 (1981): 335–50.

Kay, Lily E. *The Molecular Vision of Life: Caltech, the Rockefeller Foundation, and the Rise of the New Biology*. New York: Oxford University Press, 1997.

Kay, Lily E. *Who Wrote the Book of Life? A History of the Genetic Code*. Stanford, CA: Stanford University Press, 2000.

Keating, Jessica. *Animating Empire: Automata, the Holy Roman Empire, and the Early Modern World*. University Park: Penn State University Press, 2018.

Keller, Evelyn Fox. "Beyond the Gene but Beneath the Skin." In *Genes in Development: Re-Reading the Molecular Paradigm*, edited by Eva M. Neumann-Held and Christoph Rehmann-Sutter, 290–312. Durham, NC: Duke University Press, 2006.

Keller, Evelyn Fox. "The Biological Gaze." In *FutureNatural: Nature, Science, Culture*, edited by George Robertson, 107–21. London: Routledge, 1996.

Keller, Evelyn Fox. *The Century of the Gene*. Cambridge, MA: Harvard University Press, 2000.

Keller, Evelyn Fox. *A Feeling for the Organism: The Life and Work of Barbara McClintock*. San Francisco: W. H. Freeman, 1983.

Keller, Evelyn Fox. *Making Sense of Life: Explaining Biological Development with Models, Metaphors, and Machines*. Cambridge, MA: Harvard University Press, 2003.

325

Keller, Evelyn Fox. *Refiguring Life: Metaphors of Twentieth-Century Biology.* New York: Columbia University Press, 1995.

Keller, Evelyn Fox. *Reflections on Gender and Science.* New Haven, CT: Yale University Press, 1985.

Keller, Evelyn Fox. "Rethinking the Meaning of Biological Information." *Biological Theory* 4, no. 2 (June 2009): 159–66. https://doi.org/10.1162/biot.2009.4.2.159.

Keller, Vera. "'Everything Depends upon the Trial' ('Le tout gist à l'essay'): Four Manuscripts between the Recipe and the Experimental Essay." In *Secrets of Craft and Nature in Renaissance France: A Digital Critical Edition and English Translation of BnF Ms. Fr. 640,* edited by Making and Knowing Project, Pamela H. Smith, Naomi Rosenkranz, Tianna Helena Uchacz, Tillman Taape, Clément Godbarge, Sophie Pitman, Jenny Boulboullé, Joel Klein, Donna Bilak, Marc Smith, and Terry Catapano. New York: Making and Knowing Project, 2020. https://doi.org/10.7916/vj69-8h20.

Kircher, Athanasius. *Ars magna lucis et umbrae: In decem libros digesta; quibus admirandae lucis et umbrae in mundo, atque adeo universa natura, vires effectusq[ue] uti nova, ita varia novorum reconditiorumq[ue] speciminum exhibitione, ad varios mortalium usus, panduntur.* Rome: Scheus, 1646. http://diglib.hab.de/drucke/94-2-quod-2f/start.htm.

Klein, Ursula. "Introduction: Technoscientific Productivity." *Perspectives on Science* 13, no. 2 (2005): 139–41.

Klein, Ursula. "The Laboratory Challenge: Some Revisions of the Standard View of Early Modern Experimentation." *Isis* 99, no. 4 (December 2008): 769–82. https://doi.org/10.1086/595771.

Klein, Ursula. "Styles of Experimentation and Alchemical Matter Theory in the Scientific Revolution." *Metascience* 16, no. 2 (May 2007): 247–56. https://doi.org/10.1007/s11016-007-9095-8.

Klingenberg, Martin. "When a Common Problem Meets an Ingenious Mind: The Invention of the Modern Micropipette." *EMBO Reports* 6, no. 9 (2005): 797–800. https://doi.org/10.1038/sj.embor.7400520.

Klotz, Sebastian. "Vibration und Vernunft: Zur Experimentellen Agenda in Martin Mersennes *Harmonie Universelle* (Paris, 1636)." In *Spektakuläre Experimente: Praktiken der Evidenzproduktion im 17. Jahrhundert,* edited by Helmar Schramm, Ludger Schwarte, and Jan Lazardzig, 279–94. Berlin: Walter de Gruyter, 2006.

Knorr-Cetina, Karin. *Epistemic Cultures: How the Sciences Make Knowledge.* Cambridge, MA: Harvard University Press, 1999.

Knorr-Cetina, Karin. "Objectual Practice." In *The Practice Turn in Contemporary Theory,* edited by Theodore R. Schatzki, Karin Knorr-Cetina, and Eike von Savigny, 184–97. New York: Routledge, 2001.

Kodera, Sergius. "The Laboratory as Stage: Giovan Battista Della Porta's Experiments." *Journal of Early Modern Studies* 3, no. 1 (2014): 15–38. https://doi.org/10.7761/JEMS.3.1.15.

Kofoid, Charles A. Review of *Recent Advances in Cytology*, by C. D. Darlington. *Isis* 29, no. 2 (1938): 472–74.

Kontopodis, Michalis, Jörg Niewöhner, and Stefan Beck. "Investigating Emerging Biomedical Practices: Zones of Awkward Engagement on Different Scales." *Science, Technology, and Human Values* 36, no. 5 (September 2011): 599–615. https://doi.org/10.1177/0162243910392798.

Koyré, Alexandre. *Übersetzung von Alexandre Koyré, Descartes und die Scholastik*. Translated by Edith Stein and Hedwig Conrad-Martius. Edited by Hanna Gerl-Falkovitz. Freiburg: Herder, 2005.

Küster, Ernst. "Die keimfreie Züchtung von Säugetieren." In *Handbuch der biochemischen Arbeitsmethoden*, edited by Emil Abderhalden, 311–23. Berlin: Urban & Schwarzenberg, 1915. http://archive.org/details/handbuchderbioch08abde.

La Frenais, Rob, and Eileen Daily, eds. *Clean Rooms: Critical Art Ensemble, Gina Czarnecki, Neal White*. Exhibition catalog. London: Arts Catalyst, 2002.

Lähteenmäki, Iku, John Hodgson, and Adam Michael. "Tips for Your Pipette." *Nature Biotechnology* 15, no. 10 (1997): 1030–32. https://doi.org/10.1038/nbt1097-1030.

Landecker, Hannah. *Culturing Life: How Cells Became Technologies*. Cambridge, MA: Harvard University Press, 2009.

Landecker, Hannah. "New Times for Biology: Nerve Cultures and the Advent of Cellular Life in Vitro." *Studies in History and Philosophy of Science Part C: Studies in History and Philosophy of Biological and Biomedical Sciences* 33, no. 4 (December 1, 2002): 667–94. https://doi.org/10.1016/S1369-8486(02)00026-2.

Latour, Bruno. "How to Talk about the Body? The Normative Dimension of Science Studies." *Body and Society* 10, nos. 2–3 (June 2004): 205–29. https://doi.org/10.1177/1357034X04042943.

Latour, Bruno. "Insiders and Outsiders in the Sociology of Science; or, How Can We Foster Agnosticism?" In *Knowledge and Society: Studies in the Sociology of Culture Past and Present: A Research Annual*, edited by Henrika Kuklick and Robert Alun Jones, 3:199–216. Greenwich, CT: JAI Press, 1981.

Latour, Bruno. *Pandora's Hope: Essays on the Reality of Science Studies*. Cambridge, MA: Harvard University Press, 1999.

Latour, Bruno. *The Pasteurization of France*. Cambridge, MA: Harvard University Press, 1988.

Latour, Bruno. *Science in Action: How to Follow Scientists and Engineers through Society*. Cambridge, MA: Harvard University Press, 2003.

Latour, Bruno. "Visualisation and Cognition: Thinking with Eyes and Hands." In *Knowledge and Society Studies in the Sociology of Culture Past and Present: A Research Annual*, edited by Henrika Kuklick and Elizabeth Long, 6:1–40. Greenwich, CT: JAI Press, 1986.

Latour, Bruno, and Steve Woolgar. *Laboratory Life: The Construction of Scientific Facts*. 2nd ed. Princeton, NJ: Princeton University Press, 1986.

Lawler, Mark. "Rosalind Franklin Still Doesn't Get the Recognition She Deserves for Her DNA Discovery." *The Conversation* (blog), April 24, 2018. http:// theconversation.com/rosalind-franklin-still-doesnt-get-the-recognition-she -deserves-for-her-dna-discovery-95536.

Lawrence, Christopher, and Steven Shapin. "Introduction: The Body of Knowledge." In *Science Incarnate: Historical Embodiments of Natural Knowledge*, edited by Christopher Lawrence and Steven Shapin, 1–19. Chicago: University of Chicago Press, 1998.

Lawrence, Christopher, and Steven Shapin, eds. *Science Incarnate: Historical Embodiments of Natural Knowledge*. Chicago: University of Chicago Press, 1998.

Lefèvre, Wolfgang, ed. *Inside the Camera Obscura: Optics and Art under the Spell of the Projected Image*. Berlin: Max Planck Institute for the History of Science, 2007.

Lefèvre, Wolfgang. "Science as Labor." *Perspectives on Science* 13, no. 2 (2005): 194–225.

Leisegang, Gertrud, ed. *Descartes Dioptrik*. Meisenheim am Glan: Westkulturverlag Anton Hain, 1954.

Lenoir, Tim. "Epistemology Historicized: Making Epistemic Things." In *An Epistemology of the Concrete: Twentieth-Century Histories of Life*, by Hans-Jörg Rheinberger, xi–xix. Durham, NC: Duke University Press, 2010.

Leong, Elaine Yuen Tien. *Recipes and Everyday Knowledge: Medicine, Science, and the Household in Early Modern England*. Chicago: University of Chicago Press, 2018.

Leong, Elaine Yuen Tien, and Alisha Michelle Rankin. *Secrets and Knowledge in Medicine and Science, 1500–1800*. Farnham, UK: Ashgate, 2011.

Lessing, Gotthold Ephraim. *Emilia Galotti: Ein Trauerspiel in fünf Aufzügen*. Berlin: C. F. Voss, 1772. http://www.deutschestextarchiv.de/book/view/lessing _emilia_1772?p=5.

Lock, Margaret, Allan Young, and Alberto Cambrosio, eds. *Living and Working with the New Medical Technologies: Intersections of Inquiry*. Cambridge: Cambridge University Press, 2000.

Long, Pamela O. *Artisan/Practitioners and the Rise of the New Sciences, 1400–1600*. Corvallis: Oregon State University Press, 2011.

Lubbadeh, Jens. "Synthetische Biologie: Leben aus dem Lego-Baukasten." Spiegel Online, 2010. https://www.spiegel.de/wissenschaft/natur/synthetische -biologie-leben-aus-dem-lego-baukasten-a-670081.html.

Lynch, Michael. *Art and Artifact in Laboratory Science: A Study of Shop Work and Shop Talk in a Research Laboratory*. London: Routledge and Kegan Paul, 1985.

Lynch, Michael E. "Sacrifice and the Transformation of the Animal Body into a Scientific Object: Laboratory Culture and Ritual Practice in the Neurosciences." *Social Studies of Science* 18, no. 2 (1988): 265–89.

Making and Knowing Project, Pamela H. Smith, Tianna Helena Uchacz, Tillmann Taape, Sophie Pitman, Jenny Boulboullé, Joel Klein, Donna

Bilak, Marc Smith, and Terry Catapano, eds. *Secrets of Craft and Nature in Renaissance France: A Digital Critical Edition and English Translation of BnF Ms. Fr. 640.* New York: Making and Knowing Project, 2020. https://edition640.makingandknowing.org.

Mancosu, Paolo. "Acoustics and Optics." In *The Cambridge History of Science.* Vol. 3, *Early Modern Science*, edited by Katharine Park and Lorraine Daston, 596–631. Cambridge: Cambridge University Press, 2006. https://doi.org/10.1017/CHOL9780521572446.026.

Marcus, G. E., and Michael M. J. Fischer. *Anthropology as Cultural Critique: An Experimental Moment in the Human Sciences.* Chicago: University of Chicago Press, 1999.

Margócsy, Dániel. "A Philosophy of Wax: The Anatomy of Frederik Ruysch." In *The Morbid Anatomy Anthology*, edited by Joanna Ebenstein, Colin Dickey, and Chiara Ambrosio. Brooklyn, NY: Morbid Anatomy Press, 2015.

Margócsy, Dániel, Mark Somos, and Stephen N. Joffe. *The* Fabrica *of Andreas Vesalius: A Worldwide Descriptive Census, Ownership, and Annotations of the 1543 and 1555 Editions.* Leiden: Brill, 2018.

Martin, Jay A. "The Art of the Pipet." *HMS Beagle* (blog), April 13, 2001. BioMed-Net. https://outreach.biotech.wisc.edu/wp-content/uploads/sites/1684/2021/08/pipet.pdf.

Mascia-Lees, Frances E., ed. *A Companion to the Anthropology of the Body and Embodiment.* Chichester: Wiley-Blackwell, 2011.

Maull, Nancy. "Cartesian Optics and the Geometrisation of Nature." In *René Descartes: Critical Assessments*, edited by Georges J. D. Moyal, 263–79. London: Routledge, 1991.

Mauss, Marcel. "Techniques of the Body." *Economy and Society* 2, no. 1 (1973): 70–88. https://doi.org/10.1080/03085147300000003.

McAllister, James W. "Die Rhetorik der Mühelosigkeit in der Wissenschaft und ihre barocken Ursprünge." In *Spektakuläre Experimente: Praktiken der Evidenzproduktion im 17. Jahrhundert*, edited by Helmar Schramm, Ludger Schwarte, and Jan Lazardzig, 154–75. Berlin: Walter de Gruyter, 2006. https://doi.org/10.1515/9783110201970.154.

McAllister, James W. "The Virtual Laboratory: Thought Experiments in Seventeenth-Century Mechanics." In *Collection, Laboratory, Theater: Scenes of Knowledge in the 17th Century*, edited by Helmar Schramm, Ludger Schwarte, and Jan Lazardzig, 35–56. Berlin: Walter de Gruyter, 2005.

Meinel, Christoph, ed. *Instrument—Experiment: Historische Studien.* Berlin: Verlag für Geschichte der Naturwissenschaften und der Technik, 2000.

Merleau-Ponty, Maurice. *Phenomenology of Perception.* London: Routledge, 2002.

Merleau-Ponty, Maurice. "The Philosopher and His Shadow." In *Signs*, translated by Richard C. McCleary, 159–81. Evanston, IL: Northwestern University Press, 1982.

Mersenne, Marin. *Harmonie universelle, contenant la théorie et la pratique de la musique.* Paris: S. Cramoisy, 1636. https://gallica.bnf.fr/ark:/12148/bpt6k5471093v.

Merz, Martina. Review of *Epistemic Cultures: How the Sciences Make Knowledge,* by Karin Knorr-Cetina. *History and Philosophy of the Life Sciences* 24, no. 1 (2002): 122–24.

Mesman, Jessica. *Uncertainty in Medical Innovation: Experienced Pioneers in Neonatal Care.* Basingstoke: Palgrave Macmillan, 2008.

Miert, Dirk van, ed. *Communicating Observations in Early Modern Letters (1500–1675): Epistolography and Epistemology in the Age of the Scientific Revolution.* London: Warburg Institute, 2013.

Miert, Dirk van. "What Was the Republic of Letters? A Brief Introduction to a Long History (1417–2008)." *Groniek,* nos. 204–5 (2016): 269–87.

Mody, Cyrus C. M. "A Little Dirt Never Hurt Anyone: Knowledge-Making and Contamination in Materials Science." *Social Studies of Science* 31, no. 1 (2001): 7–36.

Mody, Cyrus C. M. "The Sounds of Science: Listening to Laboratory Practice." *Science, Technology, and Human Values* 30, no. 2 (April 1, 2005): 175–98. https://doi.org/10.1177/0162243903261951.

Mody, Cyrus C. M., and David Kaiser. "Scientific Training and the Creation of Scientific Knowledge." In *The Handbook of Science and Technology Studies,* 3rd ed., edited by Edward J. Hackett, Olga Amsterdamska, Michael E. Lynch, and Judy Wajcman, 377–402. Cambridge, MA: MIT Press, 2008.

Mol, Annemarie. *The Body Multiple: Ontology in Medical Practice.* Durham, NC: Duke University Press, 2002.

Mol, Annemarie. "Pathology and the Clinic: An Ethnographic Presentation of Two Atheroscleroses." In *Living and Working with the New Medical Technologies,* edited by Margaret Lock, Allan Young, and Alberto Cambrosio, 82–102. Cambridge: Cambridge University Press, 2000. https://doi.org/10.1017/CBO9780511621765.005.

Moyer, Ann E. *Musica Scientia: Musical Scholarship in the Italian Renaissance.* Ithaca, NY: Cornell University Press, 1992.

Mülhardt, Cornel. *Der Experimentator Molekularbiologie/Genomics.* Heidelberg: Springer Spektrum, 2013. https://doi.org/10.1007/978-3-642-34636-1.

Mulligan, Tim. "Uplands Farm." *Origins* (blog). *Unwinding DNA: Life at Cold Spring Harbor Laboratory.* Accessed April 29, 2019. http://www.exploratorium.edu/origins/coldspring/place/farm.html.

Myers, Natasha. "Molecular Embodiments and the Body-Work of Modeling in Protein Crystallography." *Social Studies of Science* 38, no. 2 (April 1, 2008): 163–99. https://doi.org/10.1177/0306312707082969.

Myers, Natasha. *Rendering Life Molecular: Models, Modelers, and Excitable Matter.* Durham, NC: Duke University Press, 2015.

Myers, Natasha, and Joe Dumit. "Haptics: Haptic Creativity and the Mid-embodiments of Experimental Life." In *A Companion to the Anthropology of the Body and Embodiment*, edited by Frances E. Mascia-Lees, 239–61. New York: Wiley, 2011.

Nancy, Jean-Luc. "Dum Scribo." In *Ego Sum: Corpus, Anima, Fabula*, translated by Marie-Eve Morin, 20–38. New York: Fordham University Press, 2016. https://doi.org/10.2307/j.ctt1b3t7tv.6.

Neher, André. *David Gans, 1541–1613: Disciple du Maharal, assistant de Tycho Brahe et de Jean Kepler*. Paris: Klincksieck, 1974.

Nanney, David L. "The Role of the Cytoplasm in Heredity." In *The Chemical Basis of Heredity*, edited by William D. McElroy and Bentley Glass, 134. Baltimore: Johns Hopkins University Press, 1957.

Neher, André. *Jewish Thought and the Scientific Revolution of the Sixteenth Century: David Gans (1541–1613) and His Times*. New York: Published for the Littman Library by Oxford University Press, 1986.

Nelle, Florian. "Eucharist and Experiment: Spaces of Certainty in the 17th Century." In *Collection, Laboratory, Theater: Scenes of Knowledge in the 17th Century*, edited by Helmar Schramm, Ludger Schwarte, and Jan Lazardzig, 316–38. Berlin: Walter de Gruyter, 2005.

Newen, Albert, Leon de Bruin, and Shaun Gallagher, eds. *The Oxford Handbook of 4E Cognition*. Oxford: Oxford University Press, 2018.

Newman, Lex. "Descartes' Epistemology" In *The Stanford Encyclopedia of Philosophy (Spring 2019 Edition)*, edited Edward N. Zalta. https://plato.stanford.edu/archives/spr2019/entries/descartes-epistemology/.

Nishikawa, Mitsuko. "Japan's Latest Nobel Laureate." NHK WORLD, October 6, 2015.

Nobel Foundation. "Elizabeth H. Blackburn." The Nobel Prize, accessed April 28, 2024. https://www.nobelprize.org/prizes/medicine/2009/blackburn/facts/.

Nobel Foundation. "Press Release: The Nobel Prize in Physiology or Medicine 1983, Barbara McClintock." The Nobel Prize, accessed May 16, 2024. https://www.nobelprize.org/prizes/medicine/1983/press-release/.

Nuttall, George H. F., and Hans Thierfelder. "Thierisches Leben ohne Bakterien im Verdauungskanal (mit einer Tafel)." *Hoppe-Seyler's Zeitschrift für Physiologische Chemie* 21 (1895): 109–21.

O'Connor, Clare M. *Investigations in Molecular Cell Biology*. LibreTexts: Biology, accessed April 27, 2024. https://bio.libretexts.org/Bookshelves/Cell_and_Molecular_Biology/Book%3A_Investigations_in_Molecular_Cell_Biology_(O'Connor).

Olesen, Finn. "Technological Mediation and Embodied Health-Care Practices." In *Postphenomenology: A Critical Companion to Ihde*, edited by Evan Selinger, 231–46. Albany: State University of New York Press, 2006. http://site.ebrary.com/id/10579116.

Ōmura, Satoshi. "The Ivermectin Story (1973–2008)." Slide show. Accessed May 1, 2019. http://www.satoshi-omura.info/ivermectin/ivermectin.html.

Ōmura, Satoshi. "Philosophy of New Drug Discovery." *Microbiological Reviews* 50, no. 3 (1986): 259–79.

Ophir, Adi, and Steven Shapin. "The Place of Knowledge A Methodological Survey." *Science in Context* 4, no. 1 (April 1991): 3–22.

Paradis, James. "Montaigne, Boyle, and the Essay of Experience." In *One Culture: Essays in Science and Literature*, edited by George Lewis Levine and Alan Rauch, 59–91. Madison: University of Wisconsin Press, 1987.

Petrescu, Lucian. "Descartes on the Heartbeat: The Leuven Affair." *Perspectives on Science* 21, no. 4 (December 2013): 397–428. https://doi.org/10.1162/POSC_a_00110.

Pfeiffer, Birgit. "Die 'Marburg Pipette': Die Geschichte und Entstehung der Kolbenhub-Pipette." PhD diss., Philipps-Universität Marburg, 2003.

Pickering, Andrew. *Constructing Quarks: A Sociological History of Particle Physics*. Chicago: University of Chicago Press, 1999.

Pickering, Andrew. *The Mangle of Practice: Time, Agency, and Science*. Chicago: University of Chicago Press, 1995.

Pickering, Andrew, ed. *Science as Practice and Culture*. Chicago: University of Chicago Press, 1992.

Pickering, Andrew. "Space: The Final Frontier." In *Collection, Laboratory, Theater: Scenes of Knowledge in the 17th Century*, edited by Helmar Schramm, Ludger Schwarte, and Jan Lazardzig, 1–8. Berlin: Walter de Gruyter, 2005.

Pickstone, John V. *Ways of Knowing: A New History of Science, Technology and Medicine*. Manchester: Manchester University Press, 2000.

Pietzsch, Joachim. "What Is Life?" Perspectives, Nobel Media AB, 1962. https://www.nobelprize.org/prizes/medicine/1962/perspectives/.

Polanyi, Michael. *Personal Knowledge: Towards a Post-Critical Philosophy*. London: Routledge and Kegan Paul, 1958.

Popkin, Richard H., ed. *The Pimlico History of Western Philosophy*. London: Pimlico, 1999.

Porter, Theodore M. "Thin Description: Surface and Depth in Science and Science Studies." *Osiris* 27, no. 1 (2012): 209–26. https://doi.org/10.1086/667828.

Prentice, Rachel. *Bodies in Formation: An Ethnography of Anatomy and Surgery Education*. Durham, NC: Duke University Press, 2013.

Rabinow, Paul. "Epochs, Presents, Events." In *Living and Working with the New Medical Technologies: Intersections of Inquiry*, edited by Alberto Cambrosio, Allan Young, and Margaret Lock, 31–46. Cambridge: Cambridge University Press, 2000. https://doi.org/10.1017/CBO9780511621765.003.

Rabinow, Paul. *Making PCR: A Story of Biotechnology*. Chicago: University of Chicago Press, 1996.

Ramstorp, Matts. *Introduction to Contamination Control and Cleanroom Technology*. Weinheim: Wiley-VCH, 2000.

Ranea, Alberto Guillermo. "A 'Science for *Honnêtes hommes*': *La recherche de la vérité* and the Deconstruction of Experimental Knowledge." In *Descartes' Natural Philosophy*, edited by Stephen Gaukroger, John Schuster, and John Sutton, 325–41. London: Routledge, 2000.

Rankin, Alisha. "How to Cure the Golden Vein: Medical Remedies as Wissenschaft in Early Modern Germany." In *Ways of Making and Knowing: The Material Culture of Empirical Knowledge*, edited by Pamela H. Smith, Amy R. W. Meyers, and Harold J. Cook, 113–37. Ann Arbor: University of Michigan Press, 2014.

Ravetz, Jerome R. *Scientific Knowledge and Its Social Problems*. New Brunswick, NJ: Transaction Publishers, 1996.

Reichle, Ingeborg. "The Art of Making Science." In *Blindspot*, edited by Herwig Turk and Paulo de Carvalho Pereira, 14–19. Exhibition catalog. N.p.: Virose, 2007.

Reichle, Ingeborg. "Taube Bilder und sehende Hände—Strategien visueller Transgression im Werk von Herwig Turk." In *Masslose Bilder—Visuelle Ästhetik der Transgression*, edited by Ingeborg Reichle and Steffen Siegel, 165–87. Munich: Wilhelm Fink, 2009.

Reichle, Ingeborg. "Unter Beobachtung: Die Kunst Schaut ins Labor." *Gegenworte—Hefte für den Disput über Wissen* 20 (Autumn 2008): 63–66.

Rheinberger, Hans-Jörg. "Cytoplasmic Particles." In *Biographies of Scientific Objects*, edited by Lorraine Daston, 270–94. Chicago: University of Chicago Press, 2000.

Rheinberger, Hans-Jörg. "Die Evidenz des Präparates." In *Spektakuläre Experimente: Praktiken der Evidenzproduktion im 17. Jahrhundert*, edited by Helmar Schramm, Ludger Schwarte, and Jan Lazardzig, 1–17. Berlin: Walter de Gruyter, 2006.

Rheinberger, Hans-Jörg. *An Epistemology of the Concrete: Twentieth-Century Histories of Life*. Durham, NC: Duke University Press, 2010.

Rheinberger, Hans-Jörg. "Gene Concepts: Fragments from the Perspective of Molecular Biology." In *The Concept of the Gene in Development and Evolution: Historical and Epistemological Perspectives*, edited by Peter J. Beurton, Raphael Falk, and Hans-Jörg Rheinberger, 219–39. Cambridge: Cambridge University Press, 2000.

Rheinberger, Hans-Jörg. *Historische Epistemologie zur Einführung*. Hamburg: Junius, 2008.

Rheinberger, Hans-Jörg. *On Historicizing Epistemology: An Essay*. Stanford, CA: Stanford University Press, 2010.

Rheinberger, Hans-Jörg. *Toward a History of Epistemic Things: Synthesizing Proteins in the Test Tube*. Stanford, CA: Stanford University Press, 1997.

Rheinberger, Hans-Jörg, and Michael Hagner, eds. *Die Experimentalisierung des Lebens: Experimentalsysteme in den biologischen Wissenschaften 1850/1950*. Berlin: Akademie Verlag, 1993.

Riess, Falk, Peter Heering, and Dennis Nawrath. "Reconstructing Galileo's In-clined Plane Experiments for Teaching Purposes." In *Online-Proceedings of the 8th International History and Philosophy of Science and Science Teaching (IHPST) Conference.* Leeds: International History and Philosophy of Science and Science Teaching, 2005.

Rietveld, Erik. "McDowell and Dreyfus on Unreflective Action." *Inquiry: An Interdisciplinary Journal of Philosophy* 53, no. 2 (2010): 183–207.

Rietveld, Erik. "Situated Normativity: The Normative Aspect of Embodied Cog-nition in Unreflective Action." *Mind* 117, no. 468 (2008): 973–1001.

Rijcke, Sarah de. "Regarding the Brain: Practices of Objectivity in Cerebral Imaging, 17th Century–Present." PhD diss., University of Groningen, 2010. https://pure.rug.nl/ws/portalfiles/portal/31628353/11complete.pdf.

Ringertz, Nils. "Award Ceremony Speech: The Nobel Prize in Physiology or Medicine 1983." The Nobel Prize, 1983. https://www.nobelprize.org/prizes/medicine/1983/ceremony-speech/.

Roberts, Lissa, Simon Schaffer, and Peter Dear, eds. *The Mindful Hand: Inquiry and Invention from the Late Renaissance to Early Industrialisation.* Amster-dam: Koninkliijke Nederlandse Akademie van Wetenschappen, 2007.

Robins, Richard, and David Trigger. "A Recent Phase of Aboriginal Occupation in Lawn Hill Gorge: A Case Study in Ethnoarchaeology." *Australian Archae-ology*, no. 29 (1989): 39–51.

Rogers, Kara, ed. *The Endocrine System.* New York: Britannica Educational Publishing, 2012.

Rorty, Amélie Oksenberg. "Descartes on Thinking with the Body." In *The Cam-bridge Companion to Descartes*, edited by John Cottingham, 371–92. Cam-bridge: Cambridge University Press, 1992.

Rorty, Amélie Oksenberg, ed. *Essays on Descartes' "Meditations."* Berkeley: University of California Press, 1986.

Rorty, Amélie Oksenberg. "The Structure of Descartes' *Meditations.*" In *Essays on Descartes' "Meditations,"* edited by Amélie Oksenberg Rorty, 3–20. Berke-ley: University of California Press, 1986.

Rose, Nikolas. "The Politics of Life Itself." *Theory, Culture and Society* 18, no. 6 (2001): 1–30.

Rose, Nikolas S. *Politics of Life Itself: Biomedicine, Power, and Subjectivity in the Twenty-First Century.* Princeton, NJ: Princeton University Press, 2007.

Sandkühler, Hans Jörg, Detlev Pätzold, and Arnim Regenbogen, eds. "Phän-omenologie." In *Enzyklopädie Philosophie*, 1013–16. Hamburg: Meiner, 1999.

Schaffer, Simon. "Glass Works, Newton's Prisms and the Uses of Experiment." In *The Uses of Experiment: Studies in the Natural Sciences*, edited by David Gooding, Trevor Pinch, and Simon Schaffer, 67–103. Cambridge: Cambridge University Press, 1989.

Schatzki, Theodore R., Karin Knorr-Cetina, and Eike von Savigni, eds. *The Prac-tice Turn in Contemporary Theory.* London: Routledge, 2001.

334

Schramm, Helmar, Ludger Schwarte, and Jan Lazardzig, eds. *Collection, Laboratory, Theater: Scenes of Knowledge in the 17th Century*. Berlin: Walter de Gruyter, 2005.

Seed, Patricia. "Contesting Possession." In *Early Modern Europe: Issues and Interpretations*, edited by James B. Collins and Karen L. Taylor, 197–206. Malden, MA: Blackwell, 2006.

Sepper, Dennis L. "The Texture of Thought: Why Descartes' *Meditationes* Is Meditational, and Why It Matters." In *Descartes' Natural Philosophy*, edited by Stephen Gaukroger, John Andrew Schuster, and John Sutton, 736–50. London: Routledge, 2000.

Shapin, Steven. "The House of Experiment in Seventeenth-Century England." *Isis* 79, no. 3 (September 1988): 373–404. https://doi.org/10.1086/354773.

Shapin, Steven. "The Invisible Technician." *American Scientist* 77, no. 6 (1989): 554–63.

Shapin, Steven. "'The Mind Is Its Own Place': Science and Solitude in Seventeenth-Century England." *Science in Context* 4, no. 1 (1991): 191–218. https://doi.org/10.1017/S026988970000020X.

Shapin, Steven. *The Scientific Revolution*. Chicago: University of Chicago Press, 1996.

Shapin, Steven, and Simon Schaffer. *Leviathan and the Air-Pump: Hobbes, Boyle, and the Experimental Life*. Princeton, NJ: Princeton University Press, 2017.

Shapiro, Lawrence A. *The Routledge Handbook of Embodied Cognition*. New York: Routledge, Taylor and Francis Group, 2014.

Shapiro, Lawrence, and Shannon Spaulding. "Embodied Cognition." In *The Stanford Encyclopedia of Philosophy (Winter 2021 edition)*, edited by Edward N. Zalta. https://plato.stanford.edu/archives/win2021/entries/embodied-cognition/.

Shostak, Sara. "The Emergence of Toxicogenomics: A Case Study of Molecularization." *Social Studies of Science* 35, no. 3 (June 1, 2005): 367–403. https://doi.org/10.1177/0306312705049882.

Shusterman, Richard. *Body Consciousness: A Philosophy of Mindfulness and Somaesthetics*. Cambridge: Cambridge University Press, 2008. http://archive.org/details/bodyconsciousnesooooshus.

Sibum, Heinz Otto. "Experimental History of Science." In *Museums of Modern Science: Nobel Symposium 112*, edited by Svante Lindqvist, Marika Hedin, and Ulf Larsson, 77–86. Canton, MA: Science History Publications/USA, 2000.

Slatman, Jenny. *L'expression au-delà de la représentation: Sur l'aisthêsis et l'esthétique chez Merleau-Ponty*. Leuven: Peeters, 2003.

Slatman, Jenny. "On the (Im)Possibility of Immediate Bodily Experience." Paper presented at the conference "Mediated Bodies," Maastricht University, Netherlands, September 14–16, 2006.

Slatman, Jenny. *Our Strange Body: Philosophical Reflections on Identity and Medical Interventions*. Amsterdam: Amsterdam University Press, 2014.

Slatman, Jenny. "The Sense of Life: Husserl and Merleau-Ponty Touching and Being Touched." *Chiasmi International* 7 (2005): 305–24. http://doi.org/10.5840/chiasmi2005749.

Slatman, Jenny. *Vreemd lichaam: Over medisch ingrijpen en persoonlijke identiteit*. Amsterdam: Ambo, 2011.

Smith, Kurt. *The Descartes Dictionary*. New York: Bloomsbury Academic, 2015.

Smith, Pamela H. *The Body of the Artisan: Art and Experience in the Scientific Revolution*. Chicago: University of Chicago Press, 2004.

Smith, Pamela H. *From Lived Experience to the Written Word: Reconstructing Practical Knowledge in the Early Modern World*. Chicago: University of Chicago Press, 2022.

Smith, Pamela H. "Laboratories." In *The Cambridge History of Science*. Vol. 3, *Early Modern Science*, edited by Katharine Park and Lorraine Daston, 290–305. Cambridge: Cambridge University Press, 2006.

Smith, Pamela H., Amy R. W. Meyers, and Harold J. Cook, eds. *Ways of Making and Knowing: The Material Culture of Empirical Knowledge*. Ann Arbor: University of Michigan Press, 2014.

Smith, Pamela H., and Benjamin Schmidt, eds. *Making Knowledge in Early Modern Europe: Practices, Objects, and Texts, 1400–1800*. Chicago: University of Chicago Press, 2007.

Snow, C. P. *The Two Cultures*. London: Cambridge University Press, 1993.

Snyder, Laura J. "William Whewell." In *The Stanford Encyclopedia of Philosophy (Summer 2022 Edition)*, edited by Edward N. Zalta. https://plato.stanford.edu/archives/sum2022/entries/whewell/.

Spedding, James, Robert Leslie Ellis, and Douglas Denon Heath. "*The Great Instauration*: The Plan of the Work." In *The Works of Francis Bacon*, Vol. 4, edited by James Spedding, Robert Leslie Ellis, and Douglas Denon Heath, 22–33. London: Longman and Co., 1858. http://archive.org/details/worksfrancisbacooheatgoog.

Strauss, Bernard S. "A Physicist's Quest in Biology: Max Delbrück and 'Complementarity.'" *Genetics* 206, no. 2 (June 2017): 641–50. https://doi.org/10.1534/genetics.117.201517.

Taylor, Charles. *Sources of the Self: The Making of the Modern Identity*. Cambridge, MA: Harvard University Press, 1989.

Teich, Mikuláš. *The Scientific Revolution Revisited*. Cambridge: Open Book Publishers, 2015. https://doi.org/10.11647/OBP.0054.

Traweek, Sharon. *Beamtimes and Lifetimes: The World of High Energy Physicists*. Cambridge, MA: Harvard University Press, 1992.

Turk, Herwig, and Paulo de Carvalho Pereira, eds. *Blindspot*. Exhibition catalog. N.p.: Virose, 2007.

Vanagt, Katrien. "Early Modern Medical Thinking on Vision and the Camera Obscura: V. F. Plempius' *Ophthalmographia*." In *Blood, Sweat and Tears: The Changing Concepts of Physiology from Antiquity into Early Modern Europe*, edited by Stephen Voss, 569–94. Leiden: Brill, 2012.

Van Epps, Heather L. "René Dubos: Unearthing Antibiotics." *Journal of Experimental Medicine* 203, no. 2 (2006): 259. https://doi.org/10.1084/jem.2032fta.

Verbeek, Erik-Jan, and Theo Bos. "Conceiving the Invisible: The Role of Observation and Experiment in Descartes's Correspondence, 1630–50." In *Communicating Observations in Early Modern Letters (1500–1675): Epistolography and Epistemology in the Age of the Scientific Revolution*, edited by Dirk van Miert, 161–78. London: Warburg Institute, 2013.

Verbeek, Peter-Paul. "Don Ihde: The Technological Lifeworld." In *American Philosophy of Technology: The Empirical Turn*, edited by Hans Achterhuis, 119–46. Bloomington: Indiana University Press, 2001.

Verbeek, Peter-Paul. *What Things Do: Philosophical Reflections on Technology, Agency, and Design*. University Park: Penn State University Press, 2005.

Verburgt, Lukas M. "The History of Knowledge and the Future History of Ignorance." *KNOW: A Journal on the Formation of Knowledge* 4, no. 1 (2020): 1–24. https://doi.org/10.1086/708341.

Vijgenboom, Erik, et al. "Handleiding Introductie & Biochemie Practicum I, Sept–Okt 2005." Faculty of Mathematics and Natural Sciences, Leiden University, 2005.

Vogt, Katja. "Ancient Skepticism." In *The Stanford Encyclopedia of Philosophy (Winter 2022 Edition)*, edited by Edward N. Zalta and Uri Nodelman. https://plato.stanford.edu/cgi-bin/encyclopedia/archinfo.cgi?entry=skepticism-ancient.

Voss, Stephen, ed. *Essays on the Philosophy and Science of René Descartes*. New York: Oxford University Press, 1993.

Waldenfels, Bernhard. *Phänomenologie der Aufmerksamkeit*. Frankfurt: Suhrkamp, 2004.

Walls, Matthew. "Kayak Games and Hunting Enskilment: An Archaeological Consideration of Sports and the Situated Learning of Technical Skills." *World Archaeology* 44, no. 2 (2012): 175–88. https://doi.org/10.1080/00438243.2012.669604.

Ward, Dave, and Mog Stapleton. "Es Are Good: Cognition as Enacted, Embodied, Embedded, Affective and Extended." In *Consciousness in Interaction: The Role of the Natural and Social Context in Shaping Consciousness*, edited by Fabio Paglieri, 89–104. Amsterdam: John Benjamins, 2012. https://doi.org/10.1075/aicr.86.

Wardhaugh, Benjamin. *Music, Experiment and Mathematics in England, 1653–1705*. Farnham, UK: Routledge, 2017.

Watson, J. D. *Molecular Biology of the Gene*. 2nd ed. New York: W. A. Benjamin, 1970.

Watson, J. D., and F. H. C. Crick. "Molecular Structure of Nucleic Acids: A Structure for Deoxyribose Nucleic Acid." *Nature* 171, no. 4356 (1953): 737–38. https://doi.org/10.1038/171737a0.

"Wax Argument." *Wikipedia*. Accessed April 25, 2024. https://en.wikipedia.org /wiki/Wax_argument.

White, Pepper. *The Idea Factory: Learning to Think at MIT*. Cambridge, MA: MIT Press, 2001.

Whyte, William. *Cleanroom Design*. Chichester: Wiley, 2001.

Whyte, William. *Cleanroom Technology: Fundamentals of Design, Testing and Operation*. 2nd ed. Chichester: Wiley, 2010.

Willet, Jennifer. "Bodies in Biotechnology: Embodied Models for Understanding Biotechnology in Contemporary Art." *Leonardo Electronic Almanac* 14, nos. 7–8 (October 2006): 1–9.

Willet, Jennifer. "(RE)Embodying Biotechnology: Towards the Democratization of Biotechnology through Embodied Art Practices." PhD diss., Concordia University, 2009.

Williams, Bernard. "Introductory Essay on Descartes' *Meditations*." In *The Sense of the Past: Essays in the History of Philosophy*, edited by Myles Burnyeat, 246–56. Princeton, NJ: Princeton University Press, 2006.

Witkin, Evelyn M. "Remembering Rollin Hotchkiss (1911–2004)." *Genetics* 170, no. 4 (August 2005): 1443–47.

Wolfe, Charles T., and Ofer Gal, eds. *The Body as Object and Instrument of Knowledge: Embodied Empiricism in Early Modern Science*. Dordrecht: Springer Netherlands, 2010.

Wolfe, Charles T., and Philippe Huneman. "Man-Machines and Embodiment: From Cartesian Physiology to Claude Bernard's 'Living Machine.'" In *Embodiment: A History*, edited by Justin E. H. Smith, 257–97. New York: Oxford University Press, 2017.

Zilsel, Edgar. "The Genesis of the Concept of Scientific Progress." *Journal of the History of Ideas* 6, no. 3 (June 1945): 325–49. https://doi.org/10.2307/2707296.

Zurr, Ionat. "Growing Semi-Living Art." PhD diss., University of Western Australia, 2008. https://research-repository.uwa.edu.au/files/3225984/Zurr _Ionat_2008.pdf.

Zurr, Ionat, and Oron Catts. "The Ethical Claims of Bio-Art: Killing the Other or Self-Cannibalism." *Australian and New Zealand Journal of Art* 5, no. 1 (January 1, 2004): 167–88. https://doi.org/10.1080/14434318.2004.11432737.

Zwijnenberg, Robert. "BIOPLAY: Clandestine Appropriation as an Artistic Strategy." In *Naturally Postnatural—Catalyst: Jennifer Willet*, edited by Ted Hiebert, 35–49. Victoria, Canada: Noxious Sector Press, 2017.

Zwijnenberg, Robert. *The Writings and Drawings of Leonardo da Vinci: Order and Chaos in Early Modern Thought*. New York: Cambridge University Press, 1999.

anatomy studies (continued)
self-study, 58–59, 70; skilled vision
in, 62; of Vesalius, 57, 62; vivisection in, 63, 64–65, 66, 67–68, 103;
and wax passage, 31, 71, 88, 94, 95,
101–2, 117, 120
Ancarani, Yuri, 215, 305n85
animals: anatomy studies on (See
anatomy studies); pathogen-free,
isolators for, 241, 242–43, 261n1
anthropological studies, 160; of Good,
239; of Latour and Woolgar,
146–47, 150, 154, 160
antiparasitic drug development,
Ōmura in, 24, 25
archaeology and ethnography, 124, 125
Aristotelian traditions, 39, 40; Descartes education on, 51; and new
sciences, 33–36, 48, 55; sense-based
epistemology in, 78, 84, 106, 178
Aristotle, 21, 106
art: anachronism in studies of,
120–22; of anatomy, 61, 70; of bio
artists, 30, 243–47; and impact
of photographs on surgeons, 215,
239; manual labor in, 22–23; and
"painter with no hands" image,
22; practical (ars), 106
artificial operation or dissection in
anatomy studies, 60, 62, 67–68
artisanal knowledge, 106, 144;
Mersenne on, 48; Smith on, 92,
93, 106, 139; Zilsel on, 93, 139
astonishment in scientific inquiry,
150–52
attitude, natural and phenomenological, Husserl on, 116
Augustine, 98–99
Augustinian traditions, 82, 97, 98–99,
100
autoethnographic accounts, 142,
187
automatons, body as, 58, 63, 178
avermectin, 25
Avery, Eric, 215
Avery, Oswald, 10
awareness of body, 26–27, 222; in
incorporation process, 222–33;

in sterile working procedures, 27,
217–18, 233
awkwardness in hands-on practices,
184–88, 244, 248
"Awkward Student, The" (Collins),
186–87

Bachelard, Gaston, 138, 139, 205
Bacon, Francis, 35, 48
bacteria: contamination with, 27, 167,
174, 199, 255; Hotchkiss research
on, 14; odor of, 4; Ōmura cultures
of, 24–26; virus infection of, 10,
11; in yeast cloning experiment,
173–74
Barker, Kathy, 216
Beadle, George, 14
Ben-Ary, Guy, 168–71, 297n86
benchwork: body awareness in, 26–27,
222–33; as embodied practice,
145–53, 155, 156; in first biochemistry practicum, 184–85, 191–93;
incorporation process in, 222–31;
in molecular biology, 175, 182–84,
190, 197; pipettes in, 1–4, 183, 191–93
(See also pipettes); sterile procedures in, 217–18 (See also sterile
working procedures); training on,
182–88, 299n139; video analysis
of, 208–17; in workshop model of
laboratories, 175
Bender, W., 200–201
Bergmann, Wilhelm, 201, 203
Bernard, Claude, 165
Binford, Lewis, 125
bio artists, 30; cleanroom
performances of, 243–47, 248
"Biological Gaze, The" (Keller), 164,
172–73
biology: gaze in, 164, 172–73, 190; interventionist approach in, 172, 173;
manual training in, 173; new intellectual approach to, 10, 13
Bioplay (Willet), 244, 309n21
black boxes: scientists as, 179, 180;
technologies as, 206, 232
Blackburn, Elizabeth, 19
Blum, Andreas, 158

58–59, 62, 69–70, 74, 79, 102–3, 278n152; on mind-body dualism, 28–29 (*See also* mind-body dualism); and new sciences, 29, 48, 51, 72, 83, 91, 94; optical studies of, 58, 72–80; on ordinary perception, 95; page count and quantity of writings, 49–50, 52; on philosophical introspection, 55, 274–75n99; as rationalist, 45, 46, 83, 84, 117; *res extensa* concept of, 128–29, 143 (See also *res extensa*); thought experiment of, 100, 104, 116, 125, 135; and *Unterschiebung* concept, 138–39; wax passage of (*See* wax passage of Descartes)

Description of the Human Body (La description du corps humain) (Descartes), 63, 64, 67, 89

Didi-Huberman, Georges, 112, 113–14; on anachronism, 120–22, 125; on displaced resemblance, 123–24

Dijksterhuis, Fokko Jan, 72–73, 78

dioptric studies of Descartes, 72, 73

Discourse on the Method (Discours de la méthode) (Descartes), 49, 54, 55, 66, 89, 94

disegno (drawing) as mental achievement, 22

disembodied conception of science, 20, 155, 265n77

disembodied epistemology, 111, 113, 127–34; and hands-off approach (*See* hands-off approach); and *Unterschiebung*, 113, 114, 135–40, 147, 148, 155

disembodied ethnography, 155

disembodied life, 244; in vitro experimentation, 164–71

disembodied observer and knower, 21–22, 23, 45, 62; in life sciences laboratory, 155, 156; and sterile body suit image, 236–37; and thinker without hands image, 21–22, 45, 47, 112, 113, 134

displaced resemblance, 123–27

dissection in anatomy studies, 57–63, 65–66, 101; of eye, 74–80

DNA, 10, 12, 204; contamination of sample with, 174, 220; damage recognition and repair process, 26, 173–74; electrophoresis of, 27; in gene technology course, 220, 222; rapid replication of, 42; in yeast cloning experiment, 173–74

Doing, Park, 185, 200

doubt: hyperbolic, 104, 115; in *Meditations*, 45, 46, 104, 114–16, 135; skeptical, of Augustine, 98

drug research and development: of antiparasitic drugs by Ōmura, 24, 25; cleanroom facilities in, 240, 241, 249, 250, 254

electrophoresis, 27, 150

Elisabeth of Bohemia, Descartes correspondence with, 82, 283n1

embodied cognition, 23–24, 127, 222, 296n59

embodiment, 163; and disembodied epistemology, 127–34; and disembodied life, 164–73; in experimentation, 28, 127, 167; in hands-on practices (*See* hands-on practices); Ihde on, 225–31; instrumental, 227; in learning, 231; in modern epistemology, 128; in pipette calibration, 5; in sciences, 127, 145–53, 156, 163, 184–88, 218; in writing, 68, 279n166

embryology studies of Descartes, 65–66

Emergent Forms of Life and the Anthropological Voice (Fischer), 215

Emerson, Ralph Waldo, 20

Emerson, Rollins A., 18

Emilia Galotti (Lessing), 22

empiricism, 34; Aristotelian, 34, 35, 36, 40; experimental (*See* experimental empiricism)

endocrinology, 160–61

engendering, 40, 41–42, 68

Enlightenment era, 22, 40, 135

enskillment, 5. *See also* training, hands-on

enzymology, 198, 205

182; on laboratory as training site, 182, 183–84; and practical turn in science, 146; on storytelling method, 149

knower: disembodied (*See* disembodied observer and knower); and known, relationship between, 105

Körper- and *Leib*-experiences, 129–31, 133

Körpererfahrung, 131

Körperlichkeit, 130

Koyré, Alexandre, 51

Küster, Ernst, 242

"Lab Hands" (Doing), 185

laboratory ethnography, 145–46, 182, 184. *See also* ethnography

Laboratory Life (Latour and Woolgar), 145–48, 153–63; as anthropological study, 146–47, 150, 154; as anti-Cartesian project, 31, 146–48; fictional observer in, 150, 153–56, 157; geometric viewpoint in, 157, 158; hands-on practices in, 159–63; historical approach in, 158, 182; in situ method in, 157–58, 161–62; organizing principle in, 157; on scientific daily life, 183; stranger device in, 150, 162; technician role in, 156, 157, 158, 159, 160; wonder and curiosity in, 150

laminar flow cabinets: bone marrow cells for transplantation in, 255; stem cell isolation from umbilical blood in, 170–71, 217–18; tumor cell handling in, 251, 252, 253, 255, 256

Landecker, Hannah, 164, 165, 166–67

Latour, Bruno, 142, 145–47, 150, 153–63; on alienation process, 206; anti-Cartesian framework of, 31, 146–48; on blackboxing of technologies, 206; on epistemology as dead discipline, 29; *Laboratory Life,* 153–63 (See also *Laboratory Life*); on materiality of instruments, 297n78; on role of writing in science, 160, 161, 253

Lawler, Mark, 10

learning process, Gallagher on, 231, 232

Lefèvre, Wolfgang, 41, 42

Leib- and *Körper*-experiences, 129–31, 133

Leiberfahrung, 131

Leiden University laboratory work, 261n1; in cleanroom facility, 43, 248–56; Eppendorf tubes in, 150, 151, 183; in first biochemistry practicum, 2, 184–85, 191–93; in gene technology course, 219–22; hand-on practices in, 2, 156–57, 174, 181, 184–85; in molecular genetics, 150, 181; participant observation of, 149; pipette calibration and use in, 1–4, 5, 6–7, 191–93; snapshot stories on (*See* snapshot stories); on yeast cloning and complementation, 173–74

Lenoir, Tim, 194

Lessing, Gotthold Ephraim, 22

life: at cellular and molecular level, 4, 6, 8, 11, 15, 27, 42, 190; existential and ethical questions in definition of, 42; in vitro, 42, 164–71; and touch, 173

life-world, 140; Husserl concept of, 139

logico-mathematical concepts, 129, 134, 139

Lynch, Michael, 162, 232

maize genetics: Emerson research on, 18; McClintock research on, 9–19, 26

"Making Clinical Sense" (Harris), 215–16

"Manufacture of Bodies in Surgery, The" (Hirschauer), 238, 308n3

manufacturing, cleanroom facilities in, 240, 241, 249

Marburg Institute of Physiological Chemistry, 197, 198, 201

Marburg pipettes, 195, 196; historical development of, 197–207

Massachusetts Institute of Technology (MIT), 126

mathematics: Collins on awkward student in, 187; Descartes on, 72, 128, 131, 144; in genetics, 12, 13; geometry in (*See* geometry); Husserl on, 134–35, 139; Kant on, 135; and logico-mathematical concepts, 129, 134, 139; in Mersenne study of acoustics, 38, 39; and metaphysics, 134, 135; in new sciences, 134; and *res extensa*, 128, 129, 132, 188; and *Unterschiebung*, 135

mathematization, 134–35

McAllister, James, 102

McClintock, Barbara, 8–21, 138; on body as nuisance, 9, 20, 21, 23, 235; genetic research of, 8–21, 26; intimate knowledge of organisms studied, 14, 15, 17, 19; observational skills of, 13–16, 17, 19, 20; virtuoso technique of, 14, 18, 20

mechanistic view, 132; of Descartes, 58, 63, 64, 67, 144, 179; in genetics, 12, 13

meditation: in Augustinian and Ignatian traditions, 98–99; *Meditations* as manual on, 96–101

Meditations (Les méditations métaphysiques) (Descartes), 6, 45–49, 68; and anachronism, 121; as antagonist of experimental approach, 44, 56; as cognitive exercise, 99, 100; doubt in, 45, 46, 104, 114–16, 135; English translation of, 271n60; and *epoché* method, 114–18; exemplifications in, 98; experience in, 98; First, 29, 114; French edition of, 49, 82, 83, 94, 288–89n1; generative repetition in reading of, 119, 125; hands-on perspective in, 137; Latin edition of, 48, 82, 94, 134, 288n1; as literary form, 96–101; as meditational manual, 96–101; meditator in (*See* meditator in *Meditations*); on perception and introspection, 274–75n99; performative or experiential thrust of, 87, 97, 98, 99, 100, 101; on philosophical exertions, 55; Second (*See* Second

Meditation); as thought experiment, 100, 104, 116, 125, 135; and *Unterschiebung* concept, 135–36; wax passage in (*See* wax passage of Descartes)

meditator in *Meditations,* 83, 84, 85–95, 102, 104–5, 119; doubt of, 114, 115; and *epoché* method, 114–16, 117; as experimenter, 83, 88–95, 96, 101, 105, 109, 133; gender of, 85–86; on sense-based knowledge, 45–46

Mendelian genetics, 10

Merleau-Ponty, Maurice, 31, 118, 128, 157, 230; on body schema, 222–24, 225, 232; as phenomenologist, 31, 128, 157, 158, 222, 229, 232

Mersenne, Marin, 30, 37–40, 48; correspondence with Descartes, 58–59, 62, 69–70, 74, 79, 102–3, 278n152; music and acoustic studies of, 38–39, 48, 49

Mesman, Jessica, 239

metaphysics, 134

microbiology: bacterial culture in, 24–26; benchwork in, 5; cleanrooms in, 29, 31–32, 140, 236, 240 (*See also* cleanroom facilities)

microcentrifuge tubes, 174, 190, 194–95, 196, 197; in gene technology coursework, 220–21, 222; incorporation of, 223

microliter technologies, 194–95, 221; historical development of, 194–207; importance of, 197

micropipettes. *See* pipettes and micropipettes

mind, and hands-on practices, 5–6, 133; and embodied cognition, 23–24, 127, 158, 222

"mind and hand" motto of MIT, 126

mind-body dualism: and disembodied knower, 21–22; and embodied cognition, 158; in genetics, 12; in Keller account on McClintock, 12; Knorr-Cetina on, 176–77, 178, 179, 180; in *Meditations,* 28–29, 84, 109; *res cogitans* and *res extensa* concepts in, 111, 127

mind's eye, 11, 12, 19, 20
model organisms, 5, 16, 20, 173–74
Mody, Cyrus, 184, 185–88, 232, 239
Mol, Annemarie, 29, 247, 256
molecular biology laboratories,
    149–53, 182; benchwork in, 175,
    182–84, 190, 197; body awareness
    in, 26–27; contamination and ste-
    rility in, 27, 43, 176; engendering
    concept in, 42; Eppendorf tubes
    in, 42, 150, 151, 183, 191; internship
    in, 26, 119–20, 156–57, 176, 181, 184,
    189–90; Knorr-Cetina on, 149,
    174–80, 182; manual character of
    epistemic practices in, 23; micro-
    liter technologies in, 194–95, 221;
    model organisms in, 5, 20, 173–74;
    object-centered interactive quality
    of practice in, 179; pipettes in, 1–4,
    5, 6–7, 183, 190, 191–93; sensory
    perception in, 4–5; touch tech-
    niques and technologies in, 190,
    191; visualization technologies in,
    42, 190; workshop model of, 175;
    yeast cloning and complementa-
    tion experiment in, 173–74
molecularization, 11–12, 42, 195, 222,
    270n47
Mühlhardt, Cornel, 189
Mulligan, Tim, 16, 18
multisited or multilocale ethnogra-
    phies, 30, 140, 149
music and acoustic studies of
    Mersenne, 38–39, 48, 49
musicology, 272n81
mutual constitution, Verbeek on,
    227–28
Mydorge, Claude, 72

naked touch, Ihde on, 228, 229, 230
narrator of *Meditations. See* meditator
    in *Meditations*
natural attitude, Husserl on, 116
natural inquiries, 34, 61
naturalists, 152–53, 154
natural philosophy, 33, 44, 48;
    Aristotelian, 35–36, 40; bias
    toward rare phenomena in, 152;

of Boyle, 43; of Descartes, 49–50,
    89; *experiences* in, 54, 55; hands-
    on practices in, 92, 93, 106–7, 108;
    Mersenne in development of, 37,
    38, 48; modes of knowledge for-
    mation in, 107; and new sciences,
    34, 35–36, 37, 48, 55; sensual per-
    ception in, 55
natural sciences, Husserl on, 116
nature, sources of knowledge on:
    Descartes on, 54; experiments as
    source of, 36, 40, 41, 78, 104
Nelle, Florian, 96
neonatal care, 239
nerve cell cultures, 164–65
Netherlands Organisation for Ap-
    plied Scientific Research, 248,
    310n34
Neurospora, 13, 14
new sciences, 33–37, 43–44; craftsmen
    and artisans in, 93, 106, 172; crisis
    of perception in, 34–37; Descartes
    in, 29, 48, 51, 72, 83, 91, 94; epis-
    temic practices in, 48; experimen-
    tation in, 33–37, 43–44, 96, 101–4,
    105, 110; hands-on practices in, 44,
    54, 84, 106–7; literary innovations
    in reporting on, 83; mathematics
    in, 134; Mersenne in, 37; sen-
    sual perception in, 34–37, 55., 91;
    thought experiment in, 178; use of
    term, 268n2
"New Times for Biology: Nerve Cul-
    tures and the Advent of Cellular
    Life in Vitro" (Landecker), 164
Newton, Isaac, 48, 150
Nishikawa, Mitsuko, 26
Nobel Prize Committee, 16–17, 18,
    264n60
Nobel Prize recipients: McClintock,
    8, 16, 20, 138, 235; Ōmura and
    Campbell, 24, 25; Warburg, 200;
    Watson, 10
nucleotide chromatography, 198, 202
Nuttall, George H. F., 241–42

object, in wax passage, 91
objectification of body, 132, 136

phenomenotechnique concept of Bachelard, 139, 205

"The Philosopher and His Shadow, The" (Merleau-Ponty), 118

philosophical introspection, Descartes on, 55, 274–75n99

Philosophical Writings of Descartes, The (Cottingham et al.), x, 53, 271n60

philosophy: experimental, 43, 44, 145; natural (*See* natural philosophy); of science, 23, 145, 194, 205, 233

"Philosophy of New Drug Discovery" (Ōmura), 25

photometry, 199, 200

physico-mathematical account of refraction, by Descartes, 72

physics, 144; Delbrück imaginary vision of, 11

physiology studies, 58, 60–61, 69, 70–71, 108

Picot, Abbé Claude, 54

pineal body, 65, 278n152

pipettes and micropipettes: automatic, 195, 196, 197, 199, 203, 204; calibration of, 1–4, 5, 6–7, 191–93; cleaning and sterilization of, 198–99; as epistemic thing, 205; Eppendorf, 183, 191, 195–96, 203, 206; in gene technology coursework, 219, 220, 222; glass, 198–99, 202; historical development of, 194–207, 301nn27–30; importance of skill with, 4; and instrument-body relations, 223, 227–28; invisible skill in use of, 159; and microcentrifuge tubes, 190, 194–95, 223; mouth used in working with, 198, 199; piston strike (Marburg), 195, 196, 197–204; with polypropylene tips, 203; as precision instrument, 42, 195; with removable tip, 202, 203; in stem cell isolation from umbilical blood, 169, 170, 217; with Teflon tips, 203; and touch in research, 150; working mechanism of, 202–3, 303n44; in yeast cloning experiment, 174

pipettor, as multivalent term, 7

piston strike (Marburg) pipettes, 195, 196, 197–207

plant genetics, McClintock research on, 8–21

Plato, 21, 133

Platonic theories, 35

Plempius, Vopiscus Fortunatus, 65, 67, 73, 103

pluralism, epistemological, 140, 175

Polanyi, Michael, 211–12

Pollock, Jackson, 124

polymerase chain reaction, 42, 203, 220, 221

polypropylene pipette tips, 203

practical turn in science, 31, 143, 145, 146, 147, 148, 150, 194

practiography, 32, 247–56

pragmatists, 293n3

Prentice, Rachel, 215

proprioception, Gallagher on, 231–32

protocols: in pipette calibration, 3; scientific writing on, 75, 83, 91, 120, 141; in stem cell isolation from umbilical blood, 168–71; in sterile working techniques, 176; in tumor cell therapy, 254; use of term, 91, 285n18; video study of hands-on practices in, 212–15; wax passage as, 125

"Pure Visitor's" viewpoint, 154

purification process, 136, 138–39, 219, 292n72

Pyrrhonism, 114

Rajewsky, Klaus, 236, 237

"Raphael with no hands," 22

rationalism, 40; of Descartes, 45, 46, 83, 84, 117; and experimental empiricism, 43–45, 49; hands-off approach in, 138; "thinkers without hands" image in, 21, 45; wax argument in, 45, 46, 83

ratio (reason), 40

reason, 84

Recent Advances in Cytology (Darlington), 12

recessive reflexive awareness, Gallagher on, 231, 233

recipe and how-to literature, 100; on eye experiment of Descartes, 75; in history of science, 100, 282n199, 284n11; wax passage as, 88, 90, 91, 100, 120

reductionist approaches, 10–11, 12, 92

refraction, Descartes study of, 72

Reichle, Ingeborg, 211–13

Rembrandt van Rijn, 57

Renaissance era, 22, 57

Reodica, Julia, 244

repetition: generative, 118–19, 125, 126, 141, 142, 256; reproducibility of experiments in, 103–4

reproducibility, in repetition of experiments, 103–4

res cogitans, 111, 126, 127, 128

research methods: on cleanroom facilities, 32; ethnographic (See ethnography); fieldwork sites in, 149, 261–62n1, 294n24; as multisited or multilocale, 30, 140, 149; participant observations in, 30 (See also participant observations)

resemblance: direct, 124; displaced, 123–27

res extensa, 131–34, 135, 143–44, 188; and alienation, 128–29, 131, 132; and anatomy studies, 131–32, 144; and devaluation of sensory experience, 117; and hands-off practices, 46; and mathematics, 128, 129, 132, 188; and mind-body dualism, 111, 127; and wax passage, 111, 117, 126, 135

Rheinberger, Hans-Jörg, 190, 233, 256, 293n3; on clarity, 138, 219; on epistemic objects, 205, 206, 218; on experimental systems, 218–19; on historicization of scientific knowledge, 143, 144, 147, 207; on instrument use, 193, 194, 207–8, 218; on intersections, 207–8, 218

rhetoric strategies: curiosity in, 152; of Descartes in Meditations, 82–83, 96–101; of effortlessness, 233; of

Latour and Woolgar, 150, 153–56; on new scientists, 44; of self-effacement, 242, 243; storytelling in, 150

Rhoades, Marcus, 15

Right Tools for the Job, The (Clarke and Fujimura), 216

Robins, Richard, 124–25

Rorty, Amélie Oksenberg, 97

Royal Academy, 48

Royal Society of London for Improving Natural Knowledge, 43, 48

Royal Society of Science, 91

Rules for the Direction of the Mind (Regulae ad directionem ingenii) (Descartes), 49, 55

Ruysch, Frederik, 60

Saccharomyces cerevisiae, 173–74

Salk Institute, 146

salon culture of France, 73, 103, 104

Say it isn't so exhibition, 209

"Scarlet O: Epistemic Politics and (Scientific) Labor" (Doing), 185

Schaffer, Simon, 41, 43, 103, 206

Schlick, Moritz, 293n3

Schnitger, Heinrich, 195, 197–98, 200, 201, 202, 203, 205

scholasticism: education of Descartes on, 51; experimentation in, 36, 77; natural philosophy in, 51, 54; scientia concept in, 34, 108, 109; as text- and theory-laden work, 44, 48, 58, 71, 108, 109, 127

science: astonishment, curiosity, and wonder in, 150–52; disembodied and immaterial concept of, 20, 265n77; as hands-on practice (See hands-on practices); history and philosophy of, 23, 27–30, 145, 194, 205, 233; as material practice, 27, 28, 143, 145; practical turn in, 31, 143, 145, 146, 147, 148, 150, 194; as social and cultural practice, 143, 145, 211

science and technology studies (STS), 145, 146, 293n10; as academic field, rise of, 194; embodied work in,

144, 185–86; generative repetition in, 142; instruments in, 204; manual tinkering in, 126; Mody on, 185–86; philosophical foundations of, 140

science wars, 153, 295n38

scientia, 43, 93, 106, 288n90; hands-off practices in, 107; physiology in, 61, 69, 71; Scholastic concept of, 34, 108, 109

Scientific Revolution, 23, 33, 268n1; hands-on practices in, 93, 139, 181

screening systems in new drug discovery, 25

SDS-PAGE protocol, video study of, 213–14

Second Meditation (Descartes), 29; doubt in, 114; and *epoché* method, 114–18; exemplification in, 98; on perception and introspection, 274–75n99; and *Unterschiebung* concept, 135–36; wax passage in (*See* wax passage of Descartes)

self-effacement, rhetoric of, 242, 243

semi-living sculptures, 245

"Senses and the Fleshless Eye, The" (Hatfield), 98

sensory body, Knorr-Cetina on, 176, 177, 178, 180

sensual perception: in acoustic studies of Mersenne, 39–40; in Aristotelian traditions, 78, 84; in cleanroom facilities, 248; deceitfulness of, 95; devaluation of, 117, 118; in *experiences,* 55; instrument-mediated, 228–29; Knorr-Cetina on, 177–78, 180; in new sciences, 34–37, 55, 91; in optic studies of Descartes, 58, 72–78; ordinary, 95; touch in (*See* touch); in wax passage, 45–48, 71–72, 84, 86–92, 95–96, 99, 105, 117

setting 04_0006 (video) (Turk and Stöger), 208, 209, 211, 212, 213, 214, 216, 217

Shapin, Steven: on Boyle experimentation, 41, 43, 106; on early history of experimentation, 103; on

hands-on practices, 44, 106–7, 181–82; on invisible technicians, 44, 45, 159–60; on new sciences, 43, 44, 106; on solitude and isolation, 101–2, 103, 104; on workshop model of laboratories, 175

Shusterman, Richard, 224, 225

silent body, Knorr-Cetina on, 179, 180

situation concept of Clarke and Fujimura, 216

skilled vision, 27, 267n107; in anatomy studies, 62; of McClintock, 13–16, 19, 20

Slatman, Jenny: on alienated body experience, 128–29, 163; on generative repetition, 118; on *Leib-* and *Körper-*experiences, 129, 130, 131; *Our Strange Body,* 128–29, 163, 223; on phenomenology, 113, 129, 163; and *res extensa* concept, 131, 133; on touch and life, 173; on two touching hands experience, 129, 229, 291n54

Smith, Pamela, 92, 93, 106, 107, 139, 144–45

snapshot stories, 142, 149, 154; on cleanroom visit, 32, 247–56; on cloning and complementation in yeast, 173–74; on first biochemistry practicum, 184–85, 191–93; on gene technology course, 219–22; on "In Touch with Life" corporate slogan, 150; on pipette calibration, 191–93; on research internship with molecular genetics laboratory, 156–57, 181; on stem cell isolation from umbilical blood, 168–71

Snowian culture shock, 157

solitary investigators, 104; Cartesian ideal of, 102; Descartes viewed as, 50, 103, 104; McClintock portrayed as, 16

solitude, Shapin on, 101–2

somaesthetics, 225

soul and soulless body, 132, 133

sound, Mersenne theory of, 38–39

Sources of the Self (Taylor), 132

355

358

www.ingramcontent.com/pod-product-compliance
Lightning Source LLC
Chambersburg PA
CBHW032342280326
41935CB00008B/416